T0271069

Right By Design

Product design is becoming increasingly challenging as product complexity increases dramatically with the advent of autonomous control and the need to achieve zero emissions. Companies continue to have poor product launches with significant numbers of recall campaigns and high after-sales warranties. It is important that potential product failures are identified and fixed during the design of a product. Failure modes found after the design has matured are normally easy to find, with some being identified by the customer, but are often difficult and expensive to fix; modifying one part will often have a knock-on effect on other parts, causing other problems. Discovering failure modes early in the design process is often difficult – requiring rigorous and comprehensive analysis – but once found, such failure modes are usually easy and cheap to fix.

This book presents an approach to product design based on Failure Mode Avoidance that utilises a series of strongly interrelated engineering tools and interpersonal skills that can be used to discover failure modes early in the design process. The tools can be used across engineering disciplines.

Despite engineering being largely a team activity, it is often the case that little attention is paid to the team process after the team membership has been identified, with membership normally being based on technical expertise. In addition to technical expertise, an effective engineering team requires individual engineers to work together efficiently. Good leadership is also required, with the leader able to both manage change and encourage individual team members to work to the best of their ability. The book interweaves technical skills, team skills and team leadership in a way that reflects their real-life interrelationship.

The book tells the fictional story of a small engineering team and its leader as they implement the skills introduced in the book and follows their experiences reflecting individual difficulties, enthusiasm, humour and scepticism in applying the methodologies and tools for the first time. In addition, the story tells of team members' interactions with their management and peers within a company that, having been very successful, finds itself in financial difficulties. It promotes constructivist learning through the reader empathising with the characters in the book. These characters ask questions that are typical of those that learners will ask about the subject matter. Learning reinforcement is also integrated into the storyline as a natural and unobtrusive feature.

The book is intended to be read like a novel from cover to cover with a storyline that motivates the reader to read on. While including in-depth technical examples, the book is not intended as a seminal text on Failure Mode Avoidance or team skills; it is intended to give the reader an understanding such that they are motivated to learn more. Having read the book, it can be treated more typically as a textbook by returning to some of the technical detail or looking to further reading such as that identified in the book.

Right By Design

A Novel Approach to Failure Mode Avoidance

Ed Henshall

CRC Press is an imprint of the
Taylor & Francis Group, an **informa** business

First published 2023
by Routledge
605 Third Avenue, New York, NY 10158

and by Routledge
4 Park Square, Milton Park, Abingdon, Oxon, OX14 4RN

Routledge is an imprint of the Taylor & Francis Group, an informa business

ISBN: 978-1-032-26008-2 (hbk)
ISBN: 978-1-032-26006-8 (pbk)
ISBN: 978-1-003-28606-6 (ebk)

DOI: 10.4324/9781003286066

Typeset in Garamond
by SPi Technologies India Pvt Ltd (Straive)

To my wife Yvonne and our family.

Contents

Preface

Engineering is a team sport. Very few product design engineers have the opportunity, or would wish, to work alone. While technical expertise is indispensable in an engineering team, it is only a part of what is needed to make an engineering project successful with the ability of individual engineers to work together effectively being equally important. Another important aspect of product design success is the ability of engineering team leaders to manage change and take full advantage of the talents of individual team members.

The environment in which a product design team works is also becoming increasingly challenging as product complexity increases dramatically with the push towards both zero emissions and autonomous control. In addition, companies continue to suffer financially because of product-launch problems as well as significant after-sales warranty and recalls. It is important that potential product failures are identified and fixed during the design of a product. The book presents an approach to product design based on Failure Mode Avoidance which utilises a series of strongly interrelated engineering tools and interpersonal skills that can be used across engineering disciplines to facilitate the design of failure-free complex products.

Aimed at practicing engineers and engineering management involved in product design as well as undergraduate and postgraduate engineering students along with university engineering staff, the book uniquely brings the subject matter to life by interweaving the technical and team skills into a fictitious storyline. The addition of realistic but fictional characters and real-life situations that involve emotions and personal reactions, that many readers will recognise, adds humour to what otherwise might appear a dry topic.

The goal of the book is to stimulate interest in the technical and team skills content such that a reader will want to learn more. The author advises the reader not to get distracted by the technical detail at the expense of losing the thread of both the fictitious storyline and the overall use of the tools and methods within the Failure Mode Avoidance framework. Having read the book, it can be treated more typically as a textbook by returning to some of the technical detail or looking to further reading such as that detailed in the book.

The application of the technical and team skills considered in the book is demonstrated through a case study. Being illustrative, the case study examples are not exhaustive.

The references given at the end of each chapter are for material that the reader might find useful in providing background information to the topics contained in that chapter. The relevant pages of a reference are given where appropriate. An additional list of more general references to material that helped to shape the content of the book is given in the bibliography.

Ed Henshall

Acknowledgements

I am indebted to Rob Herrick, Ian Morris, Nathan Soderborg and Mark Stevens who read draft versions of the book and provided constructive feedback. Mark's support gave me the impetus to continue writing the book, subsequently sustained by Ian's feedback. Rob's eagle eye for spotting typos and technical inconsistency and Nathan's discerning feedback proved invaluable.

Professor Felician Campean, who's training material was the inspiration behind the case study subject, is due thanks, and I appreciate the time we spent together, with others, in evolving the system model which forms the foundation of the technical content of the case study in the book. I also recognise the creative work done by Unal Yildirim, Professor Campean and others in extending the system modelling processes and thank Aleksandr Korsunovs for his support during the development of the case study software model.

I recognise the ground breaking work done by Tim Davis in developing and promoting the approach of Failure Mode Avoidance. I also look back fondly on the interesting times I spent with Christopher Connolly and the late John Syer discussing the integration of technical and team skills.

About the Author

Ed Henshall's career has been an academic sandwich with a substantial industrial filling. For the first decade of his career, Ed lectured in Physics within Further Education in the UK before moving into the automotive industry. He worked for the next two decades within both Manufacturing and Product Design in a multinational automotive company being based in both Europe and the United States before returning to Higher Education on a part-time basis for the next decade and a half. Ed is a former UK Royal Academy of Engineering Visiting Professor in Integrated System Design at the University of Bradford, UK.

Ed holds a PhD in Physics and is a qualified adult teacher.

Chapter 1

Beginning Anew

"Oxton Bikes Freewheeling Downhill", the headline caught Jim Eccleston's eye as he scanned the cycling enthusiast magazine while munching his toast. Putting his breakfast aside, Jim read on. The article explained that Oxton Bikes, the UK company founded by Jim Eccleston, formerly one of Britain's most successful professional cyclists, was running into difficulties despite its previous stellar performance due to several high visibility product recall campaigns over the past couple of years. Of course, this was not news to Oxton's CEO who was all too aware of the situation, but it was a stark reminder of the task he had before him in restoring his beloved company to its former glory. Putting the magazine down, Jim returned to his breakfast thankful that he had met John Perry at the engineering conference that he had attended a couple of months before.

John Perry was feeling apprehensive as he walked through the doors of Oxton Bikes reception area for the first time as an Oxton employee. He felt he was walking into the unknown and was hoping that he would not regret his decision to give up his secure job at Blade Motors where he had made a bit of a name for himself in the UK engineering headquarters of the global automotive company. John had joined Blade as an engineering apprentice almost 20 years before and worked his way up to a management role. Along the way, he had become very interested in the importance of paying attention to the people side of engineering; engineering is a team sport he would tell the teams he worked with and like any sport you neglect the importance of how people interact at your peril. He walked up to the reception desk and announced his arrival saying he had been told to report to Jim Eccleston's office.

John had first met Jim Eccleston at the Engineering Design conference in the Northern England town of Harrogate. Jim had approached John to say how much he enjoyed John's paper on the integration of the technical and people sides of engineering design and had also found the design methodology of FTED (first-time engineering design) that John had described fascinating. While being pleased to get the accolade from someone with Jim's standing, John thought little further about their encounter and was surprised to receive a call at work a week later from Jim's PA asking if Jim could call him at home that evening. John was intrigued and readily

DOI: 10.4324/9781003286066-1

agreed to the request supplying his home phone number. John spent the rest of the day wondering what Jim might say and arrived home excited to explain to his wife Jane what was happening. Jim duly rang that evening and asked John if they might meet on the following Saturday morning to talk over a proposition that Jim wanted to put to him. This left John not much the wiser and, finding himself beginning to speculate widely as to what Jim wanted, he decided that it would be best if he tried not to think too much about things before Saturday.

John met Jim in the relaxed surrounding of a local hotel. Jim got straight down to business by saying that he would like to give John some historical background to Oxton. Jim explained that after he started Oxton Bikes 15 years ago when he retired from professional cycling, the company had been more successful than he had ever imagined in selling the product that the company designed and manufactured. He described how the company had aimed its product at the higher end of the market with keen amateur and semi-professional riders being their core customers and had gained a reputation for high-quality, value-for-money products. Jim went on to say that unfortunately things started to deteriorate a few years earlier after he and the board had decided to diversify into electric bicycles and lower-end products aimed at more occasional cyclists. They also decided to expand their global markets and had held discussions with a Taiwan bicycle manufacturer with a view to setting up a distribution network in the far east and in the longer term establishing a manufacturing base there. Jim said that John might be aware of the widely published recall campaigns that had ensued from the launch of lower-end products and that these represented the tip of the iceberg with numerous problems surfacing late in the design and development process at, or even after, product launch. John said that the latter situation sounded a familiar tale and the very reason he had introduced the FTED enhancements to the engineering design process that he had spoken about at Harrogate.

Jim asked John if he saw any reason why FTED could not be applied effectively within the bicycle industry, and John said that he was sure that the approach was applicable to most, if not all, product design, including bicycles. Jim said that while he was not asking John to divulge any proprietary information he wondered if John could give him any concrete examples of the successful use of the methodology. John explained that while the FTED application in Blade had been limited, the early indicators were encouraging. John said that after some initial resistance, the technical and team skills methodologies had been well received within the team that had applied the approach, with individual engineers recognising the benefits both to themselves and the corporation. Jim asked John if he could give specific examples of such benefits and John said that several engineers had commented that they enjoyed the fact that they were doing what they called "real engineering". In citing what he called an "in-process measure", John said that the structure and content of FMEAs (failure mode and effects analysis) had significantly improved. John asked Jim if he was familiar with FMEA, and Jim said that he was. Turning to product quality, John said that data based on customer surveys and warranty showed a significant increase in quality of the system to which FTED had been applied when compared to the outgoing system at the same time in its lifecycle. Jim asked John what he meant by "a significant increase in quality", and John quoted the fact that there had been no warranty claims in the first six months of the system being in service which he said compared very favourably with the outgoing system.

John was beginning to wonder why Jim had arranged a meeting in what he saw as a neutral venue to discuss the merits of FTED when Jim answered the question forming in his mind by telling John he'd like to offer John a position at Oxton Bikes managing the implementation of an FTED-based approach to product design. Before John could respond to his proposal, Jim said that he was concerned that FTED was proprietary to Blade and asked how it had been developed. John said while the name might be proprietary the framework was not and explained that his academic supervisor and he had developed the framework while he had been doing a part-time MSc by research at a local university. John added that it was largely based on techniques that had been widely used for some time in the automotive industry. John also said that his supervisor and he had since published a few academic papers on the subject and so the framework was in the public domain. John also said that while Blade had been very supportive of his doing the master's degree, they had not shown a great deal of interest in the methodology, believing their corporate engineering design process to be tried and tested. John said that the application that he had just spoken of was a pilot for the methodology in Blade when he had worked with an engineering team while doing the research for his MSc. Jim asked John if he would recommend further reading on FTED, and John directed him to two of the papers that he had referenced in his conference paper.

Jim said he was also very interested in the way John had spoken about the integration of technical and people skills and asked John if he could say more about this. John said his interest in the subject had been sparked when Blade in the UK had brought in a sports psychologist to improve team working in the company. John added that the sports psychologist and he had worked together in looking at the integration of technical and team skills while he was leading an engineering team on a particularly difficult project. Jim said that this chimed with his own experience in working with the team's sports psychologist during his time as a professional cyclist. He went on to say that he was certain that the things he learnt about himself, and the techniques the sports psychologist taught him, contributed directly to his three grand tour podium finishes and his Olympic silver medal.

John managed to steer the conversation back to the proposal that Jim had made earlier by asking Jim what he was expecting him to do in Oxton. Jim responded that John knew better than him how he was going to do it but in essence he wanted him to improve the Oxton product design and development process, so it achieved right first-time designs with problems surfacing just before or after product launch becoming a thing of the past. Jim went on to say that he realised that this was not going to happen overnight as it related to the length of product development programmes. He committed to shielding John from problem-solving find-and-fix work and added that he would provide him with the resources that he needed for what he understood to be a rigorous and hence initially time-consuming front-loaded product design process. This was music to John's ears; most engineering managers in Blade would agree that resources should ideally be directed towards upfront product design but somehow it never happened as the company seemed to thrive on down-stream crisis management.

Jim and John then moved on to the subject of remuneration, and Jim offered John a package that he found attractive. The two men then concluded the meeting with John saying he was interested in what Jim had spoken about, but he would need to think it over and discuss it with his wife. Jim said that was to be expected and asked John to call him in the week to let him know his decision.

John spent the weekend mulling things over with Jane even though he knew immediately after the meeting with Jim that he wanted the job. He rang Jim on the Tuesday and said while he was very interested in Jim's proposal, he would like to better understand Oxton Bikes as a company, the facilities they had to offer and the way in which he might conduct his proposed role. Jim invited John over to Oxton after work one day and they settled on Thursday of that week. John thought that Jim seemed genuinely pleased to see him on his arrival and after a quick chat Jim took him on a tour of the facilities. John was surprised both by the size of the company and impressed by the state-of-the-art research and design engineering facilities. After the tour, John sat down with Jim and over a cup of coffee, he started to explain his concerns.

John began by asking Jim about the makeup of a typical product design and development team. Jim explained that typically the team would be led by an experienced engineer, have a technical specialist depending on the nature of the product being developed, component engineers again depending on the nature of the programme and a manufacturing engineer, co-opting others such as a warranty specialist as and when required. John asked Jim whether suppliers were included as team members. Jim explained that his company had good day-to-day relationships with its suppliers but did not include them in a product development team, rather contacting them on an as-and-when-required basis, adding that this was particularly true of overseas suppliers. John asked Jim if he would object to suppliers being included in future teams, and Jim reassured him that John had the freedom to do what he thought best. John said he assumed that Oxton product development teams were all based on site and Jim said that they were. John said that in co-opting team members from other locations he would need to rely on good video and audio internet connections and that he had been impressed by the Oxton video conference facilities. John then said that his experience of working with teams where not all members were on the same site was that team members tended to talk to each other through their laptops. John added that this even happened when team members were sitting next to each other in the same room. John explained that while he felt that this was fine at times, particularly when discussing technical matters, it did little for team building. Jim asked John what he would do about this, to which John replied that he would like the ability to move furniture freely within the room and the use of a freestanding conference room video camera. Jim said that given the price of audio-visual equipment that should not be a problem, but he might get some resistance if team members were asked to become furniture removers to which John laughed and said that this would be nothing new.

Jim said that it looked as if John was going to make changes at Oxton and that clearly change was required. Jim told John that he would be reporting directly to him, and he was more than happy to discuss John's needs on an as-and-when basis. Jim said he would like regular contact with John and said that John's office would be in the same area as his own. Jim said that he would like to discuss and agree any major proposed changes with John and added that once they agreed a course of action John could expect his full explicit backing. Jim then asked John if he was with Oxton Bikes to which John replied with a smile on his face that it depended on his job title. Jim asked John if Engineering Teams Manager sounded sufficiently prestigious and John said that it did. The two men concluded the meeting by discussing and agreeing the logistics of John's move from Blade.

Chapter 2

Meeting the Team

John Perry had been with Oxton Bikes for two weeks during which he had spent most of his time shadowing senior managers. While this had been both very useful and enlightening, he was glad that it was time for him to do some real work. The two things that struck him most about bicycle design, development and manufacture compared to his experience in Blade was firstly that none of the components that made up an Oxton bicycle were fabricated by Oxton and so to speak of Oxton as a bicycle manufacturer was to his mind a misnomer. In making a mental comparison between Oxton and Blade, John had seen that while Oxton did a lot of design work on the frame of the higher-end bikes and were involved in the design of other components on an occasional basis, they did not make any bicycle parts themselves. In comparing this with Blade, John conceded that while the vast majority of the components that made up a Blade vehicle were manufactured by parts suppliers, at least the body, the most recognisable part of the vehicle was fabricated in-house as were major components of the powertrain. The second major difference between automotive and bicycle manufacture, he had realised, was that the final product in Oxton was pre-assembled and shipped as a partially assembled kit to the seller, or increasingly more often with the advent of internet sales, directly to the customer to complete the assembly.

While John had met and been introduced to numerous people at Oxton, so many that he could not remember most of their names, three of his new colleagues stood out in his mind: Mo Lakhani, the Quality Manager; Paul Clampin, Design Engineering Director; and Numa McGovern, HR (Human Resources) Manager. John really liked Mo, who had joined Oxton 5 years ago from a washing machine manufacturer. He had an easy-going manner and ready smile, although he found his views on design verification and testing rather old fashioned. His mantra, which he seemed to repeat at every opportunity, was "You can do all the computer modelling and rig testing you like but there's no substitute for real-world testing with the final product". Paul Clampin reminded John of a couple of former colleagues at Blade whose main aim in life was to further their career either within the company or elsewhere. Paul had joined the company about the same time as Mo after progressing through several manufacturing companies. He learnt from others that Paul had persuaded the Board that, since the lower-end

DOI: 10.4324/9781003286066-2

offerings and eBikes were to be built from tried and tested off-the-shelf components, the budget and timing of the development programmes for these offerings could be significantly reduced over that for the premium product. Paul had spent most of the time in speaking to John lauding his own achievements while blaming the product problems Oxton had experienced on others, particularly suppliers. On the other hand, Numa was a very different character altogether to Paul. She had been with the company since its inception, and it appeared to John that she had a deep interest in the welfare of the company's employees most, if not all, of whom she was on first name terms with. He compared this to his experience in Blade where his interaction with HR was more remote and tended to focus on more negative aspects such as the occasion when he had to reprimand two of his engineers for inappropriate behaviour.

Part way through his two-week mentoring session, John had a very interesting and exciting meeting with Jim Eccleston. Jim told him that the Board of Directors had brought forward the timing for a new eBike product development programme in order that it could be used as a pilot for first-time engineering design (FTED). After outlining how the Board saw this new product in terms of the overall company portfolio, Jim said that the Board had agreed that the programme timing could be extended over the normal Oxton timing in recognition that the programme would be a pilot for FTED. Jim said that he wanted John to take responsibility for the programme, confirming that he wanted John to report to him, while recognising that this might present some difficulty since Paul Clampin would otherwise have managed the programme. Jim said that Paul had gone along with the plan when he had discussed it with him, given that it had Board backing, although Jim told John in confidence that Paul was probably not at all happy about it. John asked when the programme was due to start to which Jim responded that he could start as soon as he liked and suggested to John that he schedule the first meeting with the Product Design team soon after he finished his shadowing role. Jim added that he would get Paul to send him a list of prospective team members and said that he would make sure that Sarah Jones was one of the people Paul put forward since she was an experienced and highly capable eBike Technical Leader. John thanked Jim for this and said that he would like to have suppliers on the team to which Jim agreed, although saying that this would raise a few eyebrows. John asked which suppliers would potentially be on the programme and Jim pointed John in the direction of Margaret Burrell, the Director of Supply, saying he would give her a heads-up. John then enquired whether having suppliers work directly on the Oxton design process would raise issues of confidentiality to which Jim responded that he thought their existing NDAs (non-disclosure agreements) would cover it, although he would check it out with both Margaret and the company legal department.

Luckily, John was shortly due to shadow Margaret and during that time, he found the opportunity to sit down and speak with her. Margaret told John that she had already heard from Jim and added that she had always thought suppliers should have more direct involvement in product design but had met resistance from the engineering areas. She confirmed that it was odds-on that they would source the fabrication of the new eBike frame from HJK Bikes, the major bicycle manufacturer in Taiwan that Oxton had held discussions with about setting up a distribution network in the far east. Margaret also said that it was likely that they would source the Drivetrain from Agano, the reputed Japanese company. Margaret promised John that she would talk to her contacts in HJK and Agano and she would get back to him with the results of her enquiries. True to her word, John received an email from Margaret a couple of days after their meeting in which she confirmed that both supplier companies were keen to be represented on the Oxton Product Design (PD) team for the eBike and gave him the names and email addresses of two recommended individual team members. Eager to get on with things, John sent an email to the prospective team members introducing himself, briefly explaining why he was asking them to participate and telling them he would send them a meeting notice for the inaugural eBike design team meeting in the next few days. He was pleased to receive swift responses from both individuals thanking him for his email saying they were looking forward to joining the eBike team. It occurred to John later that day, while he was shadowing Peter Wills from Sales and Marketing, that it would be good to have someone who interfaced directly with customers of both Oxton and competitor bicycles on the eBike design team and mentioned this to Peter. Peter thought for a moment and then said that he had someone in mind at the European bicycle distributor Teutoburg whose head office was in Germany. Peter added that before going further he would like to speak to the person whom he was thinking about and said he would let John know the outcome of his conversation. The next day Peter emailed John with a name and contact details and John sent out another email in eager anticipation.

Once John had heard from Paul Clampin, and the Oxton members of the Product Design team were confirmed, he lost no time in setting up a schedule of meetings for the team with the inaugural meeting on the Tuesday of the next week. He was pleased to start to fill up his calendar since the only regular meetings were his weekly one-to-ones with Jim Eccleston that Jim's PA Vicki Harrison had arranged and the daily meetings that he had set up for himself at lunch time, which looked rather conspicuous in an otherwise empty calendar. He had learnt this trick at Blade in to trying to ensure he would be able to get something to eat, although he often had to overwrite this space in a company that seemed to him to operate purely on the basis of wall-to-wall meetings.

After his two weeks of watching others at work, John was pleased to spend what he saw as a real working day preparing for the next week's inaugural meeting of what was to be called the Oe375 programme PD team. It had been explained to him that the Oe in the code represented an Oxton eBike and 375 represented the scale of the programme, in this case that the product was an all-new, mid-price range, product composed largely of bought-in components. John had arranged a meeting that afternoon with Sarah Jones to help him prepare for the PD team meeting. Sarah seemed very interested in FTED when John gave her an overview of the framework. After discussing with John, the relative time that each of the different aspects of the framework might take, Sarah agreed to develop an overview graphic reflecting the nature of the programme timing and present this at the team meeting. John explained how the Board saw Oe375 being positioned in Oxton's portfolio and asked Sarah if she would outline this as well to which she readily agreed. Sarah and John spent the remainder of their meeting thinking of an alternative acronym for FTED given that there might be some sensitivity in using the Blade name. They quickly thought of *OFTE* for *Oxton First-Time Engineering* and then spent some time trying to think of a suitable noun beginning with *N* and, in the end, settled on *Norm* in its meaning of standard. At the end of the meeting, John phoned Vicki Harrison and asked her to run the acronym past Jim Eccleston at a suitable moment. John was pleased to receive an email from Vicki later that day saying that Jim was OK with the acronym, at least as a working title.

John felt both excited and apprehensive as he walked into the conference room where the inaugural meeting of the Oe375 PD team was to be held. John liked to arrive early for the meetings that he was facilitating to set up the audio–video (AV) equipment and check that it was working. He was reassured to find that Jim had actioned his request to replace the somewhat antiquated fixed audio equipment wired to the square table formed from nine smaller tables in the middle of the room with a ceiling microphone suspended offset from the middle of the room and in addition had added floor sockets so that room furniture could be set up more flexibly. A new large screen monitor had been mounted on the longer wall opposite the door in the rectangular room, and a free-standing conference room video camera stood in a corner of the room. These changes had involved some disruption to the use of the room, and he had heard some mutterings from people about "newcomers". Nevertheless, the speed with which the change had occurred served to reinforce his perception of Jim's reputation in the company of never letting the grass grow under his feet.

Walking over to the new control panel, John switched on the AV equipment and picked up the remote control. He was particularly impressed by the operation of the conference room video camera, the new large screen monitor on which the video camera image was displayed, the ceiling mike and the room speakers as he walked round the room counting out loud. John was also pleased by the flexibility of the new meeting software that he had recommended Oxton use which allowed him to show the conference room video camera image in addition to laptop camera images with the size of the conference room video camera image being controllable both from an app on his laptop and the remote. Having familiarised himself with the AV control, John spent a couple of minutes copying some preprepared text onto the meeting software whiteboard which he then hid. John then ensured that audio from the meeting software was outputted to the room speakers and the room mike acted as an audio input to the software without outputting directly to the room speakers. Having satisfied himself that he had set up the communications equipment as far as he could he seated himself at the table and spent a few minutes in quiet contemplation of both his surroundings and what he intended to do at the meeting.

In arranging the meeting for a two-hour duration, John intended to use the majority of the time for team building. He was expecting four people from Oxton to join him and three others to call in from remote locations. Just before 8 am, two people wandering in, one of whom was Sarah Jones and John assumed that the other was Jennifer Coulter, both had a cup of coffee and laptop in their hands. After exchanging pleasantries with John, they sat down next to each other on one side of the table adjacent to where John was sitting, and putting their coffee down, opened up their laptops and plugged them into a power socket. In preparation, John signed into the meeting software as its host as the other two locally-based team members walked in together soon after, greeting those already present. Like their colleagues before, the latest arrivals sat down next to each other, this time on the side of the table opposite to the pair who were already seated. John assumed that the older of the two men who had just arrived was Phil Eastman and the younger Sam Kumar. They also opened their laptops and started looking intently at them, typing quietly on their keyboards. John noticed from the monitor screen that the Oxton participants were signing into the meeting with their laptop audio muted.

The image displayed on the monitor changed again and now included the image of a man with greying hair sitting at what John assumed was his desk alongside the conference room video camera image and laptop camera images of the Oxton participants. John greeted the man "Guten Morgen, Wolfgang" to which Wolfgang replied, "Good Morning". Just then the screen split again, and the image of a younger man appeared and said, "Morning All" to which John and others responded, "Morning, Andy". John was glad he'd spoken to Wolfgang and Andy the previous Friday and explained the logistics of how they might attend the meeting from their locations. Wolfgang was joining from Dusseldorf and Andy from Brussels, and John hoped that Yumiko from Tokyo would soon join them. Just then, John's mobile phone rang and looking at the number John assumed that it was Yumiko. In answering her inquiry, John directed Yumiko to the link in the meeting notice that she had accepted and shortly after her smiling face joined the other images on the monitor screen. John inwardly breathed a sigh of relief that the communications equipment was up and running. John apologised to Yumiko for the lateness of the hour, to which she responded that he should not worry as she was well used to meetings at this kind of time, adding that she was working in her "Home Office".

John stood up and walked to a position in the room at which he could see both the Oxton group and the wall monitor. John checked that he was clearly visible on the video camera image and then took command of the meeting by welcoming everyone and introduced himself and explained how he came to be at Oxton, including a brief synopsis of his career to date. He then went on to say that since this was the first meeting of the Oe375 PD team, he was going to spend some time today considering team matters. He said that everyone would shortly get a chance to introduce themselves but first he wanted to discuss something. He asked if anyone played sport regularly, to which Andy replied that he played football for a local side. John asked him what the team did before any game, to which Andy responded that they changed into their kit causing some of those present to laugh. "Good," thought John, "people are beginning to relax." John then asked Andy what they did when they got on the pitch and Andy said that they warmed up. John asked Andy to explain what their Warm Up consisted of and Andy said running up and down and kicking and passing a ball in pairs or threes. John then asked Andy if there was any mental side to the Warm Up such as a discussion of tactics. Andy said that at the level at which he played the most important mental task was counting to see if they had eleven people on the pitch and one of them looked like a goalkeeper, which again prompted laughter. As the laughter subsided, John continued by saying that most people would associate Warm Up with the physical side of stretching and warming up muscles. John added that the cognitive aspect of Warm Up was also important particularly for professional sportspeople who would spend a considerable time thinking and talking about how they were going to enact their next sporting encounter.

While recognising that some people might be sceptical, John explained that his experience was that Warm Up was equally important in business teams like the Oe375 PD team and such teams perform significantly better having warmed up than not. John said that team members might be relieved to know that he was not going to ask them to get up and run round the room several times, adding that their

Warm Up would be predominantly cognitive. John then walked over to and sat at the table with his laptop and shared a slide (Figure 2.1) with the meeting participants and the monitor screen through the meeting software. John was pleased to see the small laptop camera images of all the meeting Oxton participants along with the conference room video camera image above the shared slide.

Phase	Purpose, to allow team members to ...
Place	Communicate effectively with all co-located and remote team members
Self	Transfer attention to this meeting from what they were doing previously
Others	Establish contact with another
Team	Reaffirm membership with this team
Task	Confirm the meeting purpose and agenda

Figure 2.1 Five phases of Warm Up.

Having checked that team members in the remote location could see the slide, John said that the Warm Up that they would be doing comprised of the five stages or phases described on the slide. John told team members that he would give them a minute or two to digest what was written on the slide. After a pause, John said that they were going to go through all five phases today and added that they were currently in Phase 1 "*Place*" which he said they had not quite finished yet. John said that, just as in any team sport, a business team could not perform to the best of its ability unless all individual members performed to the best of their abilities. John added that he had often been in team meetings where some people hardly contributed, and others did not contribute at all. John said that in such meetings a minority of people, quite often the more senior members, do all the talking.

John explained that if people were to contribute effectively within a team, they needed to communicate effectively with all other members of the team, and this was best done when everyone could see everyone else face to face. John said that this was of course difficult when team members were in different locations and added that people in Oxton would have noticed the use of the large monitor screen and everyone could see the image of the conference room video camera to help facilitate virtual face-to-face communication. John pointed out that currently the image of the Oxton group showed two participants with their backs to the conference room video camera. John asked the Oxton team members if they could see the faces of the other Oxton members without moving their head. There was agreement that each team member had to turn their head sideways to see the person that they were sitting next to. The person that John assumed was Phil said that he could see everyone's face on the meeting software display both on his laptop screen and duplicated on the large monitor. While John agreed with Phil that what he was saying was correct, he added that in his opinion the gallery of disembodied heads and torsos in a typical meeting software display of itself did little to engender the spirit of a cohesive team of people. John added that in his experience in using meeting software, where some people were in the same location, there was a tendency for everyone to talk to each other through their laptop screens even if they were sitting next to, or opposite, the person they were talking to.

John explained that while virtual communication was very efficient and effective, he had found that it was also beneficial to communication if everyone in the same location could see everyone else face to face. Making it clear he was just speaking to those in the room with him, John said that they were going to do something about this situation and shared a slide (Figure 2.2).

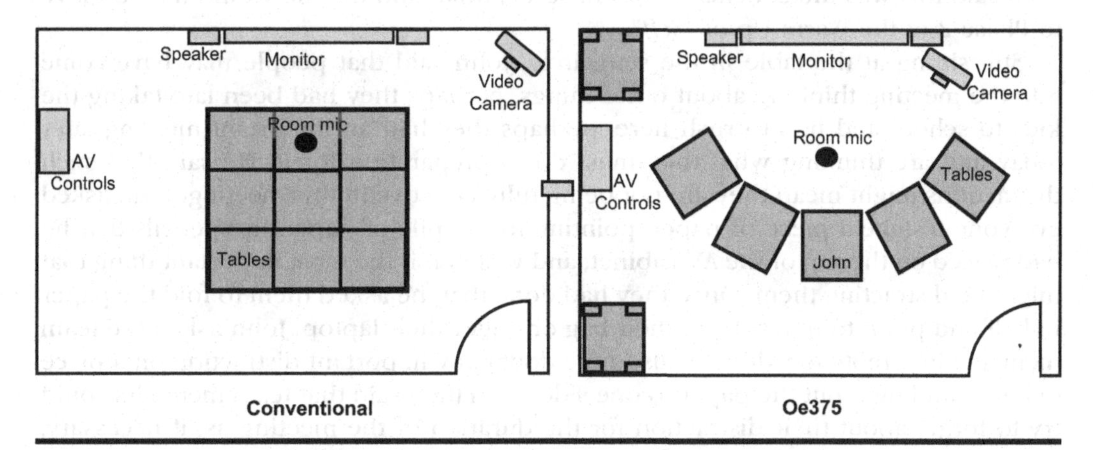

Figure 2.2 Oxton Oe375 PD team meeting room furniture configuration.

Mainly for non-Oxton team members' benefit, John explained the furniture and AV equipment arrangement shown in the sketch on the left-hand side of the slide represented the current arrangement in the Oxton meeting room that the Oxton team members were occupying which had a large central square table made up of nine free-standing smaller tables. John said that he would like to rearrange the tables in the Oxton room to form a semicircle as illustrated in the right-hand side sketch and added that he would like to sit at the base of the semicircle. Sarah said that John seemed to have lost two tables in the Oe375 configuration, and John responded that his diagram was meant to show two tables in each of the two left-hand corners with one table stacked upside down on the other. John pointed out that there were several mains sockets under covers on the floor that could be used to plug in laptop chargers. John noticed what he took to a look of either disbelief or exasperation on a couple of Oxton participants' faces with what he was requesting. After some noisy banging of furniture, some of which John thought was deliberate, the semicircle of team members sitting at tables was formed. John then sat down with his laptop at the table at the base of the semicircle.

After thanking Oxton team members for their help, John told the team that in addition to all the Oxton members being able to see each other face to face physically another purpose of what they had just been doing was to help give the remote participants an idea of the context of the meeting by seeing the faces of the Oxton participants in a single group. John explained that his experience of participating in virtual teams, where a number of team members were co-located, was that the co-located part of the team often became what he called the centre of gravity of the team with the remote members feeling like more isolated satellite members. John added that when this happened the remote participants tended not to participate fully in the meeting to the extent that it was not unusual for such members to blank their video camera and mute their audio and get on with other work while a meeting was in progress. John said that this sometimes led to a somewhat embarrassing pause when a remote member was asked a question only for that person then to ask for the question to be repeated. John noticed smiles on a couple of the Oxton team members and he took this to mean that they could associate with such a situation. John said this was the conclusion of Phase 1 "*Place*" and that he would like to move to Phase 2 of the Warm Up of "*Self*".

Still sitting at the table in the semicircle, John said that people may have come into the meeting thinking about other things, perhaps they had been late taking the kids to school and had to rush here, perhaps they had an important meeting later today and are thinking what they must do in preparation for it. He said that such distractions might mean they are not being fully effective in this meeting. John asked everyone to take a piece of paper, pointing to the pile of paper and pencils that he had placed on the top of the AV cabinet, and write on it the most important thing that might be distracting them. Once they had done that, he asked them to fold the paper in half and put it to one side, in their bag or under their laptop. John asked the team members in remote locations to also note down any important distraction on a piece of paper and then put the paper to one side. John then said that team members could try to forget about their distraction for the duration of the meeting as, if necessary, their note would remind them of it at the end of the meeting.

In glancing at his watch, John noticed that just over 15 minutes had elapsed and so moved on to Phase 3 of the Warm Up of "*Others*". John said that while he was new to the company, and meeting most of them properly for the first time, some team members had probably known each other for some time, years in some cases. John added that while some people had met before there were other team members, particularly the external members, that had not met before today. John then asked team members rhetorically how well they really knew their work colleagues and continued by saying that they were going to do an exercise called *Pairs Fantasy* to find out. John asked the team members in the room with him to move their chairs and sit opposite another team member with the two pairs spaced out, and while the Oxton team members were arranging their seating, John spoke to those in the remote locations and said he would pair them off in a minute after he had explained what he wanted everyone to do. In speaking mainly to those in the room with him, he told them that they should allocate each other as A and B with A then imagining what B is like. John shared the meeting software whiteboard on which he explained he had listed some of the things that A might imagine about B (Figure 2.3).

Things to imagine:
- Car they drive
- Favourite food
- Favourite TV/ Radio programme
- Pets
- House they live in
- Hobbies/spare time interests

At the end of the exercise find out the name of your partner, their current job function and how long they have been with their company. Identify one thing that surprised you about what they said about themselves.

Figure 2.3 Whiteboard text – Pairs Fantasy.

John continued by saying that B should remain silent as A speaks, preferably with a poker face so as to neither affirm nor deny anything A says through body language. John explained that A had 2 minutes to speak about B, then A and B should swap over, and it was B's turn to imagine what A is like for 2 minutes. John told them that he would keep the time. John then told team members that when they had finished the imagining part of the exercise, they should check the accuracy of what each had imagined about the other. John explained that having finished their discussion, he would like them to introduce their partner to the rest of the team by telling the team their partner's name, their current job function, how long they had been with their current company and one thing that surprised them about their partner. He said that team members might like to write this information down at the end of the exercise. John then requested that the Oxton team members wait for a short time before starting the exercise while he set up the logistics for the remote team members.

In looking at his laptop video camera, John told Andy, Yumiko and Wolfgang that he would like them to work in pairs along with himself by using the meeting software *Breakout Room* facility that he would manage. John said that he would allocate Andy and Yumiko to one breakout room and Wolfgang and himself to another. John said that they would find the whiteboard instructions were still accessible from the breakout room. John added that he would notify Andy and Yumiko of the timings during the exercise and asked the three of them if they were happy with what they had to do. After they affirmed that they were all happy to continue, John created the breakout rooms in the meeting software and then blanked the large monitor and muted the room microphone and speakers. John put on his headset and told all participants to start the exercise. John quickly agreed with Wolfgang that Wolfgang would be person B and John started telling Wolfgang what he imagined about him.

While interacting with Wolfgang, John managed the occasional glance at the two pairs in the Oxton room who seemed engaged. Keeping an eye on the time, he called out for the pairs to switch after a couple of minutes and sent a notification to Andy and Yumiko to do the same. Wolfgang then took this as a cue to start telling John what he imagined about him. After a further 2 minutes, when Wolfgang appeared to be struggling to say anything else about him, John asked the pairs in Oxton to stop what they were doing and advised them to spend a while feeding back to each other on the accuracy of what had been said about them. John reminded team members also to make sure they had the information they would need to introduce each other and then sent a notification to Andy and Yumiko to the same effect. After Wolfgang and he had feedback to each other and exchanged and noted the information they required to introduce the other, John joined Andy and Yumiko's breakout room and asked if they were happy to conclude the "*Others*" phase of the Warm Up which they said that they were. John then closed the breakout rooms and brought everyone back to the meeting asking the Oxton participants to take their chairs back to the semicircle of tables.

John unblanked and unmuted the room monitor, microphone and speakers. John then asked all team members as a part of the "*Team*" phase of the Warm Up to introduce their fellow pair member and to say the one thing that they discovered that surprised them about their partner. John noted that, as this was the first meeting of the team, this would help members to start to affirm their membership of this team. The verbatim introductions are shown in Table 2.1 alongside the person to whom they refer.

Table 2.1 Team Member Introduction Verbatims

Team Member	*Verbatims*
Sarah	*This is Sarah Jones. She has been with Oxton for 12 years and is currently the Technical Leader for eBikes. The thing that surprised me about Sarah is that she's a keen triathlete.*
Jennifer	*Jennifer Coulter. Jennifer joined the company as a graduate recruit last year. She works for Paul Clampin as a Technical Specialist in Drivetrain. I knew she played hockey but was surprised to learn that she plays regularly for the County as a midfielder.*
Phil	*Let me introduce Phil Eastman. Phil joined Oxton when it started and has 40 years of experience in the bicycle industry. He is a systems engineer. I was surprised to learn that he was a keen gardener and has two allotments and a large polytunnel.*
Sam	*This is Samama Hasan, known as Sam. He has been with Oxton for 2 years after graduating and is a computer modelling specialist. He breeds budgerigars which surprised me.*
Andy	*Andy Khu. He has worked for HJK Bikes for 10 years and is currently based in their manufacturing facility in Brussels working as a manufacturing engineer. The thing that surprised me about Andy is that he has a Taiwanese father and British mother and speaks 4 languages fluently.*
Yumiko	*Yumiko Izumi. She is a component engineer, she left Uni 4 years ago and started working for Agano. One thing that surprised me is that she has a PhD in Aerospace Engineering.*
Wolfgang	*Wolfgang Hurst. He is the Engineering Manager at Teutoburg who sell our bikes throughout Europe. He has been with Teutoburg for 3 years prior to which he was the Quality Manager for a lawn mower manufacturer. One thing that surprised me is that he is a keen Manchester United supporter.*
John	*John Perry. John has recently joined Oxton as the Engineering Team Manager after spending 20 years at Blade Motors. One thing that I learnt about John is that he loves going on long walks with his partner and their two dogs.*

Looking at his watch, John saw that an hour had elapsed which was as he had expected. John thanked everyone and then said that it was important within a team that people's roles were clarified, and with this in mind, he said that he would like to start allocating roles. He said that by default he was the Team Facilitator and, as they will see in the next meeting, this would include both the facilitation of the team process and the use of the overall technical design framework they would be using. He then confirmed with Sarah, again by default because of her role in the company, that she would be the technical leader in the team when it came to all things eBike. John said that they also required someone to act as a scribe and keep notes during their meetings to record the key decisions that the team made. John said that while he was not expecting formal meeting minutes, his experience was that it was important that all significant decisions were recorded and added that without this happening a team would get into what he described as the "Oh yes we did", "Oh no we didn't" pantomime routine, which prompted smiles and nods of affirmation, although John thought he saw what he took to be a puzzled look on Yumiko's face. Jennifer volunteered her services and John thanked her. John continued by saying that once they got into using the technical design framework, they would generate a considerable number of engineering documents which will need to be managed, and added that he would leave that for another day. John then asked Yumiko if she was familiar with the British tradition of pantomime, and she said that she was not. John explained that a pantomime was theatrical family entertainment based on a fairy tale which took place around the Christmas holiday. John said the performance relied on audience participation led primarily by the children with several well-known routines, one of which was that the villain will deny doing something and the audience will reply that he or she did which is repeated loudly several times. Yumiko thanked John for his explanation and said that it sounded fun.

John then told the team that they had nearly finished Warm Up and continued by explaining that the last of the five phases was "*Task*" describing this as where a team agrees what it is going to do in the meeting. Phil interjected, enunciating his words, "About time to". John asked Phil if he would like to say more. Phil explained that he came to this morning's meeting to do engineering design work and not all this team stuff. He continued by saying that if he wanted some soft skills training, he would have gone on a soft skills course. John asked if anyone else wanted to say anything and Sarah said that she had found the last hour interesting and was happy to go along with things for now based on what John had said about the importance of teamwork. Leaving space for others to speak, John said nothing but no one else spoke, with Jennifer and Sam seemingly avoiding his gaze. John then thanked Phil and Sarah for their feedback and explained that while he intended to do a Warm Up at the start of every meeting, they would be much shorter and more focused in future, adding that he had deliberately left any mention of task until now for reasons he would explain in their next meeting. John asked if it was all right to conclude the Warm Up by looking at "*Task*" and there were nods of agreement. John explained that the purpose of the "*Task*" phase of the Warm Up was to agree on both the purpose of the meeting and the agenda and added that this was the "What" and the high-level "How" of the meeting. John shared a slide showing the purpose and agenda (Figure 2.4) for the meeting that was currently taking place.

Purpose: To introduce Oe375 team members to each other, consider the programme timing and begin to establish system-level requirements.

Agenda	Time/mins	Lead
Warm Up	75	JP
Break	5	
Oe375 and Programme Timing	15	SJ
Oe375 Competitive Features	15	SJ
Warm Down	10	JP

Figure 2.4 Introductory meeting purpose and agenda.

After allowing what he felt was sufficient time for everyone to read the slide, John told the team members that he had developed the agenda for the meeting but in future he would like to agree a draft agenda with one or two of them before each meeting which would then be confirmed as a team in the meeting. John told Sarah that he would like to agree a draft agenda for the next meeting with her and asked her if that was all right. Sarah said that it was fine, and John told her he would send her a meeting notice. John asked the team members to note that the meeting would conclude with Warm Down, as every good sports team and also business team should. He told them that they had now finished Warm Up at around the projected time. John explained that he was a great believer in taking breaks even in a relatively short meeting. He explained that some people would say that they had so much work to do in meetings that they did not have time for a break. John continued by saying that his response to this was that team members could not afford not to have a break since with a break team members would be more efficient overall in the meeting. In requesting that team members took a 5-minute break, John asked them to return promptly.

During the break, John reflected on how things had gone so far. He was generally pleased but reminded himself that teams with members in remote locations require more of his attention than co-located teams as had been the case in setting up the Pairs Fantasy exercise. John noticed that they were running a few minutes late by the agenda timings, but he tended to look at agenda timings as a guide, although he recognised that he had always been keen to complete an agenda within the allotted meeting time. John reflected on Phil's interjection which to him was not unexpected since he knew that people often felt uncomfortable when being asked to do things differently. He was pleased that Phil had felt able to say what he did. Looking up from his thoughts, he noticed everyone had returned both in Oxton and in the remote locations.

In referring to the agenda, John said that he was handing over to Sarah to take the team through the high-level timings of the Oe375 programme that she and he had discussed. Sarah asked John if he would allow her to share her screen so that she could show the programme timing slide (Figure 2.5) which at this time did not have the product givens displayed.

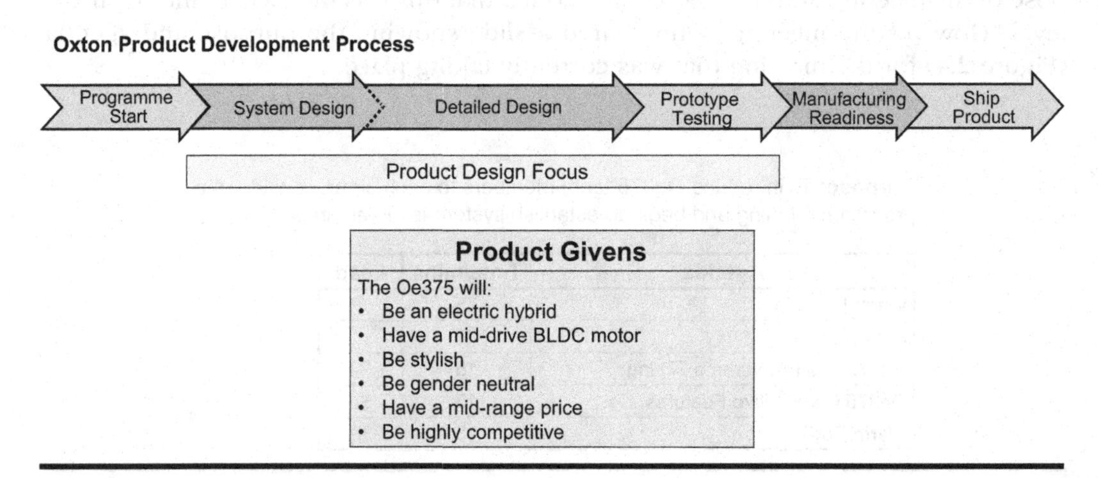

Figure 2.5 High-level Oe375 programme timing.

In referring to the slide, Sarah explained that the programme would follow the Oxton Product Development Process (OPDP). Sarah explained that the Oe375 team would focus their attention on Product Design, and since the eBike would be of a conventional design concept, this would mean the Oe375 team going through the System Design, Detailed Design and Prototype Testing phases of the OPDP. Sarah explained that the relative lengths of the phase arrows depicting the OPDP indicated the relative timing of the phases. Phil asked what the overall programme timing was, and Sarah said that this was one of the things still to be decided. John intervened and said that Jim Eccleston had got the agreement of the Board that the programme could extend significantly beyond the current timing of 12 months for budget bike programmes. Phil said that he was relieved firstly because Oe375 was not a budget eBike and secondly because the 12-month timing was proving ridiculously short even for a budget bike. Jennifer asked why the System Design and Detailed Design had the same shading with a dotted line arrow head separating them, and John said that was because of the approach that he intended the team to take, adding that he recommended that they focus on one system initially and take this through to the detailed design. John said that he would talk in more detail about this in the team's next meeting.

Sarah asked if there were anymore comments or questions, and Phil said that the relative timings were back to front. John asked Phil if he would say more about what he meant. Phil explained that in what he described as a normal Oxton programme, the time that was spent on testing would be greater than that spent on detailed design which was the reverse of the relative timings for these phases shown on the slide. Phil continued by saying that the real engineering took place during testing and launch preparation based on the issues that are identified. John thanked Phil for his explanation and said that unfortunately they could not go into this now and added that they would look at Phil's concern in the next meeting when they considered the design process that they would be using. Phil responded, in what seemed to John to be rather sarcastic terms, by saying that he could not wait and that he looked forward-with interest to what John would have to say. Sarah continued presenting the programme timing by saying that she was also intrigued by what appeared to be back-to-front timings, although she had been somewhat reassured by a discussion that she had with John in preparing the meeting. Sarah then displayed the product givens overlay and summarised those that the Board had identified in, order that the Oe375 was consistent with both the Oxton product cycle plan and the other products that the company offered. Sam asked Sarah what a BLDC motor was, to which she replied that it was a brushless direct current motor and explained that they were popular, reliable and cost effective. Sarah then handed the meeting facilitation back to John.

Having thanked Sarah, John explained that while this meeting had deliberately focused on team skills, he wanted the team to make a start on programme work. Turning to the programme timing slide, he said that he'd like the team to start looking at the last item on the Board's product givens list of ensuring that the Oe375 was highly competitive. John continued by saying that this brought them on to the next agenda item of "Oe375 Competitive Features" and explained that the Kano Model of Quality was a good place to start with this. John said that he was sure that some of them had come across the model before, asking Wolfgang given his experience as a Quality Manager to confirm his assumption, which Wolfgang did. Sharing the Kano Model slide (Figure 2.6), John said that although the model was simple, he believed

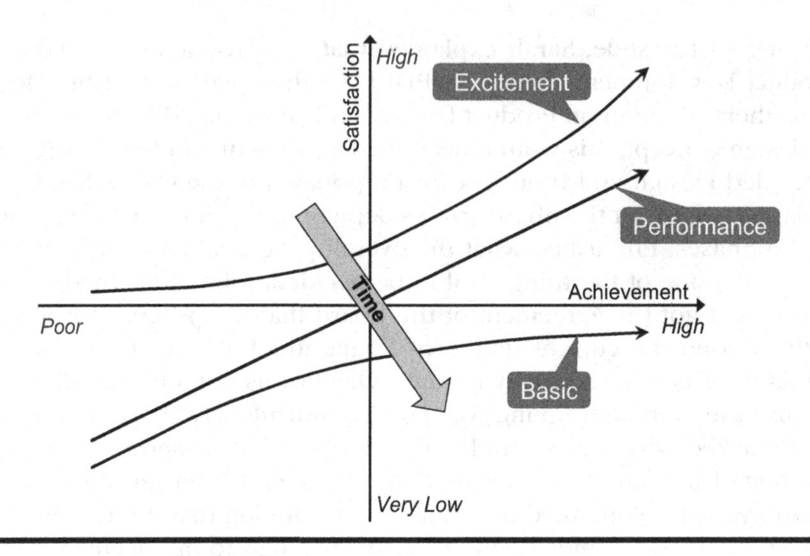

Figure 2.6 Kano model of quality.

it to be powerful and added that despite it being developed in the early 1980s he saw it as still being very relevant today.

John explained that the model distinguished three types of quality from the customers' perspective of *basic*, *performance* and *excitement*. John said that basic quality related to those features of a product that customers take for granted and will not ask about before purchase and, by way of example, he explained that a customer purchasing a bicycle would be unlikely to ask if the chain engaged smoothly with the cassette during gear change. John then said that if the chain were to slip on the cassette teeth during a gear change on the first occasion that a customer rides a bicycle that they have just bought, basic quality is not met, and the customer is likely to be very dissatisfied with their purchase. John continued by saying that customers would talk about performance quality before purchase and, by way of example of this type of quality, explained that the satisfaction of those customers looking for comfort in a saddle would be likely to increase as their perception of the comfort of a saddle increased. John explained the third category of excitement quality, related to features of a product that customers did not expect but were pleased with, and gave the example of automatic warning of low tyre pressure. John explained that since excitement quality features were not expected, like basic quality features, customers would not speak about these prior to purchase.

Jennifer asked how they would know which category a particular feature was in. The muffled sound of a mobile phone ringing broke John's concentration and he paused in answering Jennifer's question. John noticed Sam pick up his bag and take out his phone. After Sam had looked at the screen, he touched it with a finger and the phone stopped ringing. Sam apologised to team members and explained that he had forgotten to mute his phone before the meeting started. John asked Sam if it was the norm in Oxton to mute phones in meetings, and Sam replied that muting a phone was normal practice in most meetings. In asking Sam a follow-up question, John ascertained that the Oe375 PD team meeting was a meeting where phones should be muted. John decided not to pursue the matter further thinking that the incident was sufficient of itself to remind team members to mute their phones in future meetings and returned to Jennifer's question.

John said that while there were formal methods for categorising features, as far as Oe375 was concerned, he would leave it to the judgement of the team. John added that generally a new feature may well be an excitement feature if customers did not expect or ask for it but that over time a feature will move from excitement, through performance to basic indicated by the time arrow on the slide. John gave the example of disc brakes which he said were an excitement feature when they were first introduced, although they quickly moved to the performance category as initially they performed less well than calliper brakes. He continued by saying that it was not until disc brakes began to outperform calliper brakes thats they gained widespread acceptance and today many mountain-bike riders would treat them as basic quality. John concluded by saying that in addition to migrating over time, excitement and performance features will also migrate from premium to budget products and gave the example of carbon fibre wheels which he said were now a basic feature for pro-riders' bikes having migrated from such bikes to high-end amateurs' bikes as an excitement feature.

In summarising the relevance of the Kano model to the Oe375 programme, John said that if the Oe375 was to do well in a very competitive market, it was important that it did not have failures of basic quality features in the field. He continued by saying that the Oe375 will also need to match or outperform competitor bikes in terms of performance features, and it should have at least a sprinkling of excitement features. He then asked the team if before the next meeting they would compile a list of potential excitement features for Oe375. He said that in view of the Kano model, they would do well to look at existing competitor premium bikes in identifying potential excitement features for Oe375. John suggested that they should not exclude mid- and budget-range bikes and look across road and mountain as well as hybrid bikes in their search, citing the budget hybrid bike with cruise control that he had come across recently. John also advised team members to look outside of the bicycle industry, particularly to automotive, quoting his example of automatic low tyre pressure measurement. John asked Jennifer if she would be happy to collate a list of excitement feature suggestions that team members might identify and present it at the next meeting. Jennifer said that she would be very happy to do this. With this, John switched back to the agenda slide and said that it was time to look at the last agenda item of Warm Down.

John explained that just as he was going to start each meeting with a Warm Up, he would conclude each Oe375 PD team meeting with a Warm Down. John said that he believed that Warm Down in business meetings was just as important as in sport in bringing the meeting to closure and added that he would show what he meant by the Warm Down process in bringing the current meeting to closure. John asked team members if anyone remembered the five Warm Up phases, and not unexpectedly, no one could remember them all, so he displayed a slide listing the five phases in the context of Warm Down noting that the phases in Warm Down were the reverse of those in Warm Up (Figure 2.7). John then paused giving team members time to read the slide.

Phase	Purpose, to allow team members to ...
Task	Assess if the meeting met its purpose and all agenda items were covered
Team	Recognise the identity of the team including those that have had to leave early
Others	Air and resolve any personal disagreements between team members
Self	Allow individuals to reflect on how far their personal objectives have been met
Place	Return any room furniture and electronic equipment to its former state

Figure 2.7 Five phases of Warm Down.

Reminding people that the stated purpose of the meeting was to "To introduce Oe375 team members to each other, consider the programme timing and begin to establish system-level requirements", he asked if everyone felt that the meeting had achieved its purpose. Several people said yes, and Jennifer asked if John could send everyone a copy of the slides. John said that was a good point and said that he would, adding that in the next meeting they needed to discuss the management and sharing of all team documents. Moving to "*Team*", John said that because of the nature of this, their first meeting, he had done most of the talking and he was looking forward to future meetings when meeting input would be more of a team effort. He asked if people had learnt something about one another and there were nods and murmuring of affirmation. He said that clearly no one had left early today but inevitably that would happen and, when this happened, he felt it helped if this was acknowledged by the team rather than people leaving without saying anything. John continued that in terms of *Others*, personal disagreements happen in all teams and in his experience, it was far better to address such occurrences openly within the team. He told team members that he would discuss a way in which this might be done objectively to take the heat out of any disagreement in a future meeting.

In moving on to "*Self*", John said that he felt it was important that at the end of a meeting people had the chance to reflect on, and express their feelings about, the meeting. He said this was particularly important when people's expectations had not been met in a meeting and this allowed the issue to be addressed either at or before the next meeting. John stressed that, because Warm Down was a closure event, it was important that any airing of people's feelings at the end of a meeting should not be a catalyst for opening up discussion within the team. He explained that the process that they would use to achieve this was what he called "*Pass the Pen*", picking up a felt tip marker pen. He described how the pen is passed from team member to team member with only the person holding the pen able to speak, adding that they would have to use what he called "a digital pen" for the members in the remote locations. John then asked team members to make a short statement on their reflections on the meeting that was just ending and having made their statement to say how they felt emotionally, adding that emotions are described by one word, giving the examples, *happy*, *sad*, *pleased*, *disappointed*. John threw the pen to Sam and asked him to start and when finished pass the pen to his left and added that once all the Oxton team members had spoken he would say where the digital pen was. The verbatim statements along with individual's stated emotions are shown in Table 2.2.

Table 2.2 Warm Down Verbatim Comments

Team Member	*Verbatims*	*Emotion*
Sam	***I didn't really know what to expect of the meeting. Generally, I found it interesting and thought the Pairs Fantasy exercise was revealing.***	*Happy*
Jennifer	***Yes, I thought it was good. It was very different to our other meetings in Oxton, and I found it a bit strange we did not do much technical work. It was nice to meet the external people.***	*Good*
Phil	***As I said before I thought the programme timing was back to front, in fact picking up on Jennifer's point, I felt the whole meeting was back to front with far too much team stuff.***	*Frustrated*
Sarah	***I enjoyed the meeting and found the teamwork a refreshing change, although I'm looking forward to getting my teeth into some technical work.***	*Content*
Yumiko	***I found the meeting interesting, and it was nice to meet everyone. I am looking forward to doing the excitement features with the team.***	*Happy*
Wolfgang	***In my time working with Teutoburg, I have never been asked to participate in a manufacturer's design team and I found the meeting interesting and am looking forward to continuing working with everyone.***	*Pleased*
Andy	***Like Wolfgang, I haven't been in a team involving a customer company. I found the meeting OK although being in a remote location doesn't help. I'm hoping we get on with things next time.***	*Impatient*
John	***I was pleased to meet everyone and thank you all for your inputs in the meeting. I am a bit uncomfortable that most of the meeting was me talking but I know this will change. I am also looking forward to getting on with things.***	*Excited*

After the pass-the-pen exercise, John thanked team members for their input to the meetings and reminded them both of the timing of their next meeting and the piece of paper on which they had written their distraction during the Warm Up. John asked the Oxton team members if they would bring their headsets to future meetings to facilitate, any virtual small group work that they might do. John then asked those at Oxton if they would help to put the tables and chairs back to where they were at the start of the meeting and added that he would shut down the AV equipment. Having done this, John collected his laptop that was still on the table he had sat at, which by this time had been moved, along with the rest of the furniture, back to its former position.

When John got back to his office, he sat down with a cup of coffee and reflected on his first meeting with the Oe375 team. He was generally pleased with how the meeting had gone and felt that it had achieved its dual purposes of initial team building and signalling a different way of approaching meetings. In reflecting on team members feedback in Warm Down, John was pleased by the more positive comments and thought to himself that the negative comments from a few people were useful in the sense that, while he did not deliberately set out to make people feel uncomfortable, some level of discomfort indicated that the people had detected a different way of doing things. John realised, however, that he would need to get into some extended technical discussion in the next meeting if he were to carry the team with him.

Later that afternoon, John was in his office preparing for his next meeting with the Oe375 team when Paul Clampin burst in. John looked up somewhat surprised since, although his door was open, he expected at least a cursory knock. Paul, almost shouting, told John that he had been talking to Sarah about the Oe375 team meeting and added that he wanted to know what he called "all this bloody team nonsense" was about. Paul continued by saying that not only was a very long and totally inappropriate programme timing presented but apparently John spent all of a 2-hour meeting this morning talking about team stuff. John, realising that Paul had contradicted himself in a single sentence, opened his mouth to reply but Paul, looking red in the face, continued by saying that the presence of suppliers in the meeting had made the matter worse. Paul carried on his critique of the meeting by saying that Oxton suppliers bloody well did what they were asked, and, apart from the usual haggling over price, they should have very little say about the way Oxton went about their business. John managed to intervene as Paul paused for breath telling Paul that he had sworn and that made him feel uncomfortable. Paul responded by swearing again at John and telling him that Sarah was one of his best engineers and he was not having her wasting her time on this soft skills nonsense. Paul also told John that next time he wanted a meeting with Sarah outside of the PD team meetings, he should ask him first before asking her. As he stormed out, Paul told John that he was going to see Jim about what had occurred.

After Paul's exit, John sat back in his chair and thought about what had just happened. He had met this kind of resistance before at Blade, although usually expressed in a more professional manner, and had put it down to managers being worried that their engineers, in not doing work directly for them, would not be helping them achieve their performance objectives and so could detrimentally affect their annual bonus. He recalled that things had improved somewhat at Blade when allowing, if not encouraging, your people to work across the company became a positive factor in management performance appraisals. He decided to let Paul's outburst blow over assuming that Jim would back him and confident of the positive effect good teamwork would have on the Oe375 PD team performance. He also made a mental note to include the words "PD team" in the title of any meeting notice that he sent to Sarah.

Background Reading and Viewing

Warm Up	Syer, J. and Connolly, C. (1996) Warming Up. In *How Teamwork Works: The Dynamics of Effective Teamwork*, McGraw-Hill Pages 165–179
Warm Down	Syer, J. and Connolly, C. (1996) Warming Down. In *How Teamwork Works: The Dynamics of Effective Teamwork*, McGraw-Hill Pages 283–295
Team Roles	Mackin, D. (2007) *The Team-Building Toolkit*, American Management Association Pages 21–26
Kano Model	Colman, L. (2015) *The Customer-Driven Organization*, CRC Press Pages 1–9 Dave Verduyn. (2017) *The Kano Model*, You Tube, https://www.youtube.com/watch?v=bS2HMX0Ki54, Accessed December 18, 2022

Chapter 3

Framing Actions

John brought Jim up to speed with what had happened in the first Oe375 Product Design (PD) team meeting in his first weekly one-to-one meeting with him. After explaining to Jim the extended nature of the Warm Up in the inaugural meeting, John described the reaction of team members to both the Warm Up and Warm Down, describing it as generally positive with one exception without mentioning the individual's name. Jim seemed pleased and asked John how the external members had reacted, and John said that he had appreciated their input and support. Towards the end of their meeting, Jim said he had had a short discussion with Paul who had expressed some disquiet over what he called the almost total soft skills content of the first PD team meeting. Jim continued by saying that had reassured Paul by telling him that his days as a professional cyclist had taught him about the importance of directing attention to the team and asked him to reserve judgement on the way John were going about things. Jim said that he had asked Paul to continue to support John by making his people available to the Oe375 PD team. John thought that this was useful, although not wholehearted, support but acknowledged the difficulty Jim had in handling his management team. John also recognised that what he was doing was new to Jim and supporting him was an act of faith on Jim's behalf.

John had set up a series of regular meetings with Sarah with the intent of reviewing the last Oe375 team meeting that had taken place and preparing for the next meeting. In meeting with Sarah, in what became known as the Review/Preview meeting, John began by asking Sarah for her reflections on the first Oe375 meeting. Sarah said her reaction to the meeting was still much as she had said in her Warm Down feedback in that she enjoyed the meeting and found the teamwork a refreshing change and was looking forward to the technical work. As if reading John's thoughts, Sarah added that John should not worry too much about Phil's reaction, explaining that he was a typical Yorkshireman who always spoke as he thought which she said could be a bit off-putting until you get to know him.

DOI: 10.4324/9781003286066-3

In preparing for the second PD team meeting, John said he wanted to explain the concept of the teamwork skills he wished the team to experience and to look at a team-based process to make their meeting more effective. He told her that he also wanted to introduce the Oxton first-time engineering norm (OFTEN) framework. John continued by saying that he was conscious of the feedback from some team members in the first Oe375 meeting concerning the need to get into technical work and while it made sense to look at effective meetings first, since it was relevant to the meeting itself, he wondered if he should look at the OFTEN framework first. He added that, of course, he intended to conduct a Warm Up and explained that he was expecting this not to take more than 10 minutes. John was pleased that Sarah advised him to do what he felt was best since he did not wish to devalue the team skills in any way. Sarah reminded him that they also needed to review the Oe375 excitement features, and they agreed to do this after looking at the OFTEN framework as this would put the list into the context of how John intended to approach the programme.

Having arrived early to the Oe375 PD team's second meeting, John was pleased to find everything in order and awaited the arrival, and signing-in, of team members. Everyone arrived or signed in to the meeting on time except for Wolfgang. John was determined to set a team norm of starting their meetings on time, and after welcoming team members to the meeting, he mentioned Wolfgang's absence without any further comment. John said that they would start the meeting with a short Warm Up at which point Phil stood up and started to move the table that he was sitting at back from the square that it formed with the other tables. John thanked Phil and said that as in the last meeting he would like to arrange the tables in a semicircle facing the wall monitor with each team member sitting at a separate table so that they could see everyone else. John asked for volunteers to move the furniture and the other three Oxton members got up and quickly rearranged the tables according to John's wishes in a way which seemed to John to be significantly quicker and less noisy than the first time that this exercise had been completed. As in the previous meeting, John then moved the free-standing video camera so that everyone would be in shot as they sat at their tables. Once everyone was settled, John continued by saying that the Warm Up would focus on team meetings and asked team members to think of a short statement that described one aspect of engineering team meetings that they had attended that made them efficient and another aspect of the meetings that could be improved. After a few minutes, John said that he would combine the "*Others*" and "*Team*" Warm Up phases and asked team members to go around in turn making their statements (Table 3.1).

Table 3.1 Warm Up Verbatim Comments on Team Meetings

Team Member	*Verbatims*	
	Good Aspect	*Aspect to Improve*
Phil	*Getting to know everyone and working with the team*	*Time keeping, we never started meetings on time*
Sarah	*Having clear goals and objectives*	*Meetings were long, some seemed to go on forever*
Sam	*Strong facilitation by the team leader*	*Often not been able to express my opinion on things*
Jennifer	*It was an exciting project with everyone involved*	*The team did not stick to the agenda and so some urgent things weren't covered in a meeting*
Andy	*We had a good leader who kept meetings on track*	*The internet communication was bad, we always started late because of setting things up, and the link dropped on occasions.*
Yumiko	*Most of the work took place outside of the formal meetings*	*I was lost at times in the meetings not knowing what was going on*

John thanked the team for their statements which he said that he found interesting. In the final phase of Warm Up, John referred team members to the meeting purpose statement and agenda he had sent out the day before and asked if anybody had a comment. Nobody said anything, so John asked the remaining Oxton team members who had not already done so to join the meeting using the link in the meeting notice that he had distributed the day before and to mute their laptop microphones and speakers. After allowing time for this to happen, John moved on to the first agenda item entitled Effective Meetings. Just at that moment, Wolfgang's smiling face appeared on the monitor screen and he apologised for being late, saying that his previous meeting had overrun. John told him not to worry and gave him a brief summary of what had happened up to that point in the meeting. John then shared a flowchart on a slide so that it appeared on both the wall monitor screen and team members' laptop screens (Figure 3.1) and checked that all team members could see the graphic.

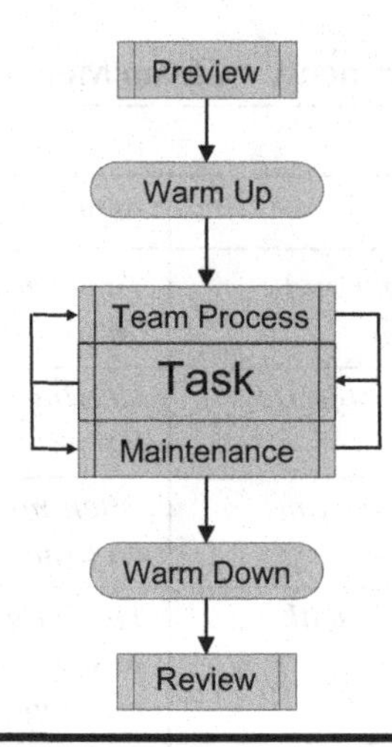

Figure 3.1 Effective Oe375 meetings.

John began by saying that as the flowchart implied the core of any engineering team meeting was the engineering *Task* that was to be completed. John added that the two key interlinked subprocesses of the *Team Process* and *Maintenance* enabled the engineering task to be achieved in an efficient and effective way.

John explained that for the team to perform to the best of its abilities, some attention needed to be paid to how team members interacted with each other as a part of the Team Process. John said that as their series of meetings progressed, he would introduce a number of interrelated team skills with the intent of helping the team to improve the way in which they worked together, adding that he would go through the list of key team skills that he wanted to share with the team progressively over future meetings in a minute. Drawing the team's attention to *Maintenance*, John said that, as engineers, they knew only too well that many engineered systems require regular maintenance and that the system that is the team was no different. John reminded them, by way of example, of what he had said in their last meeting about the benefits to what he would now call the Task and Team Process of taking breaks. John added that team members knew from personal experience that people cannot perform to their best when they are tired and hungry and explained that a maintenance break would typically be an opportunity to "rest your brain", get a drink and snack and see to any other personal needs. Some team members smiled at this phraseology.

John continued by explaining that the core of their meetings would be topped and tailed with *Warm Up* and *Warm Down* adding that team members were now familiar with these two aspects. John asked team members to note that very deliberately Task came after Warm Up in the Effective Meetings model and reminded them that he had said in their last meeting he would discuss this in this meeting. John explained that engineers and engineering managers like nothing more than talking about engineering and so tend to jump straight to task in any meeting. John then gave the example of a meeting that he had attended where an engineer had sat in a meeting for 15 minutes before realising that he was in the wrong meeting, because the meeting had not had even the most cursory Warm Up by discussing the meeting purpose and agenda. This example raised smiles in the team and John was sure that this was based, at least to some extent, on shared experience.

In moving on, John said that he thought that planning or previewing a meeting before it took place was important and to that end had set up a series of planning meetings with Sarah. John added that they would ask other members to join them on an as-and-when-required basis. Referring to the last box on the flowchart, John said in conclusion that it was important to review a meeting after it had finished to see what had been learnt particularly with respect to Task and Team Process and consider what might need to be done in subsequent meetings to make the team more effective.

John then asked team members for their comments on the model. Andy said he thought it seemed pretty straightforward. Phil said he agreed with Andy and went on to say that his experience of team skills was that there was a significant gap between theory and practice. Phil added that he, along with colleagues at Oxton, had attended several soft skills training courses and explained that his experience was that while they were usually enjoyable events they had not made an iota of difference to the way people interacted with each other on their return to the workplace. John said that his experience was similar until he came across the approach to team skills that he wanted to share with the team. In not wishing to oversell the team skills he intended to introduce to the team, John added the proviso that “the proof of the pudding was in the eating”. Realising that some team members might not be familiar with the idiom that he had just quoted, John added that the team members could make their own judgement about the usefulness of the team skills he would discuss with them.

John said that this brought him nicely to the next agenda item of what he called “People Skills”, bringing team members’ attention to the new slide that he had shared (Figure 3.2).

- Warm Up & Warm Down
- Effective Meetings
- Tacit and Explicit Knowledge
- Visualisation
- Listening Skills
- Questioning Skills
- Framing Information
- Feedback in Meetings
- Attitudes
- Attribution

Figure 3.2 Oe375 People Skills.

John told team members that his intent was to bring this list to life as their team meetings progressed in the same way in which they had participated in, and would continue to participate in, Warm Up and Warm Down. John brought team members' attention to the title of the slide by saying that while up to now he had been using the more widely accepted description of Team Skills, from now on he would speak of People Skills because many of the techniques listed were beneficial on a personal basis as well as in interaction with others. John added that he was not keen on the term "soft skills" since to him it seemed to make them less important than the macho sounding "hard skills", although he emphasised that this was his personal view. Given that the slide only contained factual information, John did not invite discussion but moved straight on to the next agenda item which he called the core of the Oe375 product design process.

John explained that it was inevitable that potential failure modes would be created during the design and development of the Oe375 eBike for a number of reasons. He cited the occurrence of mistakes, a lack of understanding of how customers will use the eBike and poor understanding of how supplier components and subsystems will integrate in the overall eBike design as three reasons. John added that what was important was how quickly after a potential failure mode was created it was discovered and a countermeasure put into place so that it did not become an actual failure mode later. Sam asked John what he meant by a failure mode, and John replied that it was the way a system might fail. John added that it was important to find out how a system failed because that would help in establishing how the failure might be avoided.

John then shared a slide (Figure 3.3) and said that to help the external team members he had defined the phases of the Oxton Product Development Process (OPDP) at the bottom right of the slide.

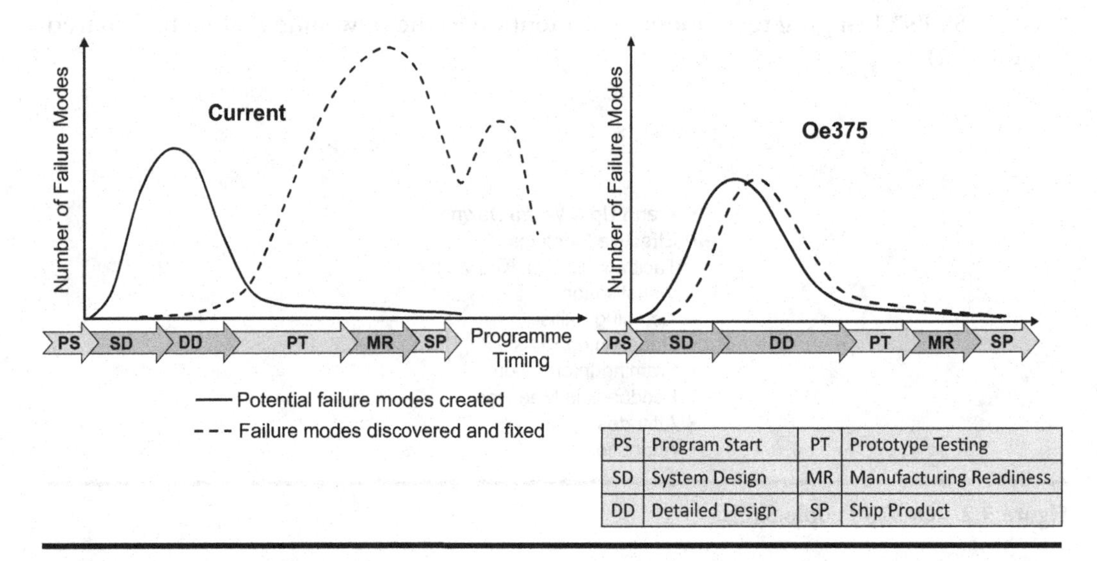

Figure 3.3 Avoiding failure modes.

John said that the curves on the left-hand side of the slide illustrated the current situation in Oxton with potential failure modes largely being created up to the Prototype Testing (PT) phase, as depicted by the solid line curve. John explained that the dotted line curve on the "Current" graphic was based on academic research and showed that product failure modes tended to be identified late in a programme typically around PT and launch. John added that he had had a chat with Mo Lakhani who had confirmed that the general positioning of this curve corresponded to the current approach to product development in Oxton. John said that his vision for Oe375 was represented by the right-hand graphic in which potential failure modes were discovered and fixed soon after they were created.

Sarah asked why the Current dotted line curve had a dip at Ship Product (SP) and was higher than the Oe375 curve. John asked Phil if he could answer Sarah's question based on his experience. Phil said that he assumed that the second hump on the dotted line curve after SP was because of the problems found by customers riding their bikes after purchase and added that he did not know why the Current dotted line curve was higher. John asked Phil what tended to happen when Oxton fixed a problem late in the programme, to which Phil replied that fixing one failure often caused other problems, and judging by his expression John assumed that a light had gone on in Phil's head.

John then reminded team members of what Phil had called the back-to-front programme timings in their last meeting and said that the Current graphic explained why a lot of time might be required later in a programme to achieve the Prototype Testing (PT), Manufacturing Readiness (MR) and Ship Product (SP) phases. John added that moving from the Current position to that represented by the Oe375 curves required significant effort in achieving the initial System Design (SD) and Detailed Design (DD) phases and this accounted for the programme timing that they had discussed in the previous meeting. John asked Phil if that clarified the situation for him, and Phil said that he now understood John's reasoning and added that for now he would reserve judgement on whether the Oe375 timings were appropriate and achievable. John thanked Phil for his response and told him that at this stage he had every right to be sceptical. Phil then said that one thing he would give John was that the last couple of phases in the current process did tend to merge into one mad rush to discover and fix problems.

John added that the aim of Failure Mode Avoidance was to find and fix potential failure modes early in the design process and that each failure mode should be fixed once. John continued by saying that finding failure modes late in a programme was relatively easy and explained that his experience was that such problems come out and bite you. John then explained that finding potential failure modes early in the design process, as team members would find out, normally took significantly more effort. John added that if this was the bad news, the good news was that fixing potential failure modes early in the design process was relatively easy, normally much less expensive with no knock-on effect in causing other failure modes. Wolfgang interjected at this point saying that not all late failures could be easy to find since some escaped to the customer, to which John responded that while this was true, customers usually find such problems quite easily, however.

John explained that a potential failure mode might occur in a product if either a mistake was made during the design of the product or the design itself was not up to the job. John said that he assumed that product design teams in Oxton did not make mistakes, and Sarah said that she wished this was the case. John asked Sarah if she could give an example of such a mistake. Sarah said that a classic case was the galvanic corrosion of an aluminium alloy seatpost mounted to a carbon frame in one of Oxton's high-end bikes. John asked Sarah if she would explain further. Sarah said that as every engineer should know that if two metals are placed in contact through an electrolyte such as moisture, there was a tendency for metal ions to pass from one metal to the other under the potential difference set up by the metals. Sarah added that this tendency was high with electrical contact between aluminium alloy and carbon because together they set up a relatively high electrical potential difference. Sarah said that she was sorry if she was boring other team members with what they already knew, and Sam said that he would appreciate Sarah continuing if the other team members did not mind. As there were murmurings of agreement with what Sam had said, Sarah continued by saying that metals are arranged in what is known as a galvanic series with those metals near together in the series less likely to cause corrosion than those some distance apart since those closer in the series set up a smaller potential difference between them than those further apart. Sarah explained that aluminium alloy and carbon were towards opposite extremes of the series, so any application where they can be connected electrically though an electrotype such as rain water or sweat was prone to corrosion. Sarah summarised her explanation by saying that using two metals that were separated by some distance on the galvanic scale in an engineering application where they were in electrical contact in the presence of moisture was a mistake.

John thanked Sarah for what he described as her enlightening example and said that we all make mistakes from time to time, and in engineering, it was important that they were seen as a learning experience. John added that making a mistake once was acceptable, although repeating the same mistake was not. Sam said that he had a couple of comments and said that Sarah had said that the galvanic series listed metals and carbon was not a metal. Sarah responded by saying that carbon was included in the galvanic series as it was an electrical conductor and so in this sense behaved as a metal. John thanked Sarah and asked Sam about his second comment. Sam pointed out to John that he had said that making mistakes was only one way in which potential failure modes could occur during product design and had also said that potential failure modes could also occur if the design was not up to the job. John thanked Sam and said that he had got rather sidetracked in talking about mistakes. John said that what he meant by a design not being up to the job was that it was not sufficiently robust and added that he was speaking of robustness here in the Taguchi sense. Sam said that he was not sure what John meant by the Taguchi sense of robustness. John asked if anyone in the team could give Sam a brief explanation of the term robustness in the sense that he had been using it.

Wolfgang said that he would like to do this and continued by explaining that the concept of robustness was developed by Dr Genichi Taguchi in the context of what is called Taguchi Methods. He added that Taguchi Methods include Parameter Design which is based on statistically designed experiments in which a design can be optimised at low, no, or reduced cost to operate when subjected to what Taguchi called "Noise" by adjusting design parameters. Wolfgang continued by explaining that Noise was things that an engineer cannot control, or chooses not to control, and gave the example of the environment in which the product operates. Wolfgang concluded his description by saying that a robust product was one which performed as expected despite the presence of Noise. Sam thanked Wolfgang saying that he remembered learning something about this at university. John also thanked Wolfgang for what he described as his clear and concise explanation.

John then said that it was also important that a design was no more complex than was necessary to achieve its requirements since the more complex a design the greater was the risk of potential failure modes occurring. John added that reusing proven components was also a good way of avoiding failure modes, although care had to be taken since a component which is proven to be robust in one application may not be as robust in a different application because, what John termed, the "Noise Space" might be different in the new application. Jennifer asked if the failure of a proven component in a different noise space should be considered as being due to a mistake or a lack of robustness. Andy said that it was clearly a lack of robustness because of the different noise effects. Phil said that he thought it was a mistake to use the proven component without investigating its robustness in the new application. John agreed with Wolfgang and Phil and said that he felt that they were both right. John added that he thought it might be helpful to define a mistake as "*An error that is made by not applying existing knowledge which, if it had been applied, would have prevented the mistake occurring*". John advised that his rather narrow definition was best applied in the context of Failure Mode Avoidance. Phil said that he agreed with John in both respects, firstly his definition was helpful in an engineering context, and secondly, if he accused his wife of not applying existing knowledge the next time she misinterpreted something that he said that would be less than helpful.

After the laughter had subsided, John said that the primary purpose of considering failure modes as being due to mistakes or a lack of robustness was to facilitate the identification of a potential failure mode by focusing thinking on the detail of a situation. John stressed that the categorisation of a particular failure mode as a mistake or a lack of robustness was of secondary importance. John said that what he called "the good news" was that the engineering framework that he was going to explain next facilitated both the identification of mistakes and the achievement of product robustness.

Jennifer then effectively asked the question that Sam had asked earlier in the meeting by asking John why he kept talking about failure modes rather than just failures. John asked her to hold the question for a while as he would answer it shortly. John was pleased by the way in which team members had entered into discussion and wondered how many of them had noticed the change in the Team Process. John said that the answer to Jennifer's question was contained within the engineering framework that he was shortly going to discuss and shared a slide (Figure 3.4).

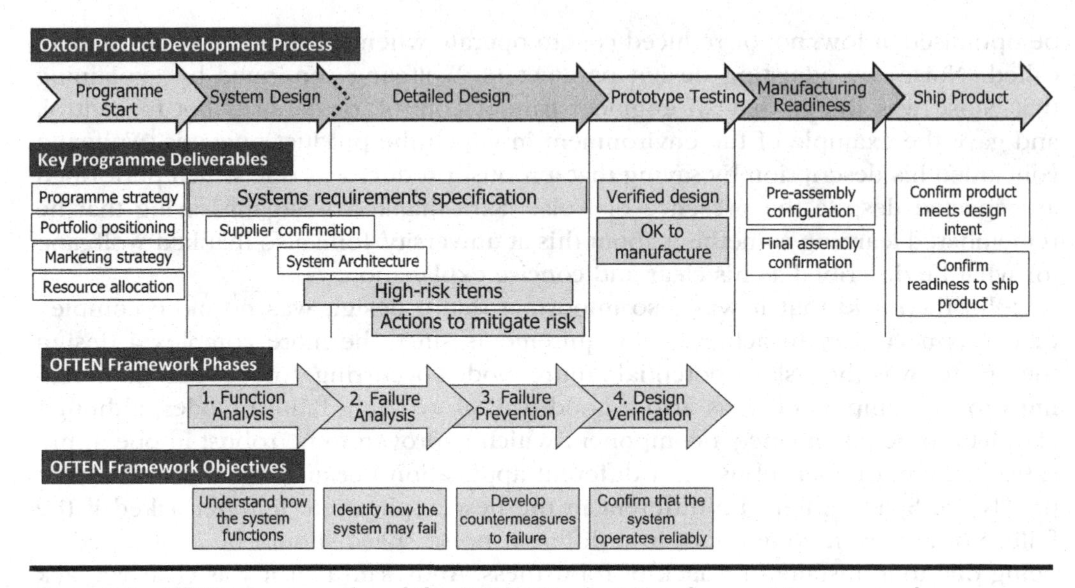

Figure 3.4 OFTEN Framework and the OFTEN Product Development Process roadmap.

John said that Oxton team members would be familiar with the top half of the slide he was sharing being what he described as the roadmap depiction of the Oxton Product Development Process (OPDP). John explained that the bottom half of the slide summarised the engineering framework that he planned to lead the team through. With some hesitancy, feeling that the acronym was somewhat contrived, John said that OFTEN which he deliberately pronounced as “Of – Ten”, stood for Oxton first-time engineering norm where norm meant *Standard*. He was pleased that there was no immediate reaction to the acronym, although Wolfgang said that if he had been given a 100 Euros for every “Right First Time” initiative he had seen, he would be a rich man. John told him that he had every right to be sceptical but hoped that it would not stop him supporting the application of Failure Mode Avoidance principles to Oe375. Wolfgang said that he would reserve judgement.

John explained that the framework was based on the four phases of Function Analysis, Failure Analysis, Countermeasure Development and Design Verification with the objectives of each phase being summarised at the bottom of the slide. John said that the OFTEN framework facilitated the achievement of the key programme deliverables in the shaded boxes within the System Design, Detailed Design and Prototype Testing phases of the OPDP. John asked if there were any questions or comments, and Jennifer said that the slide did not show all the programme deliverables usually included on the OPDP roadmap. John agreed with Jennifer and explained that he had only included what he saw as the key deliverables for the sake of clarity of the slide. Since there were no more questions, John said that he wanted to explore the OFTEN framework in a bit more detail and shared another slide (Figure 3.5).

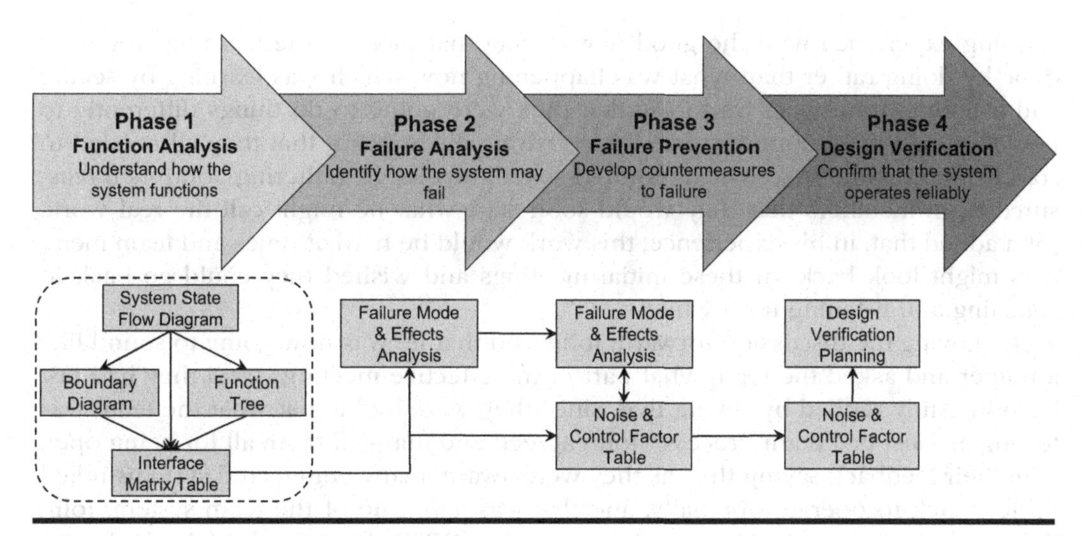

Figure 3.5 The four-phase OFTEN framework.

John said that the top of the slide illustrated the overall four-phase framework with the primary engineering methodologies used within each phase shown on the bottom half of the slide. John added that the arrows connecting one phase to the next represented the flow of coherent information between phases and added that this flow was iterative. Andy asked John what he meant by saying that the flow was iterative, and John said that even engineers are fallible and very occasionally miss an item of detail or need to go back and correct something in the light of further information. John said that, as team members would experience in using the framework, the coherence of the forward information flow resulted in backward traceability, allowing the effect of a correction in one phase to be traced back to a corresponding correction in a preceding phase. John continued by saying that although he would explain the detail of each phase as they applied it during the programme, he would like to have some discussion with team members now about the framework itself.

At this point, Sam intervened and said heatedly that what was happening seemed to him like training and added that he thought that he had been asked to join the Oe375 programme to do some digital engineering not to go on a training course about a new-fangled way of doing things based on using team skills. John closed his laptop lid, thanked Sam for his feedback and sitting back in his chair asked what other members felt. An animated discussion ensued primarily within the Oxton members of the team, resulting in the consensus that the Oxton team members were surprised with the direction that their first two meetings had gone. John asked Andy, Yumiko and Wolfgang for their views and again there was consensus, although this time it was that the three external team members did not have any preconceived ideas about what to expect but were pleased to be asked to join the team.

John said that rather than training he saw what they had been doing as learning, adding he was not overly keen on the term training seeing this term as being more to do with enhancing the physical skills of sportspeople or modifying behaviour in animals. John continued by saying that he saw the whole Oe375 programme as a learning experience with the good news being that most of the learning would be done by doing rather than what was happening now which was learning by seeing and listening. John went on to say that they were going to do things differently to how they had been done historically in Oxton which meant that they all needed to start from the same base which required some knowledge gathering. John then reassured team members that they would soon start what he might call the real work. John added that, in his experience, this work would be hard at times and team members might look back on these initial meetings and wished they could go back to watching and listening for a while.

In moving the discussion forward, John said that he was now going to sound like a trainer and asked the team what part of the effective meetings triad they had just been in. Andy replied by saying that, since they were looking at what the team was feeling, it must be "Team Process". John agreed and thanked them all for being open with their feedback saying that, as they were aware, many engineered systems relied on feedback to operate optimally, and this was also true of the team system. John then said that he would like to get back to the OFTEN framework and asked team members if it was OK to do this, to which team members responded that it was.

In turning to the primary tools shown on the slide (Figure 3.5), John explained that, as he had said, an important aspect of the OFTEN framework was the coherent flow of information from one phase to the next and asked team members to note that some tools were used in more than one phase which facilitated information flow between phases. John added that there was also significant information flow between tools. John asked if there were any tools shown that were not familiar to team members. Sam said that while he was familiar with state flow diagrams, he had not come across a tool called the system state flow diagram (SSFD). The Oxton and external team members all agreed that they had not come across this tool before. John said that this was a key tool in the OFTEN approach and was used to achieve the functional decomposition of a system by considering the transitions of the states of energy, material and information within a system. Phil said that in his experience this was called a boundary diagram. John explained that the SSFD and boundary diagram were closely linked, with the essential difference being that the SSFD identified the functions necessary for a system to convert an input to an output by considering the state transitions that happened in a system. John added that in comparison the boundary diagram considered the flow of energy, material and information through a system from input to output. Andy asked John what he meant by state transitions, and John gave the example of the state of electrical power in an eBike battery being converted to the state of torque by the motor. John said that he thought that the basis of the SSFD was best understood by team members developing a real engineering example which he intended that they would do in their next meeting. John said the methodologies used in Phase 1 of the framework were all closely interrelated, and this was why he had drawn the dotted line box around them on the slide.

In returning to John's question about unfamiliar tools, Sarah said that Oxton used the interface matrix but not the interface table. John said that in OFTEN the team would use both the interface matrix and the interface table as part of interface analysis. In response to a question from John to the external team members, Yumiko and Andy said that their companies used interface tables. John asked if there were any other unfamiliar tools, and Jennifer said that she had not come across the noise and control factor table (NCFT) and to her knowledge, this was not used in Oxton. The other Oxton and the external members also confirmed that they were not familiar with this tool. John said that the NCFT was used in improving product robustness in Phase 3 of the OFTEN framework and in verifying this robustness through Design Verification in Phase 4.

John asked if there were any other differences between the Oxton framework and what Oxton did now, to which Andy replied that while HJK did their own FMEA (failure mode and effects analysis), he had never seen an FMEA from Oxton. Sarah said Oxton did do FMEAs. John asked if other members were familiar with FMEA to which they all responded that they were. John added that, as he was sure team members knew, there were several different types of FMEA, including the Design FMEA (DFMEA) which focused on the design of a product and the Process FMEA (PFMEA) which analysed a process such as that used to manufacture a product. John added that the Oe375 team would use the DFMEA within the OFTEN framework. John asked Sarah when Oxton would typically develop a DFMEA, and she replied that they were developed once a design was complete. In following up on her reply, John asked Sarah how Oxton currently identified functions in developing a DFMEA to which she replied that they did not. John asked Sarah to describe a typical failure mode in a DFMEA that she had worked with others in developing. Sarah explained that since the DFMEA was done mainly to satisfy the Quality Audit, she usually developed the document on her own, revising where necessary the DFMEA for the outgoing product and added that a cracked chainring would be a typical failure mode. This example confirmed John's view that Sarah had been speaking of a hardware-based DFMEA so he made the point explicit to team members by telling them that, as they may be aware, there were two main standards for DFMEA, one of which is based on identifying failures in parts with the other standard being based on failures of function. John added that they would be developing function-based DFMEAs because doing this in Phase 2 of the OFTEN framework was consistent with the identification of functions in Phase 1 through the SSFD. Turning to Jennifer, John told her that the answer to the question that she had asked earlier as to why John spoke of failure modes was because when a failure of function is considered, the way it has failed is described in terms of the type of failure mode that is exhibited. John added the team would consider the detail of this in Phase 2 of the framework.

Phil then raised his head from his laptop screen to look at John who, in noticing this, remarked to him that he had a frown on his face. Phil said that something that John said earlier had struck him and that was that the boundary diagram was closely linked to the SSFD, and he was not sure what he meant by this. John said that what he meant was that the boundary diagram was developed from the SSFD. John asked Phil how they developed the system boundary diagram in a programme now, to which Phil responded that since the design concept of the bicycle had not changed much over his lifetime, the boundary diagram only needed tweaking from one programme to the next and cited the example of changing calliper brakes to disc brakes. John asked Phil what a typical Oxton boundary diagram showed, to which Phil responded by saying that it showed the links and flows between bicycle components, with the flows depicted being those of energy, material and information. John said that sounded good and then quoted the architect Louis Sullivan "*Form ever follows function*". John explained that in line with Sullivan's advice in applying the OFTEN framework the team would identify the functions responsible for energy, material and other flows before considering hardware. Phil responded by saying that sounded like another example of back-to-front engineering and asked John why they would want to do this. John returned to his laptop and searched for a file and, on finding it, shared it quickly removing the Blade logo before doing so (Figure 3.6).

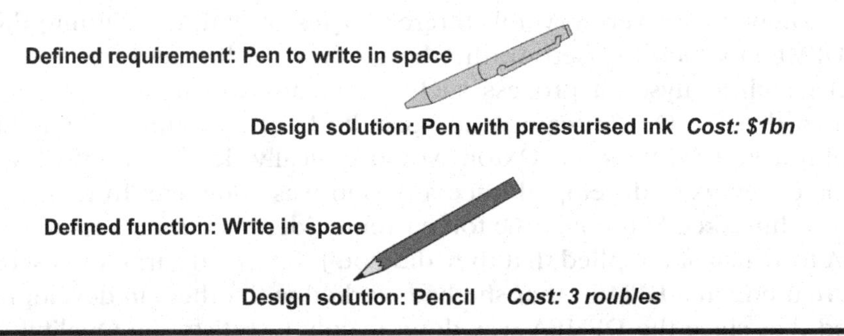

Figure 3.6 A mythical tale.

In explaining the slide, John said that while it was a myth it nevertheless promoted a strong message in that design engineers tended to focus on solution rather than first considering what a product design was expected to do. John continued by saying that the approach of focusing straight away on solution can constrain the creativity that is required in developing a product that stands out from competitors' products. John then reminded team members that the Oxton Board wanted Oe375 to be highly competitive. John explained that, as team members would see when they got into the detail of OFTEN, functional analysis is at the core of the framework and added that this approach was based on a sound academic foundation.

Jennifer then asked John what he had meant when he had said that the example on the slide was a myth. John explained, the myth is that NASA (National Aeronautics and Space Administration) spent millions of dollars developing a pen to write in space while the Russians used a pencil. He continued by saying that both NASA and the Soyuz space programmes originally used pencils with NASA using a highly expensive mechanical pencil. John said that an outcry against the cost of the mechanical pencil caused NASA to look for something cheaper and they started to use the space pen which was developed in the US independently of NASA. John added that the Soyuz programme also switched to using the space pen. Yumiko then asked John if he could send her the details of any academic papers that discussed the functional approach that he had mentioned, and John told Yumiko that he would send her and the rest of the team some references.

John returned the team's attention to the OFTEN framework by switching back to the Framework slide (Figure 3.5) and reassured Phil that the boundary diagram was an important tool in OFTEN and added that, unlike in the current Oxton approach, the boundary diagram was developed from the analysis of functions rather than parts. John explained that the team had identified two key differences in comparing Phases 1 and 2 of the OFTEN framework with the current Oxton PD process and stated these as starting with function analysis rather than hardware and developing functionally-based DFMEAs. John said that he would like to consider Phase 3 of the framework entitled Countermeasure Development. John reminded team members of what he had said earlier in the meeting that there were two principal ways in which potential failure modes might occur during product design, firstly due to mistakes and secondly due to a lack of robustness. John added that the best way of avoiding making a mistake during product design was by paying attention to detail. John said that the rigorous analysis conducted in Phase 2 based on interface analysis helped here as did the use in Phase 3 of design guidelines like engineering standards.

Andy intervened at this point to comment that care had to be taken in using engineering standards, adding that he thought that a poor standard was worse than no standard. John asked Andy if he would explain his statement. Andy said that a poor standard might miss a key failure mode and lull engineers into believing that they had captured all the significant failure modes associated with a particular aspect of a design when this might not be the case. John thanked Andy for his comment and said that he agreed with what he had said. John then asked Sarah if there was an Oxton engineering standard for the seatpost which identified galvanic corrosion as a potential failure mode. Sarah replied that there was an engineering standard for the seatpost which at the time of designing the high-end bike only included detail on its mating with the frame, the strength of the materials, their relative geometry, and the clamping force. John asked if the standard had been updated in the light of the experience with the high-end bike, and Sarah said the standard had been updated to include the issue of galvanic corrosion.

In turning attention to the second important way of developing countermeasures to potential failure, John asked how much robustness improvement work was done in Oxton, and Sarah responded by saying that they did not do much robustness work in Oxton, leaving that to suppliers. John agreed that robustness was best achieved through the design of components and asked Sarah if she had any discussion with suppliers about this. Sarah responded that as a technical specialist she kept in close contact with any suppliers designing parts for a new product to help ensure they matched Oxton's expectations and added that she answered any questions that they had. John asked Sarah what kind of questions suppliers asked her to which she replied that they were mainly about how their component would interact with other components on the bicycle. In response to prompting by John, Yumiko said that as far as she knew Agano worked largely on their own in ensuring their parts were robust but as Sarah had described they kept on close contact with their customer companies. Yumiko added that Agano had a strong reputation for research and development. Andy said that this situation was similar in HJK. John asked Sarah if, in working with suppliers, Oxton gave the suppliers a specification for the components they were to design which included a list of noise factors that the component would be subjected to in operation. Sarah replied that any specification that Oxton gave suppliers did not include any details on noise factors as such. John concluded the discussion on Phase 3 by saying that developing supplier specifications which included information on noise factors was another difference between OFTEN and the current Oxton PD process.

In looking at Phase 4 of the OFTEN framework, John then asked the Oxton team members how Oxton verified their designs. Jennifer said that as far as her area of expertise, the Drivetrain, was concerned they performed rig tests and road tests, with the rig tests using the final components that were to be used on the bicycle and the bicycle also being composed of the final components, adding that the tests were "bogey" tests. John asked Jennifer if she could explain the term bogey in case anyone was not familiar with it. Jennifer said that it meant that they test to a target, for example, the number of revolutions of the chainset with a specified pattern and frequency of gear changes in a rig test, or with road testing, the target was typically the distance covered on different on-road or off-road surfaces. Jennifer added that the ultimate aim of the bogey test was that there should be no failures. John asked Phil and Sarah if Jennifer's description of what happened to the Drivetrain applied to other bicycle systems and they both said that very little rig testing was done for other areas of the bike with the focus being on-road and off-road testing of bikes and eBikes. Phil added that they also did quite extensive wind tunnel testing with the premium bicycle designs. John then asked the Oxton team members if they included the effect of noise factors in their testing and the general agreement was that while noise factors, such as the environment and road surface variation, would be included as a matter of course they did not specify which noise factors would be included. John then asked Andy and Yumiko about what happened in their companies, and they concurred that in both companies while on-road and off-road testing was important, they mainly did rig testing. Andy and Yumiko were again in agreement that while their companies did their best to include noise in their verification testing, this was difficult as often they did not know how their components were going to be used so they did not know which noise factors were important or what the magnitude of these noise factors would be.

Phil spoke up at this point saying that he found this discussion fascinating and while he had visited supplier companies on many occasions, their testing had always seemed like a black art to him. He added that this testing seemed to work though because he was generally impressed by the quality of parts coming from companies like Agano and HJK. John thanked Phil for what he called an enlightening statement. Largely addressing the Oxton members of the team, John said he had noted that Jennifer said that in the Drivetrain testing, the aim was that none of the systems should fail and asked if it was usually the case that parts did not fail during testing. Phil laughed and said "If only" in a loud whisper as the other Oxton team members smiled. Reinforcing what he had said in the previous meeting, Phil continued by saying that things got busy during testing dealing with failures and other unexpected events. John asked Phil what he meant by unexpected events, and Phil gave the examples of excessive wear in brake pads or wheels needing realignment. John then asked the team if it would be fair to say that in Oxton verification testing was a key part of the design process and was not just a final confirmation of design robustness. After the Oxton members, except for Sam, all agreed, John said that they had identified the key difference between the current Oxton PD process and Phase 4 of the OFTEN framework. Noticing that Sam did not say anything in response to his question, John asked him if he got involved in any virtual testing work. Sam replied that while he had joined Oxton on the basis of using his skills in developing engineering computer models, the reality was that he spent the majority of his time trouble shooting problems associated with control software and added that he even helped to sort out IT issues. John thanked Sam for his feedback and said that he hoped to make better use of Sam's talents as a member of the Oe375 team which would include the development of software-based engineering models.

John reflected that his strategy of encouraging team discussion in Oe375 meetings was beginning to work well, although the flip side of this was the extra effort needed to manage the team process and ensure that the team did not lose sight of the core tasks defined by the agenda. John felt that this was one of those occasions where the meeting needed to move on and so made this explicit by suggesting that Jennifer report back on the list that the team had developed of potential excitement features for Oe375. John handed the meeting over to Jennifer, and she thanked him and shared her screen with the team (Figure 3.7).

Wireless charging	Regen braking
Ultrafast charging	Regenerative pedalling
Park, charge & go	ABS braking
Blind spot assistance	Mild hybrid commuter
Google map navigation	Fingerprint unlock
Smart phone connectivity	Antitheft alarm
80-mile range	Detachable battery
Powered gear change	Easily detachable wheels
Smart speed selection	Folding bike

Figure 3.7 Potential excitement features for Oe375.

Jennifer explained that this was the final list that the team had agreed on and that everyone had contributed with some of the features being put forward by more than one person. She added that initially the list was longer and between them they had agreed to shorten it, for example, by removing features that the team felt were too wacky. John said he was very impressed with the creativity of the team and thanked them all for their work. Andy explained that the option of the mild hybrid commuter meant an eBike with regenerative braking and pedal to charge to supplement external charging of the battery. John asked him why this feature for the eBike would be for commuting, and Andy explained that the proposal was to incorporate smart control that could learn the rider's route and the manner in which the rider interacted with the power assist system and subsequently provide assistance at the appropriate times without the need for the rider to intervene. John said that that sounded interesting and creative and had answered another of his questions. John asked the team if they had any more comments, and Phil explained that some of the features were complementary and gave the example of the mild hybrid generating its own electric power which extended the eBike range. Wolfgang then explained that, while many eBikes had removable batteries, normally this was for ease of charging or replacement and added that by detachable the team meant a battery which could be removed easily to enable the eBike to be used as a normal bicycle. After allowing time for others to say something, and when they did not, John asked that team members give him an example of a wacky feature that they had rejected. Jennifer gave the examples of a satellite TV display and a hotplate for cooking food, to which John responded by saying that he understood why these suggestions were rejected. Looking at the time, John concluded the discussion by saying he would discuss the list with Jim Eccleston in his next meeting with him and get back to the team. John then quickly moved towards Warm Down.

John said that there was an outstanding task before Warm Down as the team needed someone to manage the documentation that would be generated by the team. John suggested that the team use the *ShareDrive* file sharing facility associated with the branded software he was using to conduct their team meetings and explained that this would enable the external members to access the files. John added that the team documents would be proprietary to Oxton and said that as such the person managing the files should be one of the Oxton team members. After a short silence, Jennifer said that she would volunteer for the job, and John thanked her and suggested that she should use the programme name along with a strong password as the sign-in to access the file sharing site. In continuing to speak to Jennifer, John added that she could also upload the team decision log that they spoke about at their last meeting to the ShareDrive. Jennifer thanked John for his advice and said that she was thinking along the same lines and added that she would circulate the sign-in details.

In starting the Warm Down, the team agreed that the meeting had met its stated purpose and then moved into the pass-the-pen exercise. The consensus was that, while some frustration had been expressed that the first half of the meeting seemed like training, they had found the second half interesting and more like an engineering discussion. John thanked team members for their feedback and then reminded them that, as he had said earlier in the meeting, they would start some real work in the next meeting. John then closed the meeting and asked for volunteers to help move the tables in the Oxton room back into a single square and everyone helped to do this. After the room was returned to its former state, Sarah asked John if he had a moment to discuss something, and John said that he had. Sarah said that she was dying for a cup of coffee, so John and she collected some refreshment on the way.

Having reached John's office, John noticed that Sarah shut the door behind her before sitting down at his table with him. John asked Sarah what was on her mind, and Sarah said that it concerned the integrity of Oxton's engineering standards and other design guidelines. Sarah continued by saying that after she had heard about what she called the high-end bike seatpost debacle, she had reviewed other Oxton standards and found some of these to be wanting. Sarah cited the example of a chainset and pedal standard which made no mention of rider using the pedal to mount and dismount a bicycle and so had missed an important aspect of rider usage. John asked Sarah what proportion of the standards she had looked at needed improving. Sarah replied that about a quarter of the standards that she had reviewed would benefit from an updating and described the situation as being like rotten apples explaining that the minority of poor standards affected the perceived integrity of all standards. John asked Sarah what, if anything, she had done to rectify the situation. Sarah said that she had discussed the matter with Paul Clampin and had recommended to him that she lead a small working group with a view to resolving the situation. John asked Sarah how Paul had reacted to her suggestion, and Sarah described him as being less than enthusiastic. Sarah quickly added that she was speaking to John in confidence and said that it would not be helpful if Paul got to know that she was talking to John about him. John assured her that he would not mention their discussion to anyone. Sarah said that Paul did say that should Sarah or others come across any standards while working on a programme that they felt needed improvement that they should take what Paul described as appropriate action.

Sarah said that she did try to raise the issue of standards with Paul on a subsequent occasion, and he had told her that engineers were intelligent people and would be able to recognise what needed to be done despite any shortcomings in a standard. Sarah said that she pointed out that she thought herself to be intelligent and had a good memory but still wrote out a list when she went supermarket shopping in case she forgot something. John asked Sarah what Paul's response to her analogy was. Sarah said that he told her she was too busy working on the tasks assigned to his department to get sidetracked on what he called broader company matters and then changed the subject.

After pausing to take in what Sarah had told him, John said that they should certainly do as Paul had suggested and update any poor standards that they came across on Oe375. John added that the advantage of this strategy over the use of the working group was that the team would be coming at the standard with the knowledge they had gained from the interface analysis they would be conducting in Phase 1 of the OFTEN framework, although he conceded that they would only be updating selected standards. In smiling, Sarah said that what John had suggested also had the advantage of not invoking Paul's ire. John thanked Sarah for sharing her experiences with him and asked her how she though the Oe375 meeting that had just finished had gone.

Sarah said that despite of what she called the "interruption", she had enjoyed the meeting and, added that like the other Oxton members she had found the two Oe375 meetings very different to the usual PD meetings she had attended in Oxton. John asked Sarah what was different between the two Oe375 meetings and other PD meetings. Sarah said that normally in such meetings people get down to the engineering task in hand almost immediately as a matter of course, and any agenda, if there is one, for the meeting is verbalised at the start of the meeting by the engineering manager who is leading the meeting. John asked Sarah if there was any discussion of the process to be used in performing a task, and Sarah answered that if there was it was usually cursory. John asked what happened when people who were new to the company or people from external companies joined the meeting. Sarah said that newcomers were largely expected to pick things up as they went along, and external people such as suppliers were only asked to participate in meetings with Oxton engineers when there was a problem which needed their expertise to help solve. John observed that the Oxton team members must have found the two Oe375 meetings strange and said that it was hardly surprising that there was some resistance since change was often an uncomfortable experience. Sarah reassured John that although she had found the meetings to be different, she had found them interesting and was looking forward to applying the OFTEN framework in earnest.

Sarah then said that in her opinion Sam's outburst in the meeting was not just directed at John since, as Sam had hinted, he was upset with the way his current job was working out. John asked Sarah to elaborate. Sarah said that Sam had expressed his frustration to her on a couple of occasions and explained that as Sam had said in the meeting, he had been taken on at Oxton to support software model-based engineering and yet had found himself being, what he described to her as, an IT general dogsbody. John said that he thought that was unfortunate and asked Sarah who Sam reported to, and Sarah told him it was Joe Bower, the IT Manager. John reflected that he thought that Sam reporting to IT was a bit strange when he had been brought into the company to do engineering but decided that he had not been in the company long enough to start questioning the company organisation. John said that as he had implied in the meeting, he intended that Sam would lead the development of engineering software models for Oe375 and added that he would speak to Joe when the opportunity presented itself to get his support for this. Sarah thanked John for the opportunity to speak to him and, on getting up to leave, said that she had found the meeting useful. John thanked Sarah in return for her frankness.

John usually took a coffee break mid-morning when he was able to, and the day after the Oe375 PD team meeting, he was standing in the coffee queue in the canteen when he realised that Paul Clampin was behind him. His cheery greeting to Paul was met by what sounded to him more of a grunt than a greeting. Paul asked John if he had a few minutes and, on John acknowledging that he did, Paul suggested they should get their coffees and sit in the area in the canteen with comfortable chairs. When his turn came at the head of the queue, John asked Paul what he would like to drink and paid for the two coffees. After the two men had both sat down, John started to enquire how Paul was keeping, but Paul made it clear he wanted to talk business and interrupted him. Paul said that he had spoken to both Sarah and Jennifer yesterday afternoon about their latest Oe375 meeting and was disappointed to hear that the meeting comprised of more soft skills stuff and, what Paul described as, a cosy fireside chat about how Oxton conducted their product design. John responded by saying he preferred the term People Skills to soft skills and was about to explain why, when Paul interrupted him again this time with the one word "Whatever" spoken in a louder voice than was normal for conversations in the canteen. Paul said he wanted to cut to the chase and said that he needed Sarah to stop working on Oe375 at least temporarily and re-join the team working on O254, the budget bike programme which was running into some difficulties with suppliers. Paul quickly continued before John could speak by saying that he realised that he had agreed with Jim for his people to support Oe375 and so he was going to replace Sarah with an undergraduate on a thick sandwich degree course who was spending his industry year in Paul's department. John was taken aback and paused before saying that this was hardly a like-for-like replacement, at which Paul looked at his watch, stood up, said he had a meeting and walked away leaving John to contemplate how to best deal with the new situation that he had found himself in. Looking down at the table in front of him, John saw both coffees, untouched.

John sat back, and in sipping his coffee, thought about what had just happened, reflecting it was made all the more unpalatable by the fact that he was impressed by Sarah's positive approach and willingness to support the change that implementing the OFTEN framework entailed. He decided this was something that would not blow over and began thinking about how he might best discuss the situation with Jim Eccleston.

Background References

Effective Meetings	Connolly, C. (1996) Communication: Getting to the Heart of the Matter, ***Management Development Review***, 9(7), 37–40, https://doi.org/10.1108/09622519610153938 Mackin, D. (2007) ***The Team-Building Toolkit***, American Management Association Page 65
Failure Mode Avoidance	Lonnqvist, A. and Gremyr, I. (2008) A Note on Failure Mode Avoidance, ***11th Quality Management and Organizational Development Conference***, 33(8), 109–116, https://ep.liu.se/konferensartikel.aspx?series=ecp&issue=33&Article_No=8, Accessed 20 July, 2022 Pages 100–103
Functional Decomposition	Ulrich, K. and Eppinger, S. (2008) ***Product Design and Development***, 4th edition, International Edition, McGraw-Hill Pages 100–103
Form Follows Function	Hwang, K. (2020) Form Follows Function, Function Follows Form, ***Journal of Craniofacial Surgery***, 31(2), 335, https://doi.org/10.1097/scs.0000000000005891 Curtin, C. (2007) ***NASA spent millions on a pen able to write in space***, Scientific American, https://www.scientificamerican.com/article/nasa-spent-millions-on-a-pen-able-t/, Accessed 18 December, 2022

Chapter 4

Tracking States

In thinking about how he might best ensure that Sarah Jones remained on the Oe375 Product Design (PD) team, John decided to have a chat with the HR Manager Numa McGovern since he did not want to go straight to Jim Eccleston feeling that the onus was on himself to manage the situation. He decided to act straight away and was pleased to see Numa sitting at her desk in her office as he knocked on her open door and asked if she had a few minutes. As expected, she invited him in with a broad smile and walking out from behind her desk motioned him to sit down in an area of her office with two comfortable chairs opposite to each other. Closing the door, she asked if he would like a coffee indicating the coffee maker on a table in the far corner. John politely declined her offer saying that he had just had one.

Numa asked John how she might help him and, without naming anyone, John explained that there was pressure on him to allow a key experienced member of the Oe375 PD team that he was currently managing to withdraw from the team and be replaced by an intern. John said that Jim Eccleston had recommended this particular team member and went on to explain the impact that this would have on what he was doing. He added that there had been some resistance to what he was doing with the Oe375 team from the area in which this particular team member was based. Numa asked John if he had spoken to the team member's manager about the situation and John responded, thinking that the situation was becoming transparent, that it was the team member's manager who had requested the return of the team member. Numa asked if John had explained to this particular manager the effect that the change of team membership would have, and John replied that his discussions with this person had been rather one way and abrupt. Numa sat back in her chair thinking, and John was sure she knew who he was talking about. After a short while, Numa asked John if he had regular one-to-one meetings with Jim, and he said that he did. Numa continued by saying that firstly, she was going to talk in generalities, and secondly, offer some specific advice. She told John that she had found that standing up to what she said might be called forcefulness often produces results and went on to suggest that John ask Jim if he might invite the manager to the next one-to-one without saying why. Numa continued by suggesting that he address the issue by raising it during the course of the meeting with Jim in a matter-of-fact way in terms of overall resourcing,

DOI: 10.4324/9781003286066-4

adding that she was sure that Jim's presence would act as a catalyst to achieving a satisfactory outcome. John sat back thinking about what Numa had said and thought that he understood what she was suggesting and why. John stood up and thanked Numa for her help to which she responded by wishing him good luck as he opened her office door and left.

John checked with Vicki Harrison who said that there was no need to ask Jim as it would be perfectly fine for John to invite Paul Clampin to his one-to-one. Unfortunately, when John checked Paul's calendar, he saw that he already had a meeting at the time of his one-to-one with Jim so he decided he would let him know of his intention before sending the meeting notice and walked along to Paul's office. Paul was just finishing a meeting when John arrived and so he stood outside of his open office door until everyone but Paul had left. Putting his head round the door, John told Paul in a neutral manner that he had set up a meeting between Jim, Paul and himself for the next day and added that as he was confident that Paul would be able to postpone his existing meeting, he would send Paul a meeting notice shortly. Without allowing Paul time to respond, John walked away and went back to his office.

John arrived at his one-to-one with Jim early and, as he had hoped, was able to mention to Jim that he had invited Paul and explained to him that he thought it would be useful for the three of them to get a shared understanding of what he was doing. He added he would also like to maximise team effectiveness particularly in relation to the two Oe375 PD team members who worked for Paul. At this point, Paul breezed in, greeting Jim and John in a cheerful manner and sat down opposite to John at the table where Jim and John were already seated. Paul asked Jim what the meeting was all about, and before Jim could answer, John repeated what he had just said to Jim. For Paul's sake, John briefly described what had happened at the Oe375 PD team's first meeting during which time Paul looked to Jim several times. John thought that Paul was checking whether Jim was going to speak, but when Jim did not say anything, Paul also remained silent. John then moved on to give a summary of the team's second meeting, and Jim was sufficiently interested in the Effective Meetings process that he asked John to take them through the flowchart which John did. Jim remarked that the model was very interesting and had wider application than the Oe375 team, asking Paul for his reaction. Paul responded by agreeing with Jim and added that Oxton could do with more discipline in its meetings, although saying that he was not convinced about the use of Warm Up and Warm Down. Jim said that the power and importance of Warm Up and Warm Down should not be underestimated, and Paul said he agreed.

In continuing his summary of what had happened in the second Oe375 meeting, John introduced the first-time engineering design (FTED) framework and mentioned to Jim that he might recall that it had been rebadged as OFTEN, explaining what the acronym stood for. Jim said he recalled Vicki telling him about this and said that while he was not sure about the name on first hearing it was growing on him, to which Paul added that he quite liked the name. John took the two men through the detail of the framework during which time Jim said nothing as he had a good understanding of it by now, and Paul mirrored him by remaining silent. John then explained that after he had taken the Oe375 team through the framework they had spent some time discussing the model and comparing it to the existing Oxton PD process. Aware of

Paul's previous reaction to this discussion, John added that he was pleased that the discussion was animated since he assumed that this would help team members to internalise the framework. Jim asked Paul what he thought of the framework, and Paul replied that it looked interesting, but he did not think that Phase 3 was relevant to Oxton being a supplier responsibility. Jim retorted that, having read a couple of the academic papers that John had recommended to him, he felt that the quantification of what he called "Noise Space" was very much an Oxton responsibility and that this quantification should be apparent in the design requirements that Oxton gave suppliers. At this, Paul said that perhaps he had not made his thinking clear and continued by contradicting what he had just said by restating what Jim had said almost word for word. John thought that he detected a slight smile on Jim's face.

Having discussed the OFTEN framework, John showed the other two the list of potential excitement features for Oe375, and Jim said he was impressed by what he saw as a creative list and followed this up by asking John to explain what was meant by a mild hybrid commuter. John explained how the team saw such a feature, and Jim thanked him and asked Paul what he thought of the idea. Paul said that he found it interesting. Jim said that as far as he knew, the power generated by regenerative braking and regenerative pedalling on an eBike was minimal so the idea of the mild hybrid, while being interesting, was probably not that useful, although adding that the team should not rule anything out at this stage. Jim continued by saying that he found the idea of a detachable battery to enable the eBike to be used more as a conventional bike attractive. He said it reminded him of his pro-cycling days of throwing away water bottles towards the end of a stage to gain a small weight advantage and added that he had had a few near misses in that respect. Jim continued by saying that ideally the motor would also be detachable, adding that in his experience easily detachable wheels were a godsend and that the team might pursue that idea too. John said he would be pleased to convey Jim's feedback to the team. John then concluded his summary of the Oe375 meeting by saying that there had been some resistance in the meeting to what the participants saw as training and briefly explained to John and Paul what he had said to the team members about the need to gain common knowledge as a base to move on from. As John was mentioning the resistance in the meeting, he noticed Paul turn his head towards him so Jim could not see and smirk. Jim said that this type of resistance was to be expected, although he assumed that the team would soon move on to design analysis. John assured Jim that the design analysis work would start in earnest in the next meeting.

Jim thanked John for the summary and told him to keep up the good work and moved the meeting on by saying that he knew that John wanted to talk about maximising team resource effectiveness. Feeling that this was the key moment in the meeting, John said, what he had been rehearsing in his head, that he was very pleased with both Sarah Jones' and Jennifer Coulter's contribution to the team and that he was particularly appreciative of Sarah's help. John then said that he was aware that Paul was contemplating asking Sarah to switch her attention back to O254 because of workload difficulties. Jim asked Paul if this was correct and Paul responded that, as John had said, it was just something he had been mulling over. Jim raised his sitting position and said to them both that, because of the importance of the Oe375 programme to the company, he had decided to become the champion for the programme. Jim continued by telling Paul that, as he knew, he had specifically requested that Sarah work on the Oe375 and said that was the way, in his now official role of programme champion, he wanted it to be. Jim added that the same thing went for Jennifer. Paul smiled at Jim deferentially and said that was fine by him, adding that John could count on his support. Jim closed the meeting by thanking Paul and John and saying that he had found it a useful meeting after which Paul and John stood up and walked out of Jim's office with Paul glaring silently at John when he was out of Jim's vision. John thought to himself that he seemed to have won that battle, but the war was probably far from over.

Later that day, Numa McGovern called John and asked him how his meeting with Jim had gone to which John responded by saying that it had gone as well as he could have hoped. John thanked Numa for her help, and she replied that she was only too happy to assist. As he put the phone down, John thought to himself that now he had seen both sides of Paul Clampin's character, he felt better able to interact with him.

Following his one-to-one with Jim, John had decided that it would be useful to develop a use case diagram (UCD) to better understand how a rider might interact with the eBike and sketched out a rough draft in order to discuss his idea with Sarah at the Review/Preview meeting that he had arranged with her prior to the next Oe375 PD team meeting. In looking forward to the Review/Preview meeting, John had decided not to mention anything to Sarah about his recent conversations with Paul since he suspected that Paul may not have approached her yet with a view to recalling her from the Oe375 team. When John and Sarah duly met the following day, on being asked by John if she had anything to add to the feedback on the last Oe375 meeting that she had given him immediately after the meeting, Sarah said that she had nothing to add. In looking forward to the next Oe375 team meeting, John showed Sarah the initial draft of the UCD that he had developed, and Sarah suggested a couple of amendments. Sarah told him that she knew from a previous project that Jennifer had some expertise in use case analysis, and John asked Sarah to run their draft past her, requesting that one of them present it to the team in the Oe375 meeting. In developing the agenda for the Oe375 meeting, John said that the development of a system state flow diagram (SSFD) was to be the core task and added that he wanted to look at the context of the SSFD in Phase 1 of the OFTEN framework. John continued by saying that tacit knowledge, defining this as the knowledge gained through experience residing in team members' heads, was important in developing an SSFD and explained that he wanted to look at the difference between tacit and explicit knowledge in the meeting. John added that the technique of visualisation was key to unlocking tacit knowledge and he intended to demonstrate this to the team. John suggested they start the meeting with a Warm Up centred on visualisation and asked Sarah if she would conduct the Warm Up. Sarah responded by saying that she would be happy to do this provided that John would take her through what was expected of her, and they spent some time discussing how the Warm Up might be conducted.

Having agreed the Warm Up exercise, John asked Sarah what else they needed to do in the Oe375 meeting, and she reminded him that he was going to run the teams' excitement feature list past Jim Eccleston. John said that he obtained Jim's feedback on the excitement features and could report back to the team. Having established the content of the Oe375 meeting, John and Sarah finalised the meeting agenda. As Sarah was leaving John's office, she turned round and said that she had forgotten to mention one thing and reminded John he had said that he was going to send team members a list of references related to Failure Mode Avoidance. John thanked Sarah and said that he had completely forgotten about sending the references and added that he would do it right way. John thought that the easiest thing to do was to upload a copy of his conference paper to the team ShareDrive, which he did, and send an email directing team members to both the paper itself and the references that accompanied it.

John was looking forward keenly to the third meeting of the Oe375 programme team as he sat in the conference room awaiting team members' arrival. Phil had arrived early and together they had arranged the tables and video camera in the same configuration as the last meeting. Sarah walked in with her usual cup of coffee and after greeting John and Phil set herself up at one of the tables. Wandering over to Sarah, John asked her if she was all set for the Warm Up, and she said she was. John then asked Sarah how her triathlete training was going, and she explained that, while it was difficult to juggle the three full-time jobs of work, being a wife and a mother, with a lot of support from her husband, she was managing reasonably well to train. Sarah added that she had an important competition in a couple of weeks and John wished her luck for that. By this time, everyone had signed in, and John suggested to Sarah that she make a start. After John had taken his usual seat, Sarah stood up and greeted the team members and explained that she was taking the Warm Up.

Sarah said that she was pleased to see the room at Oxton all set up when she arrived and that she could see from the monitor screen that Andy, Yumiko and Wolfgang had joined the meeting. Sarah explained that for *Self* she would like team members to imagine that they were interacting in some way with an eBike. She continued that for *Others* they should describe what they were doing to a fellow team member who should guess what the action was. She said that she was not asking them to mime what they imagined but to describe in words and went on to say that the exercise had a serious point which John would explain during the first agenda item. Sarah said that she would describe an action she was imagining to help team members understand what she meant and, closing her eyes, she said that the picture in her head was that she was sitting on the eBike and in moving along it was slowing down, and, with her eyes still closed, she said the eBike had quickly come to a stop. On opening her eyes, she asked the team what she had described and there was a chorus of braking. Sarah then gave team members a couple of minutes to think of what action they were going to describe and then asked the three Oxton members, excluding John, to pair with a team member in one of the remote locations and paired them off by name. She said that each pair member should describe the image that was in their head using the Chat facility in the online meeting software, with the other pair member responding by Chat. After a couple of minutes, there was sporadic laughter, and following a few more minutes, in walking round the room looking at laptop screens, Sarah was able to judge that the exercise was finished.

Bringing the attention of the meeting back to her, Sarah asked for comments and there was general agreement that it was a fun exercise, although Andy said he found Sam's description too ambiguous. John asked Andy to say more, and Andy explained that he thought Sam was avoiding a pothole, but he was actually coming round a sharp bend when going down a steep hill with a strong cross wind. This caused laughter in the team, and when it had subsided, John said that was a good example, and John asked if anyone else had had difficulty in understanding what their partner was describing. Following this prompting, Jennifer said that she had found it difficult initially since from Yumiko's description of one of her legs going up and down, she had thought she was riding a bike with one leg but, with a bit more information, she gathered she was pumping up a tyre with a foot pump. Again, this prompted laughter, and John thanked Andy and Jennifer saying he would pick up the examples later in the meeting. John then apologised to Sarah for interrupting and handed the meeting back to her. Sarah then concluded the Warm Up by taking the team through the meeting purpose and agenda and handed over to John for the first agenda item which she had said was feedback on the teams' excitement features list for Oe375. John told the team that Jim Eccleston had thought that the list was creative, and that Jim particularly liked the ideas of a detachable battery and easily detachable wheels but was sceptical that the return from a mild hybrid commuter was useful, although adding that Jim had said not to rule anything out at this stage.

John began the next agenda item by explaining the difference between Explicit and Tacit Engineering Knowledge. John said that *Explicit Engineering Knowledge* was organised knowledge expressed in formal language and disseminated through textbooks, slide presentations, electronic media, mathematical relationships, graphics, software models and so on. John continued by saying that *Tacit Engineering Knowledge* was subjective and unstructured knowledge residing in people's heads that was difficult to disseminate. John then showed a slide summarising what had occurred in the Warm Up exercise (Figure 4.1).

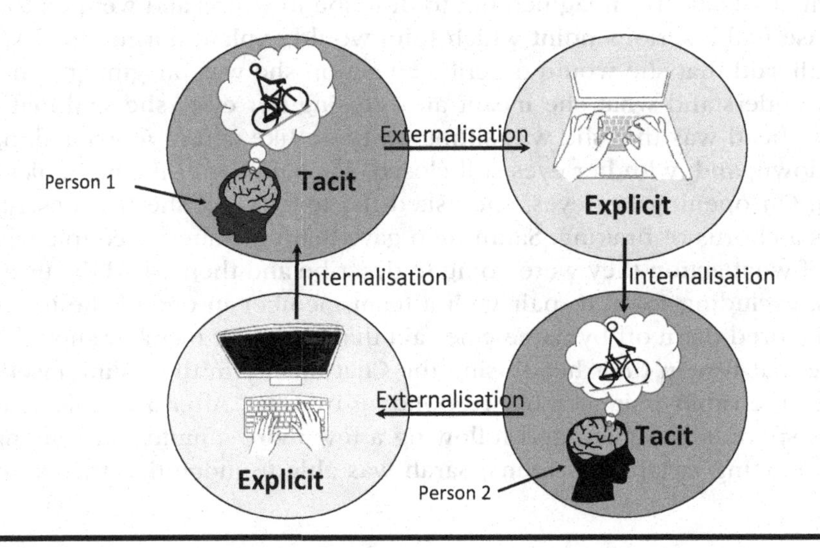

Figure 4.1 Tacit and explicit knowledge transfer.

John explained that the slide was intended to represent what happened earlier in the *Self* and *Others* phases of Warm Up with one person, whom John called "Person 1", transferring their mental model, that is their tacit knowledge, to the other person by making it explicit with the Chat facility. He continued by saying that the second person, whom John called "Person 2", then converted this explicit knowledge to their own tacit knowledge in the form of a mental model. John said that the process then reverses, like a feedback loop, with the knowledge going back from Person 2 to Person 1 through the tacit be explicit, and explicit to tacit, conversions. John continued his description of the knowledge transfer shown on the slide by saying that Person 1 ends up with a second mental model and checks this out against their first mental model. John explained that, as the graphic on the slide illustrated, tacit knowledge could only be transferred between people by making it explicit since we cannot see inside another's head, adding that this was probably just as well at times, which prompted laughter. John then brought team members' attention to the terms of *Externalisation* and *Internalisation*, as shown on the slide, to mean the transfer of tacit knowledge to explicit knowledge and explicit knowledge to tacit knowledge, respectively.

John continued by saying that the conversions of tacit to explicit and explicit to tacit made the process of knowledge transfer prone to error as illustrated in the graphic with Person 1 thinking about a person cycling up a steep incline and Person 2 interpreting Person 1's explicit description of their mental model as a person cycling up a slight incline. John asked team members to think back to their experience in Warm Up saying that at least some of them found this transfer of knowledge was subject to error which was to be expected. John added that some of them had found that they had to go round the loop depicted on the slide at least twice in order to improve the integrity of the knowledge transfer. John continued by saying that the difficulty in sharing tacit knowledge means that there is a tendency for engineering teams to focus almost exclusively on explicit knowledge in their day-to-day communication and added that this significantly diminished the efficiency of the team.

John said that the technique they had used in Warm Up of communication through the sharing of a mental model with another was called *Visualisation* and added that as a team they would be doing quite a bit of this. John explained that there were two kinds of visualisation that they would be using, with the first being the sharing of the mental images of an engineering system that individual team members formed in their heads. John added that the second type of visualisation was making the shared mental images visible as a model of the system being considered. John said that taking best advantage of both types of visualisation depended on good communication between team members. John said that he recognised that communication between Oe375 team members was not helped by the team being separated geographically. John told the team that, as at least some of them had found in the Warm Up exercise, the process of sharing tacit knowledge was normally iterative being dependent on the asking and answering of questions and that in their next meeting team members would start to look at communication skills a bit more closely. At this point, John asked the team members if they had any comments, and Sarah said that she liked to form pictures in her head of engineering concepts, and it was useful to think about how best to share them with others. Wolfgang agreed with Sarah and said that he found what John had said to be very interesting and he felt it had made him more aware of how he thought about things. John thanked Sarah and Wolfgang for their comments and said that team members would get an opportunity to use visualisation later in the meeting in the context of the OFTEN framework. In moving on to the next agenda item regarding the Use Case Diagram, John looked to Sarah who pointed to Jennifer, and John handed the meeting over to her.

Jennifer began by sharing a slide displaying the UCD that she had developed from John's and Sarah's draft (Figure 4.2).

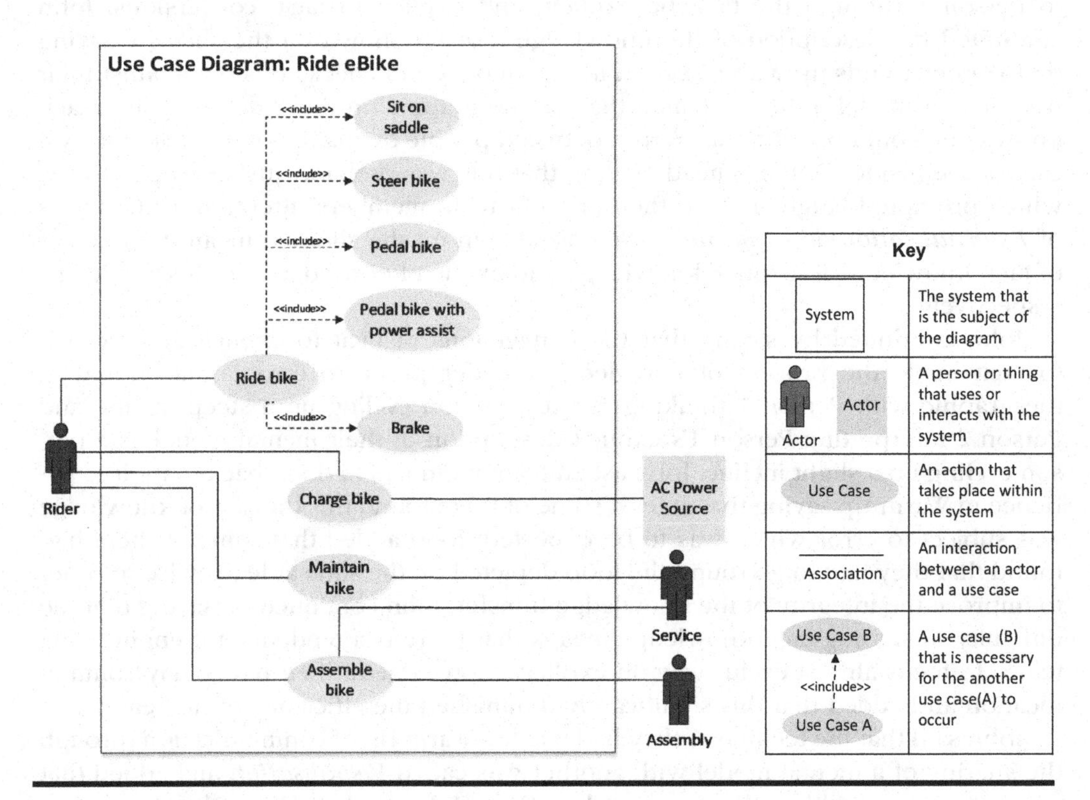

Figure 4.2 Use case diagram for Ride eBike.

Jennifer said that she had worked with Sarah and John in developing the UCD and asked if the other members of the team were familiar with the technique. Phil said while he had come across UCDs before, his understanding was a bit shaky. Jennifer explained that the UCD should help the team to get a better shared understanding of how the rider and other stakeholders interact with the way the eBike system functions, and, in doing so, it should help to identify the user and stakeholder's requirements of a system. In listening to John speak about tacit and explicit knowledge, Jennifer had recognised that, although she did not, think about it at the time, what John, Sarah, and she had done in developing the UCD was to share their mental models of system functionality and make them explicit. Pleased with this realisation, she made this point to the team. Jennifer brought team members' attention to the Key beside the diagram and after pausing to allow team members to assimilate the information it contained, she said that she thought the diagram was largely self-explanatory and asked team members for questions or comments. Phil asked Jennifer if she would expand on what the label "include" meant. Jennifer said that in riding an eBike the rider will need to *Pedal bike* and so the use case of pedalling the bike is included in the use case of *Ride bike*. Phil thanked Jennifer for her example.

Since no one else spoke, John said he was impressed by what Jennifer had done given the rather sketchy version that he had developed with Sarah's help. Jennifer thanked him and asked the team if there were any further comments on the content of the UCD. Phil said that he noticed that the UCD did not include any of the teams' excitement features such as regenerative pedalling, and, in agreeing with Phil, Jennifer said that once the team had decided which excitement feature to include, these would probably deserve their own use case diagrams. Sam observed that the powered-only mode was not shown on the UCD despite it being legal in some global markets. John intervened to explain that while the motor-only mode was legal in some parts of the world, for example, some States in the US, Oe375 was being designed, at least in the first instant, for the UK and European markets, and he recommended that the UCD be left as it was. Since there was no disagreement with John's recommendation, he handed the meeting back to Jennifer, and she asked if there, were any further comments and Yumiko said that the graphic had a different appearance to the UCDs that she had developed. In reply, Jennifer noted that there was no one accepted standard for UCDs, and Oxton used the widely accepted representation depicted on the slide. John noted that eBike assembly had been added as an actor since he had last seen the UCD and made this point to Jennifer. Jennifer explained that the people who pre-assembled and final-assembled the bike were important stakeholders because experience had shown that designing a system without thinking clearly about how it is to be pre-assembled and shipped can lead to problems late in the design process. Looking at the time, Jennifer saw that she had overrun her allotted agenda timing and thought to herself that this was as much John's fault as hers. Jennifer concluded the discussion by saying the UCD gave a good graphical representation of system-level requirements which the team would find useful and handed the meeting back to John. John said that in the next agenda item they would look at the SSFD (system state flow diagram) in beginning to design an eBike that met stakeholder requirements. John added that before beginning the SSFD the team should take a short break.

John caught Sarah on her way out of the room and told her that he liked the way she had conducted the Warm Up and added it seemed as if team members had enjoyed the exercise. Sarah thanked John and said that she had enjoyed the experience. On restarting the meeting after the break, John said he thought it is important that the OFTEN framework was set in an overall engineering design context and apologised in advance if this seemed like training. He said that his experience in working with engineering teams was that they get so involved in using engineering design tools they tend to lose sight of why they are doing, what they are doing and where it is leading to. He continued by asking team members if they were familiar with the Systems Vee model, and, since there was a murmuring of consent, he shared a slide depicting the model (Figure 4.3).

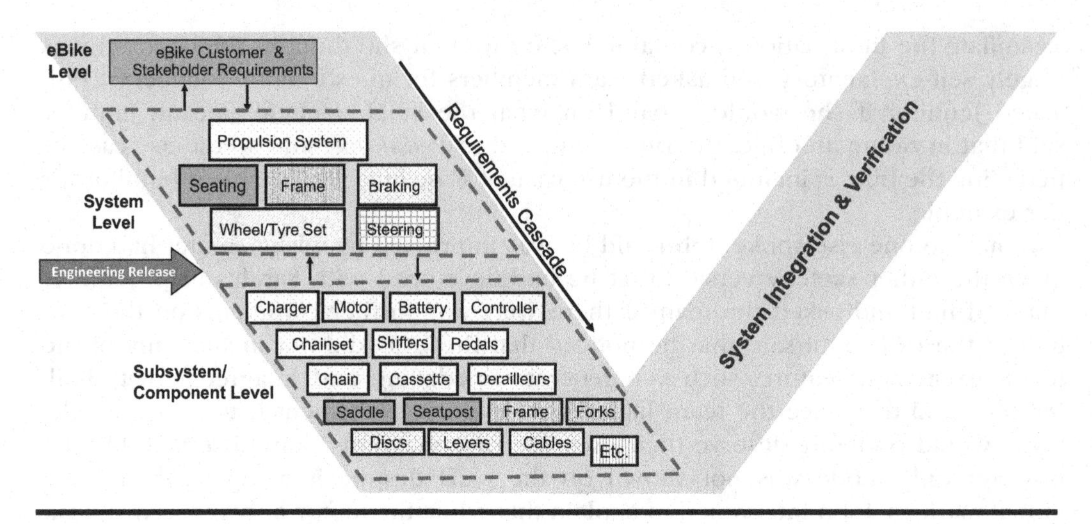

Figure 4.3 Systems Vee model – eBike partitioning and decomposition.

Before explaining the content of the slide, John mentioned that he had uploaded a PDF copy of all the slides he was going to show today to the team's ShareDrive in a file named "meeting slides" annotated with today's date. John said that, providing he remembered to, he would do this for all their future meetings.

Turning his attention to the slide, John said that the graphic depicted a top-down product design process which incorporated the OFTEN approach. In referencing the left-hand side of the Vee model, John said that the eBike-level requirements were established using techniques such as the UCD and then deployed to the systems level. John added that the design process progressed at system level to establish system-level requirements which were then deployed to the corresponding subsystems and components further down the Vee. John explained that the top-down requirements cascade allowed the components and subsystems to be designed in the context of their corresponding systems and hence of the overall eBike.

In bringing team members' attention to the right-hand side of the Vee, John reminded them that, once designed, the components and subsystems were integrated into their corresponding systems and ultimately into the eBike. John concluded his explanation of the model by saying that the systems integration process included verification that the components and subsystems met their defined requirements both in the context of the system of which they were part and in the context of the eBike. John then asked the team for their comments.

Wolfgang said that he thought John had said that OFTEN was based on a functional, rather than a hardware, approach to design and yet the Systems Vee slide seems to show hardware decomposition. John thanked Wolfgang for his comment and asked him what his mental image was if he said the words Electric Bicycle to him, to which Wolfgang replied that his mental image was of an electric bicycle which prompted laughter from others. John continued by asking Wolfgang to describe the image, and Wolfgang responded by saying that the object in his mind had two wheels attached to a frame with handlebars, a saddle, pedals, Drivetrain, battery and motor. John remarked to Wolfgang that his mental model seemed to be based on hardware and said that this was because it was based on his experiences of what an electric bicycle was. Taking the example of the eBike Propulsion System, John said that this box label could also indicate a functional model and said that the team were shortly going to look at such a functional model. John then reassured Wolfgang that he had not made a mistake and said that everyone would almost certainly have the same kind of mental model as him with the exception perhaps of an OFTEN geek who might see an eBike function tree in the Propulsion System box, which prompted smiles. John concluded his discussion with Wolfgang by saying that he had raised an important point in that engineers often have difficulty seeing past hardware and explained that the first step to doing so is for us to be aware of our preconceptions. John said that they would look further at the role preconceptions play in people's thinking in a future meeting.

Following Wolfgang's discussion with John, Phil said that, in his experience, he had not found the Systems Vee model to be very helpful explaining that, while he could not disagree with it, it had not helped him in doing systems engineering. John said that he sympathised with Phil's viewpoint and continued by saying that the Systems Vee was about the "What" and as such did not help much with the "How". John explained that the OFTEN framework should help deliver the "How" and added that he would let team members judge for themselves how well it does this after they had used it.

John moved the meeting on by saying that he wanted to take the team through Phase 1 of the OFTEN framework in a bit more detail than they did at their previous meeting, reassuring the team members that when they had done this they would, at long last, begin the Oe375 engineering design journey. John shared a slide showing the four phases of the OFTEN framework with Phase 1 *Function Analysis* highlighted (Figure 4.4).

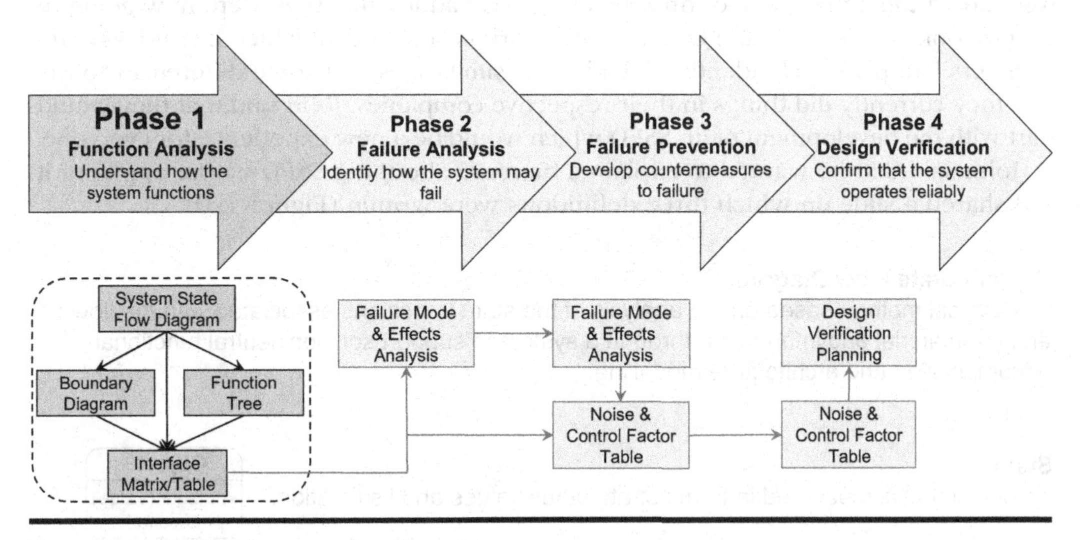

Figure 4.4 OFTEN framework – Phase 1.

John said that the objective of Phase 1 of the framework was to understand how a system functions and added that, as they had discussed in their last meeting, Phase 1 included the use of the boundary diagram, function tree, system state flow diagram and interface analysis as the primary methodologies for this phase. John explained that, as team members would shortly experience for themselves, the development of an SSFD uses both kinds of visualisation that he had spoken about before the break in that it represented a visible model of a system and was based on team members, shared mental images of how the system worked. John added that the SSFD underpinned the analysis that was conducted in this phase with both the function tree and boundary diagram being derived from the SSFD with the information gained from applying these tools leading directly into interface analysis. John explained that the output of Phase 1 was detailed functional requirements documented in interface tables which as the graphic implied facilitated the development of the failure modes and effects analysis (FMEA) in Phase 2 of the framework. John continued by saying that, as they had established in their last meeting, Oxton currently developed the boundary diagram based on existing hardware. John asked how a function tree was developed in Oxton. Sarah said that function trees were hardly ever developed in Oxton, adding that when they did develop a function tree it was developed through brainstorming. John then asked the Oxton team members if the boundary diagram was an input to the development of the interface matrix, and Sarah and Phil replied in unison that it was.

Turning his attention away from the Oxton team members, John asked Andy and Yumiko what happened in their companies, and they both agreed that while the SSFD was new to them they used function trees, boundary diagrams, interface matrices and interface tables on a regular basis. They also agreed that, like in Oxton, the functions in a function tree were brainstormed first and then arranged in the tree and the boundary diagram was developed based on existing hardware. John thanked the team members for their inputs which he said helped him to understand how they currently went about the initial stage of product design. He added that they were now going to start to look at Phase 1 of the OFTEN framework in some detail which he said, as team members had previously identified, had some similarities and some differences to the way they currently did things in their respective companies. John said that they would start with the development of an SSFD which would be a new experience for everyone.

John said that he wanted to explain a bit more about the SSFD before applying it and shared a slide on which three definitions were written (Figure 4.5).

System State Flow Diagram
A graphical method based on the analysis of the state transitions associated with the flow of energy, material and information through a system to support solution-neutral functional decomposition and architecture modelling.

State
An object that is measured in terms of attributes/values and its location.

Object
A thing that can have physical and/or cerebral existence.

Figure 4.5 Definitions – system state flow diagram.

John said that the key definition was that of the SSFD and this definition relied on the two definitions beneath it. John paused to allow team members to read the definitions before explaining that the box to the right of the State definition was a graphical depiction of a State. Sam asked what John meant in saying that an object could have cerebral existence, and John explained that he meant that an object could exist in someone's brain as a mental image. John said that to help explain this further and make the definitions more concrete he would look at the example of the household system of a kettle and shared another slide (Figure 4.6).

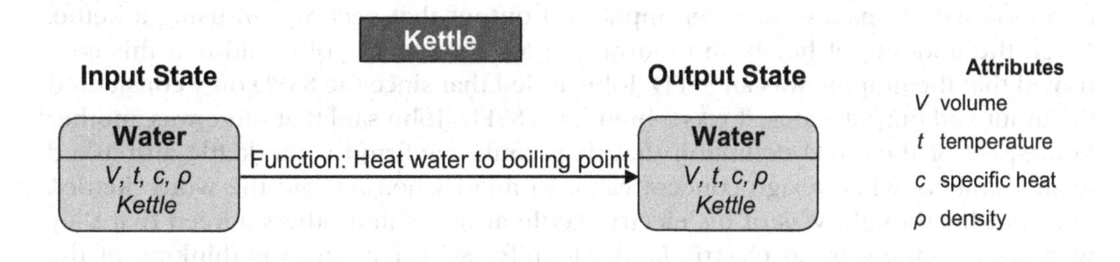

System State Flow Diagram
A graphical method based on the analysis of the state transitions associated with the flow of energy, material and information through a system to support solution-neutral functional decomposition and architecture modelling.

Figure 4.6 High-level SSFD for kettle.

John explained that the graphic depicted the principal input and output states for the kettle with both input and output states being depicted graphically as an object with attributes and location. John said that, since the main aim of the graphic was to illustrate the concept of state transition, he had simplified matters ignoring the need to have a source of heat. John said that one important aspect of the graphic was that the arrow symbolised the function required to achieve the state flow and added that he had stated the function required to achieve the transition of the input state of water to the output state of water in the middle of the graphic. John said that his experience was that familiarity with boundary diagrams acted as a conceptual block at first to many engineers in their initial interaction with SSFDs because they assumed that since the diagram considers energy and/or material and/or information and included arrows, the arrows represented the flow of these entities. John said he would introduce a simple graphic to reinforce the point that the arrows on an SSFD represented functions shortly.

John then asked team members what distinguished the input and output states, and Andy said that they were distinguished by their attributes of volume, temperature and density. John asked Andy to say more, and Andy explained that typically the input volume and temperature of water in the kettle might be 500 ml at 10 °C, while the output volume would be slightly less due to vaporisation and the water temperature would be around 100 °C. John thanked Andy and said that he was right in that people heating water in a kettle would normally only be interested to know if the water was hot and they had enough of it. John added that if someone wanted to quantify the energy transition, they would also need to know the specific heat c along with the density ρ of the water and so he had added these as attributes of each state. John asked team members to also note that in this example the location was the same for the input and output states.

John asked team members if they agreed that the graphic at the top of the slide matched with the definition at the bottom of the slide in the respect of depicting the principal state transition between input and output that occurred in using a kettle. Given the nodding of heads and murmuring of agreement, John said that this confirmed that the graphic was an SSFD. John added that since the SSFD only considered the input and output states, it was a high-level SSFD. John said that there was another key aspect of the SSFD definition that they could confirm for the kettle and asked team members what design concept came to mind when he said the word "kettle". Phil said his thought was of his electric kettle at home and others agreed that they were also thinking of an electric kettle. Jennifer said that she was thinking of the kettle she used while camping. John asked Jennifer how her kettle was heated, and Jennifer said that she placed it on a small gas camping stove. John asked Phil and Jennifer if they felt the SSFD shown on the slide was applicable to the type of kettle that they had imagined, and both agreed that it was. John said that this meant that the graphic was design solution-neutral unless any team member could think of a design concept of a kettle to which the graphic did not apply. Since no one said anything, John took the opportunity to make the point that the solution-neutral aspect of the SSFD helped creativity as the engineering design was not constrained at the beginning of product design to a particular hardware-based design concept. John added that this did not mean that the SSFD could not be used where a design concept had already been chosen.

In returning to the kettle example, John said that since the SSFD was a high-level depiction of the kettle, it could not be used for functional decomposition or architecture modelling, apart from the function stated on the graphic and the architectural description of "kettle". John added that they would analyse an eBike system next where the SSFD supported both functional decomposition and architectural modelling. Phil asked John why, since he had spoken of state transitions, the system state flow diagram was not called the system state transition diagram. John said that he saw it as a useful construct to consider the state transitions as a flow and added that this flow would become more apparent when they considered the eBike SSFD example.

Taking a deep breath, since he knew this was a key moment for the Oe375 PD team as they transitioned to the detailed engineering analysis, John said that he would like to start Phase 1 of the OFTEN framework as far as Oe375 was concerned by analysing the Drivetrain at system level. He added that, as team members knew, the Drivetrain was one of the key systems on an eBike since it needed to integrate seamlessly with the electric motor-based power source and shared a slide (Figure 4.7).

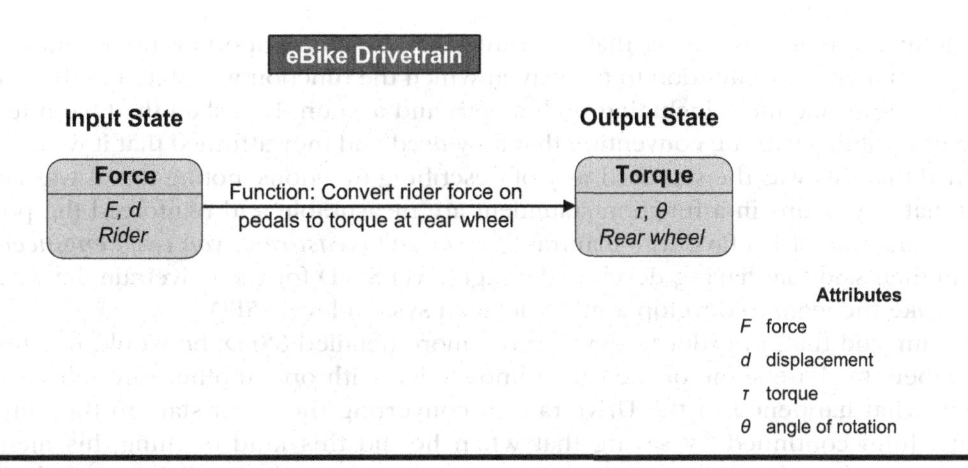

Figure 4.7 High-level system state flow diagram for Drivetrain.

John said that the graphic shown on the slide was the high-level SSFD for the Drivetrain that team members would recognise as having a similar layout to the high-level SSFD for the kettle. John explained that the input to output states shown on the slide were associated with the transition of states of energy associated with the Drivetrain and added that he wanted to explain how the name of the input and output states of *Force* and *Torque* related to this state transition. In considering the input state, John said that he was using the term *Force* to describe the input state in the SSFD as shorthand for *Energy associated with force* and had indicated this by including both force and displacement as attributes. John added that as team members would know the work done by a force was the product of the force and the displacement through which the force acted with work done being equivalent to energy.

Turning to the output state, John said that like the input state the name of the state of *Torque* was shorthand for *Energy associated with torque*, and in this case, like the input, he had quantified this energy by considering the work done. John added that the work done this time was the product of the torque and the angular displacement which as team members could see were the attributes associated with the output state.

After a pause in which no one spoke and in view of the question that Phil had posed earlier regarding the name of the SSFD, John asked what the flow of states through the eBike Drivetrain system shown on the slide was. Yumiko responded that it was the flow of the state of rider force energy on the pedals to the state of torque energy at the rear wheel. John thanked Yumiko, reinforcing to the team that Yumiko's description included the location associated with each state. John asked the team to note that, like the kettle, the Drivetrain SSFD included a statement of the function required to generate the state flow with the arrow on the graphic being associated with this function. John brought team members' attention to the fact that he was considering both pedals in stating the function since the work done by the rider in propelling a bicycle or eBike was largely done on the down part of the pedal movement around the chainset axis with each pedal having its down part of the movement as the other pedal was moving up.

John continued by saying that he wanted to stress an important point and drew the team members' attention to the way in which the function was stated in the SSFD as an action statement including both a verb and a noun. He asked the Oxton team members if this was the convention that they used and they affirmed that it was. John added that this was the standard way of describing functions, noting that it was critical that any nouns in a function statement are measurable and reinforced the point by stating one of his favourite mantras *if you can't measure it, you can't engineer it.* John then said that having developed a high-level SSFD for the Drivetrain, he would now like the team to develop a more detailed system-level SSFD.

John said that, in order to develop the more detailed SSFD, he would like team members to share some of their tacit knowledge with one another through visualising what happened in the Drivetrain in converting the input state to the output state. John continued by saying that when he did this kind of thing, his mental model was based on his experience of the system he was visualising, and closing his eyes, John said that the picture he saw inside his head in this case was of a conventional Drivetrain with a pedal, chainset, chain and cassette. Opening his eyes, John added that this was not a problem in terms of developing a solution-neutral SSFD provided that he did not allow his mental model to constrain his thinking. John asked the team members to get into pairs to do the visualisation as this was often the optimum number for sharing tacit knowledge. Given that he did not feel it appropriate to get involved and aware that there were seven team members, he asked for help in forming pairs. Andy suggested that the three people in the remote locations work together, and John thanked him for the suggestion and said he would create a breakout room in the meeting software and assign Andy, Wolfgang and Yumiko to it.

John then told the team that what he wanted each of them to do was to think of themselves as a small amount of force energy that the rider applied to the pedals to rotate the chainset through a small angle and visualise what happened to them as energy in travelling to the rear wheel in terms of their state. John continued by explaining that team members should think about what kind of energy they were at different locations on their journey, adding that he found that closing his eyes helped him in a visualisation. John asked the team to record the state flow as a list starting with "I am a small amount of force energy at the pedals" and finishing with "I am a small amount of torque energy at the rear wheel". John added that he would like each pair or three to develop a single list between them. John said that he had one piece of advice in developing such a list which was that each line in the list should correspond to a recognisably different state. In expanding on this advice, John said that if a team member got to the level at which they were describing energy being passed from one link of the chain to another, then this was at too detailed a level of analysis. John asked if there were any questions about what they should do. As there were no questions, John then set up the breakout room for the external team members and told the team members to start the visualisation in their groups.

John noticed Jennifer and Phil had formed one pair with Sam and Sarah forming the other, with both pairs seemingly engaged in the exercise. John let the exercise continue for five minutes before asking the Oxton pairs how they were getting on, to which they replied that that they had finished. John joined the breakout room and asked Andy, Wolfgang and Yumiko the same question and they said they needed a couple more minutes which John gave them after which he brought the team back together.

Believing the lists would be similar and noting that Jennifer had recorded her and Phil's list on her laptop, John asked Jennifer to share her screen with the team. John asked the other team members to identify the similarities and differences between their lists and Jennifer and Phil's list. The comparison showed the three different lists to be largely consistent apart from some slight wording differences. Team members quickly agreed on the wording of the list which allowed Jennifer, with a few edits, to convert her and Phil's list into the teams' list (Table 4.1).

Table 4.1 State Flow in the eBike Drivetrain

Drivetrain
I am a small amount of:
- force energy at the pedals
- torque energy in the chainset
- tension energy in the chain at chainset
- tension energy in the chain at cassette
- torque energy at rear wheel

John said that before going further he would just like to clarify something and said that he had noticed that the Oxton engineers referred to the *chainset* and he assumed that this was what he would call the *crankset*. Phil responded by saying that the words chainset and crankset were interchangeable, with chainset being the preferred term in the UK while the use of crankset predominated in other parts of the world such as North America. Phil added for good measure that the chainset comprised of the chainrings, the right and left crankarms and the spindle upon which the chainset rotated.

John thanked Phil for his clarification and asked team members for their reaction to the exercise. Sam said it felt a bit strange at first and added that he could not remember ever being asked to work with his eyes shut at Oxton before. Jennifer agreed with Sam and said that she found it good to think about things from first principles which was helped by closing her eyes. Wolfgang also agreed that while it seemed strange at first the process worked well. Sarah said that this visualisation was different to the earlier Warm Up exercise in that they all seemed to have very similar mental models. John said that this was almost certainly because this exercise was more focused, being directed to a specific system, whereas the Warm Up exercise was much broader. John thanked Sam, Jennifer, Sarah and Wolfgang for their feedback, adding that he would now show team members how they could convert their list into the Drivetrain SSFD and shared a slide (Figure 4.8).

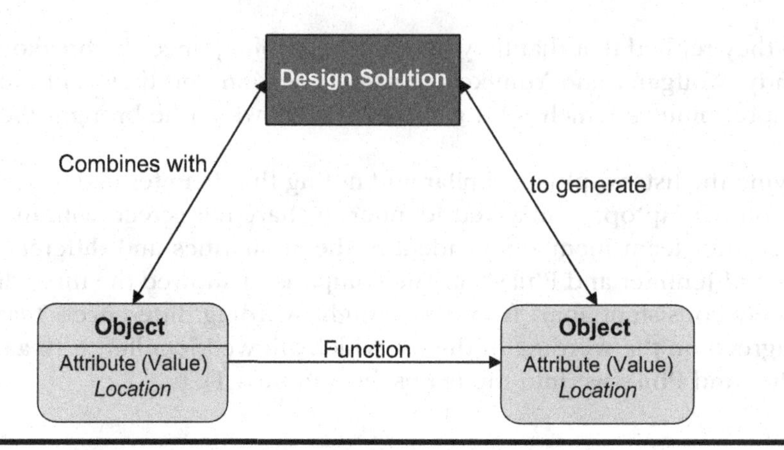

Figure 4.8 SSFD object, function and design solution triad.

John explained that he found the triad graphic to be useful in identifying the function necessary to achieve the state flow from one state to another and saw the graphic reinforcing the point that he had made earlier that the arrow on an SSFD symbolised the function required to generate the state flow. John said the triad also included what was called the *Design Solution* which was the entity that generated the function required to achieve the state flow and added that this should be design solution-neutral. John said that while the graphic probably seemed rather abstract, he hoped that it would become more tangible when the team used it in a moment.

John told the team that a triad template was available to them on the team ShareDrive in the SSFD folder and asked Jennifer if she would upload the list describing the state flows in the eBike Drivetrain that the team had just developed to the same folder. John then said that he would explain how the triad was used by looking at an example from the Drivetrain and edited and resaved the triad template slide under another name so that it included the teams' list of state flows in the Drivetrain (Figure 4.9).

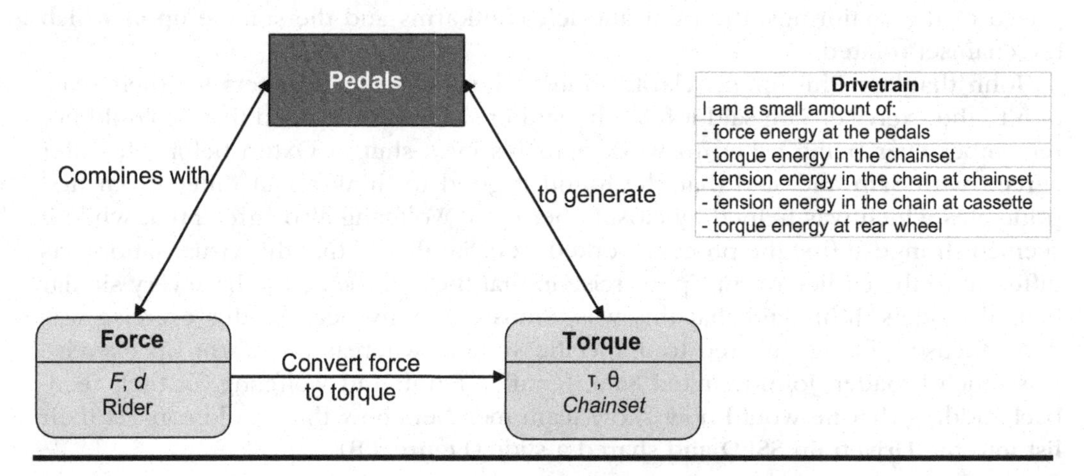

Figure 4.9 Triad for transformation of force at pedal to torque at chainset.

John said that he would consider the first state flow indicated by the first two states on the team's state flow list. John added the detail to the triad template talking the team through it as he did so and speaking slowly in saying "*The rider* ***combines with*** *the* ***pedals*** *to exert a small amount of* ***force*** *energy with attributes of* ***force*** *and* ***displacement*** *which* ***generates*** *a small amount of* ***torque*** *energy in the* ***chainset*** *with attributes of* ***torque*** *and* ***angular*** *rotation. The state flow is produced by the function* ***Convert force to torque***". John emphasised the words in his descriptive sentence as he was editing the slide, pausing between naming the Design Solution and each Object. On finishing the editing of the triad, John sat back from his laptop feeling that his performance deserved a round of applause which did not materialise, although he noticed that he had the full attention of team members.

John continued by telling the team members that it was their turn to develop triads and asked them to work in the same groups as earlier. He allocated each group one of the three remaining state flows denoted by two consecutive lines on their list so that they each had a different flow, reminding them that the triad template was on the ShareDrive and reinstated the breakout room for the external members. Keeping an eye on the Oxton team members, John noticed after a few minutes that both pairs were no longer looking at a laptop or a piece of paper but were sitting back in their chairs talking. Realising both that three people working together on a task often takes longer than two because of both the trebling of the one-to-one relationships and the fact that this more complex group was working remotely, John allowed several more minutes after it appeared that the two Oxton pairs had completed their triads. After what he judged was an appropriate time, John entered the breakout room to find that the three external team members had completed their triad and so asked everyone to return to the meeting. Each group then presented their triad to the team in turn with their results being represented in Figure 4.10.

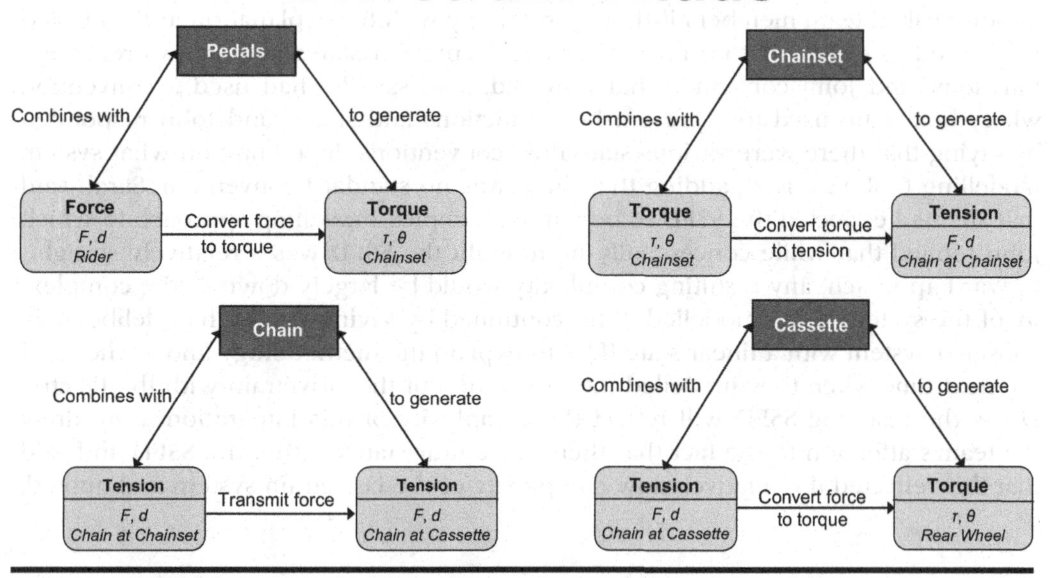

Figure 4.10 Drivetrain triads.

John thanked the team for their efforts and said that he was impressed with what they had done. He said that he had added the triad that he developed earlier to complete the set. He drew the team's attention to the fact that apart from the first and last states, which were the input and outputs to the system, all the states were duplicated on the slide, explaining that this was because each of the duplicated states was both what he described as a *to* state and a *from* state. John continued by saying that the team were now able to develop the Drivetrain SSFD, and if the team agreed, he would show them how this was done by synthesising the SSFD as they watched. As the team members readily agreed to John's suggestion, John asked Jennifer if she would help him to do this, saying that, judging by the computer skills she had already shown she was more adept at using graphic packages than he was and he wanted to complete the SSFD before the meeting ended. John posted the triads slide to the ShareDrive and sat down next to Jennifer at her laptop. As he expected, under his guidance, she edited and rearranged the triad slide in half the time that it would have taken him, with her audience watching with interest what was close to a smooth animation (Figure 4.11).

John thanked Jennifer for her skillful assistance and began summarising what Jennifer and he had done. John explained, as he said team members could see from the slide, the flow of states in the SSFD was that represented by the triads with their order being based on the state flow list that the team had developed earlier. John continued by saying that the duplication of states present on the triads had been eliminated in identifying the state flow on the SSFD since each state occurred only once as both a "to" and a "from" state. He added that the arrows along with the function statements that they represented were positioned between the same two states on the SSFD as on the corresponding triad. John pointed out that on the SSFD each Design Solution from the triads was associated with the function that it produced. John concluded his explanation by saying that finally a dotted line box was drawn such that the inputs and outputs on the high-level SSFD became the inputs and outputs to the system-level SSFD.

John asked team members if they were happy with his explanation and Sam said he wanted to clarify that on an SSFD, boxes represent states and arrows represent functions, and John confirmed that they did. Sam said he had used a convention where boxes are used to represent both functions and states, and John responded by saying that there were various state flow conventions depending on what system modelling tool was used, adding that there was no standard convention. Sarah said that she had expected the SSFD to be a more complex modelling approach to which John replied that while conceptually he thought the SSFD was a relatively straightforward approach, any resulting complexity would be largely down to the complexity of the system being modelled. John continued by saying that he had deliberately chosen a system with a linear state flow to explain the methodology and as they will see next time when they modelled the integration of the Drivetrain with the Electric Drive, the resulting SSFD will reflect the complexity of this integration. John drew the team's attention to the fact that there were four states within the SSFD and said that this reflected the relatively low complexity of the Drivetrain system and quoted,

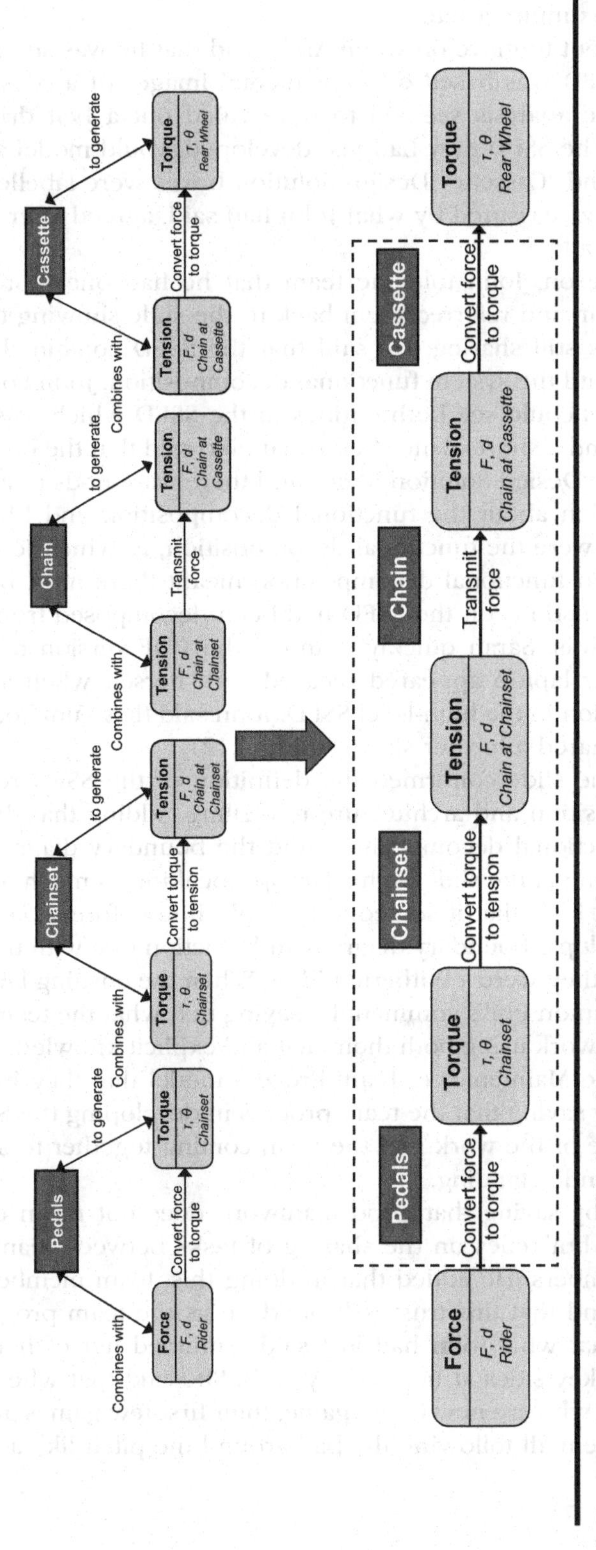

Figure 4.11 Development of Drivetrain SSFD from triads.

what he called a useful rule of thumb, that an SSFD should contain a minimum of four states and a maximum of ten.

John was just about to move on when Andy said that he was still a bit concerned that because the SSFD was based on their mental images of a conventional chain-based Drivetrain the analysis seemed to have ruled out a belt drive. John asked Andy if he thought the SSFD they had just developed would model a belt drive system if the "Chain" and "Cassette" Design Solution boxes were labelled as "Belt" and "Sprocket", and Andy, reassured by what John had said, agreed after a few moments of thought that it would.

Anxious to move on, John told the team that he had one more thing that he wanted to show them and referred them back to the slide showing the SFFD development that he was still sharing. He said that the SSFD contained both a system boundary diagram and the system functional decomposition. John continued by asking if team members could see both entities in the SSFD which was followed by a few puzzled looks and a short silence before Yumiko said that the boundary diagram was indicated by the Design Solution boxes and there were nods of agreement. John reiterated his question about the functional decomposition and Phil said that the function statements were the functional decomposition, to which John agreed. John then pointed out that functional decomposition meant there must be a function or functions that the functions on the SSFD had been decomposed from and he asked what this function was. Sarah quickly scanned the PDF version of the slides and looking up from her laptop appeared pleased with herself when she pointed out that it was the function in the high-level SSFD. John said that Yumiko, Sarah and Phil were spot on and shared a further slide (Figure 4.12).

John said that the slide confirmed the definition of the SSFD referencing both functional decomposition and architecture modelling, adding that the function tree represented the functional decomposition and the boundary diagram the architecture modelling. Staring at the slide on his laptop, speaking as much to himself as the others, Phil said out loud that if someone had told him before today's meeting that the team could develop a boundary diagram and function tree with their eyes closed, he would have said they were a blithering idiot. When the ensuing laughter had subsided, John picked up on Phil's comment by saying that what the team had done was based on good teamwork using both their tacit and explicit knowledge. He reminded the team of the Task, Maintenance, Team Process model that they had discussed in their last meeting by saying that the team process in developing the SSFD was based on pairs doing some of the work and the team coming together to share that work to gain a common understanding.

John continued by saying that good teamwork does not mean everyone doing everything together but relies on the sharing of tasks between team members and groups of team members. He added that in doing this, team members will need to trust one another, and that this trust will build up as the team progresses. Jennifer intervened to say that what John had just said reminded her of her experience in coaching junior hockey sides at her hockey club. She said that whenever you get a group of youngsters who are new to the game, their first few games are always comical to watch with them all following the ball around the pitch like a swarm of bees.

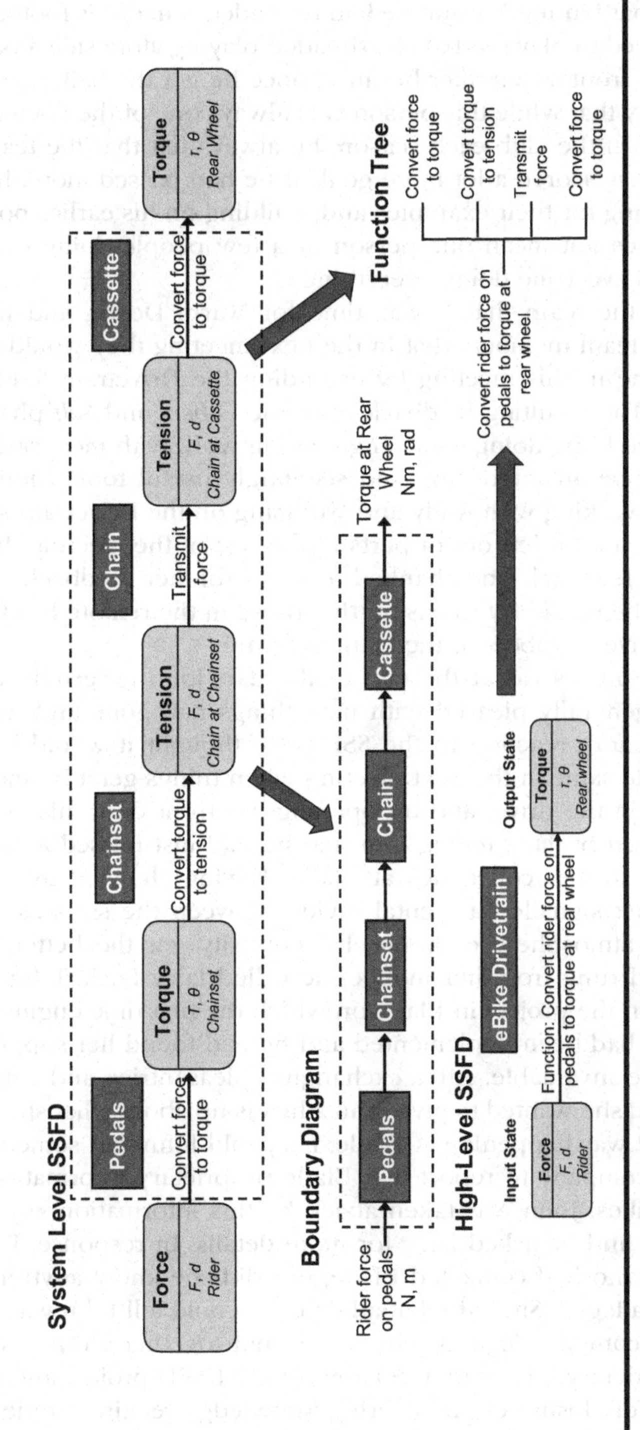

Figure 4.12 Derivation of boundary diagram and function tree from SSFD.

She continued by saying that was why they spent a quite a lot of time practicing both individual and passing skills.

Wolfgang said that Jennifer's example had reminded him of his football days, adding that when he used to play football, he dreaded playing alongside a particular person who played up front as a striker because, once he got the ball, he rarely passed it. He went on to say that while this person was always one of the contenders for the highest goal scorer in the club each season, he always felt that the teams in which he played would have scored a lot more goals if he had passed more. John thanked Jennifer and Wolfgang for their examples and, building on his earlier point, said that good teamwork does not mean one person or a few people doing everything any more than it means everyone doing everything.

John then told the team that it was time for Warm Down, and in looking at *Task*, he reminded team members that in the next meeting they would build on the work they had done in this meeting by extending the Drivetrain SSFD to include the Electric Drive. The resulting feedback from the *Others* and *Self* phases was that people were pleased to be doing some engineering work with most saying that they found the SSFD to be an interesting and seemingly useful tool. Yumiko said that while she enjoyed working with Andy and Wolfgang on the Drivetrain state flow list and the triad, she felt a bit left out of parts of the rest of the meeting. John thanked the team for their feedback and thanked Yumiko for her feedback on the Team Process telling her he would try to ensure that those in the remote locations felt that they were more active members of the team in future.

Travelling home in his car at the end of the day, John reflected on the Oe375 meeting and was generally pleased with how things had gone and was particular happy about the team's reaction to the SSFD. He thought it would be interesting to see if they felt the same in the next meeting when things get a bit more complex. He parked his car in the drive, and on opening the front door, his good frame of mind was interrupted by Jane telling him that he had just missed a call from Lucy Collins, one of his former colleagues at Blade. Having changed into loungewear, which he always felt signalled a mental divide between the stresses of work and the more relaxing atmosphere of home, his curiosity got the better of him and, noting that she had rung from her mobile, he called Lucy back. John had worked closely with Lucy on the project in Blade on which the first-time engineering design (FTED) framework had been implemented and he had found her support and positive approach to be invaluable. After exchanging pleasantries and catching up on family life, Lucy said she wanted to give John a heads-up about what she described as "a bit of a flap" that was happening at Blade. Lucy told John that someone had been in touch with the company to report that Blade proprietary information was being shared at Oxton Bikes. John was taken aback by this information assuming that it must relate to him, and he asked Lucy for more details. In response, Lucy said that she did not know who had contacted Blade, nor did she know anything about the detail of what was alleged. She said that all that she could tell John was that she had been asked by the company legal department to put together a report outlining the nature of any proprietary information relating to the FTED project and to gauge the seriousness of any disclosure of engineering knowledge relating to the two patents that had been filed as a result of the project. Lucy continued by saying that while she would have liked to have spoken at more length with him about how he was getting on at Oxton Bikes, she did not feel that to be appropriate in the circumstances. Lucy added that she was calling John in confidence because she saw him as a friend and said that she would appreciate it if John would treat their phone call as never having happened and, in wishing John a good evening, rang off.

John was flabbergasted by what Lucy had said and was angry that someone could do what they had done to demean him. Remembering that he had mentioned to Paul Clampin while he was shadowing him, that the FTED project in Blade had spawned two patents, John's immediate reaction was that it must be Paul Clampin who had contacted Blade. In thinking about this further, John realised he had mentioned the patents to several others at Oxton as an illustration of the creativity that the FTED framework promotes. John also thought that it would not take a genius to figure out the system that the patents were associated with from the engineering case study that he had used in the published papers he had co-written, including the recent conference paper. He was reassured that the papers had been sanctioned for publication by the Blade Public Relations department. John was also pleased that his MSc thesis did not contain any sensitive content and so had not required a moratorium to be placed on it restricting access. John concluded that, given the flimsy circumstantial evidence that he had, he could not accuse Paul Clampin, or anyone else for that matter, of contacting Blade.

On talking the matter over with Jane, John was comforted by the fact that she believed in his innocence without him having to tell her that there was no foundation to the accusation. Following a discussion which only served to raise more questions than answers, they decided that since there was nothing that they could do before John got into work the next morning, they had best try not to think about it anymore that evening. John found this very hard to do, however, and went to his home office to confirm what he already knew to be the case, that he had no potentially incriminating files. A check of his laptop, his external hard drive and his various flash drives only verified that, while he did have some of the slides that he had developed at Blade, these were harmless in the sense that they only related to material that was already in the public domain when he left Blade. The worse offence he thought he could be accused of, was that by retaining the Blade logo on some slides, it could be said that he had retained material that he had developed while he was employed by Blade. He thought of going through and removing the Blade logo from these slides, but he knew a record of what he had done would probably remain on the hard drive which might tend to make him look guilty when he had nothing to hide.

John felt that this was a situation where it might be good to seek advice from his union. He had joined the union as an apprentice and maintained his membership throughout his time at Blade but had let his membership lapse on joining Oxton since Oxton was not a unionised company. On reflection, John thought it better not to give the matter anymore visibility than it deserved and decided that even if he had still been a member of the union, he probably would not have raised the matter at this stage. In the end, John tried to take his mind of things by joining Jane to watch a couple of episodes of the latest season of their favourite detective series on the TV streaming service they subscribed to, but he found it difficult to concentrate. In the end, Jane and he retired to bed early to spend rather a restless wakeful night.

Background Reading and Viewing

Tacit and Explicit Knowledge	Nonaka, I. and Konno, N. (1998) The Concept of "Ba": Building a Foundation for Knowledge Creation, ***California Management Review***, 40(3), 40–54, https://doi.org/10.2307/41165942 Pages 40–45
Systems Vee Model	Siemens Software. (2015) ***System Engineering Brief: Managing Complexity with a Systems Driven Approach***, You Tube, www.youtube.com/watch?v=uEmX7rw0fKg, Accessed 18 December, 2022
Visualisation – Sharing Mental Models	Gillespie, B. and Chaboyer, B. (2009) Shared Mental Models Enhance Team Performance, ***Nursing in Critical Care***, 14(5), 222–223, https://doi.org/10.1111/j.1478-5153.2009.00357.x
System State Flow Diagram (SSFD) SSFD Object, Function, Design Solution Triad	Campean, F., Henshall, E., Yildirim, U., Uddin, A. and Williams, H. (2013) A Structured Approach for Function Based Decomposition of Complex Multi-disciplinary Systems. In: Abramovici, M. and Stark, R. (editors) ***Smart Product Engineering. Lecture Notes in Production Engineering***, Berlin Heidelberg, Springer, https://doi.org/10.1007/978-3-642-30817-8_12 Yildirim, U., Campean, F. and Williams, H. (2017) Function Modeling Using the System State Flow Diagram, ***Artificial Intelligence for Engineering Design, Analysis and Manufacturing***, 31(4), 413–435, https://doi.org/10.1017/S0890060417000294 Pages 413–422

Chapter 5

Applying Guidelines

John woke early and knowing that Jim Eccleston was normally the first person to arrive at work, as well as being the last to leave, he decided to see if he could bump into him on his way into the office from the car park. Setting off, at what was for him an hour earlier than normal, he arrived in the Oxton car park to see it practically empty apart from a couple of cars that he assumed had been there overnight. He was relieved to see that Jim's reserved space near the building entrance was empty and strategically parked his car some distance away so that as soon as he saw Jim's car enter the car park, he could leave his, hoping it would seem as if he had just arrived. John's assumption was that he could walk over to the building entrance by the time Jim was getting out of his car. John spent the time listening to his favourite early morning news and current-affairs radio programme and found himself shouting less than complimentary names at a politician successfully evading direct questions by quoting generalities. Jim duly arrived 20 minutes after John, and John's strategy worked like a treat as he strolled by Jim's car, hoping he looked untroubled, just as Jim was getting out. After exchanging greetings, John knew that he could not broach the subject of the dissemination of Blade confidential knowledge since that would compromise Lucy Collins. Instead, Jim asked John if he had seen anything of yesterday's stage of La Vuelta, the Spanish cycling grand tour currently entering its final week. Luckily John had thought that he had better show some interest in professional cycling on taking up his new job, as up to then the extent of his knowledge was if pushed being able to name a couple of the British riders who had won the Tour de France. Even though his current exposure to the sport was almost exclusively based on the evening highlights of the La Vuelta daily stages shown on TV, he had been surprised how exciting it was. He was also beginning to understand how important teamwork was in pro-cycling, being impressed by some of the team tactics that he had seen. He told Jim that he had not caught anything yesterday but said that this year was particularly exciting, basing this assessment on what the TV pundits had said.

Unfortunately, by the time they had climbed the stairs and walked down the corridor to the office area, Jim had not said anything about the alleged incident that was at the forefront of John's mind. Consequently, John was left none the wiser as Jim

DOI: 10.4324/9781003286066-5

walked past Vicki's desk and into his own office, leaving John to walk the short distance to his own office. John wondered if Jim's silence on the matter was because he had not heard from Blade or because he was still thinking about what to do about the situation. John assumed that it was probably the former since he had no idea when Blade had received the call, what action they would take and how long it would take to assess the report Lucy and the legal department were putting together. Indeed, he hoped that Blade would decide that, given the tenuous nature of both the accusation and the way in which it seemed to have been made, it was not in their interest to pursue it. However, when John thought further he knew that the reason why they had filed patents was because of the competitive edge they potentially gave Blade so it was unlikely that they would choose to ignore the situation, and so it was a case of when, not if, Jim would ask to speak to him about the matter.

John decided that he had little option but to continue working as usual until he heard otherwise and sent a text with the one word "Nothing" to Jane who he knew would be anxious to know what had transpired between himself and Jim. Looking at his calendar, he saw, as expected, his first meeting of the day was the Oe375 Review/Preview meeting. John had invited Yumiko to attend the first part of the meeting to better understand her comment in the Warm Down of the previous Oe375 meeting and to discuss with her how future meetings might be improved from her perspective. Because of Yumiko's attendance, he had scheduled the meeting early in the day which he thought was useful since this would help him not to dwell on the unfounded accusation against him. Along with Sarah, John had also invited Jennifer to the meeting having asked her if she would like to take the Warm Up in the next Oe375 meeting. Jennifer had agreed to manage the Warm Up on the understanding that John talked her through what she had to do. John was pleased to see that Yumiko had accepted the Review/Preview meeting invitation since he had only sent the meeting notice to her the day before, long after she would have left to go home.

Yumiko called in precisely on time on the web-based video meeting link John had sent her. Sarah and Jennifer had yet to arrive, so after greeting Yumiko, John said that he had invited her to the meeting both to get her reflections on yesterday's team meeting as a whole and to discuss her specific feedback during Warm Down. Yumiko said that she was sorry if John had been offended by what she had said, to which John responded by saying that he had not been offended in any way, telling her that he always welcomed feedback. He continued by saying that unless we receive feedback, we cannot be fully aware of the effect that what we say and do, or not say and not do, has on others. At this point, Sarah and Jennifer walked into John's office, obligatory cups of coffee in their hands, and, after exchanging greetings, John summarised what Yumiko and he had been discussing before their arrival. John suggested they split the Review/Preview meeting into three parts: firstly, he said he would be grateful if Sarah, Jennifer and Yumiko would give feedback on yesterday's team meeting, secondly, he proposed that they discuss the best course of action to address Yumiko's Warm Down feedback, and lastly, conduct the preview of the next Oe375 team meeting. John said that Yumiko was welcome to stay for the whole meeting and added that he did not expect her to do so because of the lateness of the hour. Sarah, Jennifer and Yumiko agreed that John's agenda seemed fine, with Yumiko adding that she would have been interested to remain for the whole meeting but, as she was meeting a friend that evening, she would leave after the second agenda item.

In giving feedback on the last Oe375 team meeting, Yumiko, Jennifer and Sarah largely reiterated the positive side of the Warm Down feedback by saying they liked the use of visualisation in sharing tacit knowledge and thought the system state flow diagram (SSFD) looked a useful tool. John then asked Yumiko to tell Sarah, Jennifer and himself more about the reasons why she said that she felt left out of parts of the last Oe375 team meeting, stressing that it would help him to understand her reasoning if she was frank in her feedback. Yumiko said that she could best describe how she felt if she gave an example, and John said that he thought that was a good approach. Yumiko said that as a representative of a supplier company, she had to be careful of what she said for fear of saying something that might upset an important customer company. She continued by saying that for this reason, she had not said anything about the choice of using a belt-driven Drivetrain even though Agano has quite a lot of experience in this area. John asked her to say more, and Yumiko said that there are clear advantages and disadvantages of a belt-driven system over a chain-driven system. Sarah intervened and asked Yumiko to give her an example of what she meant to which Yumiko responded by saying that a belt had a significantly longer life than a chain but was the more expensive option. John asked Yumiko if she would have said this in the meeting if he had specifically asked her to give the team some of her expertise in this area, and she said that of course she would.

After mulling things over, John said that he recognised that Oxton and Agano were in a commercial relationship, but he hoped that would not get in the way of Yumiko being a valued member of the Oe375 team. John told Yumiko that he very much appreciated her input to team meetings and told her that, as he saw it, her presence on the Oe375 team was to Agano's and Oxton's mutual advantage. He explained that Oxton benefitted from her expertise, and he was hopeful that by Agano and Oxton working together in partnership, Agano would benefit from having a clearer understanding of Oxton's requirements. John said that he saw Yumiko as a full member of the Oe375 team and was hopeful that her relationship with other team members would be based on mutual trust and respect, adding that, as he had said in the Oe375 team meeting, he recognised that trust between people did not happen overnight but grew over time. He said that there was a practical measure that they could take which would help him, or whoever was leading the meeting at the time, to know when team members wished to say something and that was to start using the meeting software hand gesture facility. John paused hoping this would give Yumiko the opportunity to respond, which she did by thanking John for what he had said and saying that using the hand gesture was a good idea. John said that it would be opportune if she would give a brief presentation in the next Oe375 team meeting comparing the use of chain- and belt-driven Drivetrains, since the time was rapidly approaching when the team needed to make a choice. Yumiko said that she would be pleased to do so and then said that she had to go. In wishing John, Sarah and Jennifer goodbye, Yumiko signed off from the meeting.

Once Yumiko had left the meeting, Sarah asked John if she and Jennifer could discuss something with him in confidence to which John responded that of course they could. Sarah said that she and Jennifer were both pleased to be working on the Oe375 programme and felt that they were learning a lot, both because of the Oxton first-time engineering norm (OFTEN) approach and the People Skills. Sarah continued by saying that although the Oe375 team were meeting frequently, she was attending the Review/Preview meetings and they had worked on the use case diagram (UCD), they were not fully occupied at the moment with Oe375 work. Sarah added that if it became apparent to Paul, their manager, that they had spare time, he would not hesitate to fill the gaps with other tasks. Jennifer picked up from Sarah by saying that their experience of working on other programmes was that they required a full-time commitment and that they wanted to confirm that this would be the case for Oe375 as they did not want to become overloaded in the future working on both Oe375 and tasks for Paul. John thought for a minute and then said that this was something that he had been thinking about as well. He continued by saying that while, as he had explained to the Oe375 team, they had spent a lot of time in learning mode up to now, the workload was likely to pick up dramatically soon, particularly once they had developed a comprehensive propulsion system boundary diagram and got into interface analysis. John said the workload would require team members to work full time on the programme, although that should not preclude the two of them helping Paul out with key aspects of his work occasionally. John continued by saying that he expected the workload to be such that it would need to be shared between team members, adding that he did not expect that the external members would be able to devote a lot, if any, of their time over and above the meetings. John explained that he saw that the core task in the next Oe375 meeting was to extend the Drivetrain SSFD into a propulsion system SSFD by including the Electric Drive. He suggested if Sarah and Jennifer found they had the time today, they might think about developing state flow triads for the Electric Drive. John asked Sarah and Jennifer for their reaction to what he had said, and they agreed that it seemed reasonable and thanked John for confirming what they had thought about the programme workload and said that they would be pleased to work on the Electric Drive state flow triads. John said that because electrical flow was usually considered in terms of electrical power, one tip that he had was to initially think of the state transitions through the system in terms of energy and to convert this to power when quantifying states in the triads.

In moving on to develop the agenda for the next day's Oe375 team meeting, John said that because of their discussion this morning, in addition to suggesting that the team might like to use the meeting software hand gesture, he wanted to discuss projected Oe375 workload, and ways of working, with the team. Sarah suggested that the discussion of both topics could be combined into a single agenda item entitled "Team Process", and John said that he liked Sarah's suggestion. John repeated that he also wanted to look at the integration of the Electric Drive with the Drivetrain by expanding the Drivetrain SSFD as the core task in the meeting. John continued by saying that, as he had mentioned in yesterday's meeting, he also wished to start looking at team communication skills. John was pleased to see Jennifer developing a draft agenda as they were speaking, and Sarah, looking at Jennifer's laptop screen, added that Warm Up and Warm Down were the only items missing. John checked to see if Jennifer was still happy to conduct the Oe375 meeting Warm Up, and she said that she was, although adding that she was somewhat apprehensive about it. John said that they should spend the rest of the meeting discussing what the Warm Up would look like and Jennifer and Sarah readily agreed to his suggestion.

Following the Review/Preview meeting, John spent some time preparing his materials for the Oe375 team meeting, realising as he was doing so that he could link his input on communication skills with Yumiko's presentation and sent her an email explaining how this might be done. John then turned his attention to his library of slides which he had accumulated during his time at Blade from the various presentations on the first-time engineering design (FTED) framework he had given in Blade, at the University, and the set from the Harrogate Conference. As his quick check the evening before had confirmed, none of the slides contained any proprietary information but quite a number were set in the Blade context of automotive. John had been careful to develop the slides he was using on the Oe375 programme from new, although admittedly a number were based on slides that he had developed while at Blade. He had also been at pains to go back to the original source material in developing any People Skills slides he had used with the Oe375 team to improve their clarity, having decided that the material that he had used previously tended towards being overly theoretical. While the latter task was quite time consuming, he found it therapeutic and was approaching it on a meeting-by-meeting basis. All in all, while he retained what might be seen as Blade property, he felt comfortable that he was neither plagiarising Blade materials nor divulging any proprietary information in working with the Oe375 team.

John had heard nothing from Jim Eccleston by the time he had got to the conference room the next morning for the Oe375 team meeting. He had gone straight to the room without going to his office as he would rather not get distracted from carrying out his role with the team to the best of his abilities. Even though he was a few minutes early, he was pleasantly surprised to see the room set out and Phil, Sarah and Jennifer already there and thought to himself that this was an indicator of team spirit manifesting itself. Jennifer and Sarah were deep in conversation and looked up briefly to greet him as did Phil who was looking at his laptop. John booted his laptop and asked Jennifer across the room if everything was OK for Warm Up to which she replied that Sarah and she had just run through things and she was fully prepared. Once Sam had arrived, and the remote members had signed in, John handed the meeting over to Jennifer.

Jennifer told the team that she was conducting the Warm Up that morning and that she wanted team members to consider the important People Skill of listening. She shared the meeting software whiteboard to reveal a definition (Figure 5.1).

<u>Listening</u>

Waiting impatiently for a pause or interrupting the speaker in order to express your views

Figure 5.1 Definition of listening.

After pausing, Jennifer read out the definition. She continued by saying that the definition of listening might well strike a chord with team members in that people who look as if they are listening to someone else speaking are often paying more attention to what they are going to say themselves and so they are often deaf to most, if not all, of what the other person is saying. Given that the Warm Up was presently in the *Self* phase, John had advised Jennifer not to get into a team discussion at this point, which she nimbly avoided by saying that what she had just described was certainly her own experience and moved on to tell the team members that there was a simple but powerful remedy to this situation. She said that the remedy was called *Restatement* which requires the listener to reflect back to the speaker what they had heard. She added that she found that it helped her if she began the restatement with words like "*I think what you are saying is …*" or "*What I heard you say is …*". Jennifer added that Restatement was a part of what is called *Active Listening*.

Jennifer said that she would like the team members to get into two groups of three, suggesting the Oxton team members form one group and the remote members the other in the breakout room which she asked John to create. Jennifer continued by saying that, in turns, they should have one person speaking, one person listening and restating to the speaker and the third person observing and judging how accurate the restatement was. She said that the speaker should tell the listener one thing they had learnt so far in applying the OFTEN framework and why they thought what they had learnt was important. Jennifer requested that team members made their restatements in their own words and not just act as a voice recorder. She added that the observer could give the listener marks out of 10 for the accuracy of the restatement. Jennifer asked that each team member take on the role of speaker, listener and observer in turns and added that it would be helpful if each speaker talked about a different aspect of the OFTEN framework. Finally, she asked team members to get into their groups.

As the team started the restatement exercise, Jennifer glanced across at John who gave her a thumbs up. Jennifer kept an eye on the Oxton group and noticed when they had finished and, as John had advised, gave the remote group a couple more minutes before she entered the breakout room to find that Andy, Wolfgang and Yumiko had also completed the exercise. After inviting team members to come back together, Jennifer asked them how they found the exercise. Sam said that in knowing that he had to restate, he had concentrated on what the speaker was saying, and several other team members agreed with this sentiment. John was determined not to take over the facilitation from Jennifer so resisted the urge to comment on what Sam had said. Jennifer continued by asking the team how easy or hard they had found the exercise, and Wolfgang said he found that it was rather difficult and took some effort and there was general agreement with what Wolfgang had said. Phil and Andy disclosed that they got marks deducted for missing points out of their restatement. Finally, Jennifer asked how team members felt when they were receiving the restatement of what they had said, and Andy said he was reassured that Yumiko had really listened to what he was saying. Given that no one else spoke, Jennifer thanked the team for their participation in the exercise and concluded the Warm Up by taking the team through the meeting purpose and agenda before handing over to John.

John thanked Jennifer and asked for a volunteer to take Warm Up at the next meeting, and Wolfgang said he would like to do it. John said that he would invite Wolfgang to the Review/Preview meeting so that they could agree how the Warm Up might be conducted. In moving on to the second agenda item of "Team Process", John said that there were a couple of aspects of the way the team might work together that he would like to discuss, adding that one was the delegation of tasks and the other member participation. Taking the second topic first, John said that following Yumiko's comment at Warm Down in their last meeting, he had discussed with Yumiko, Sarah and Jennifer what they might do in order that everyone might participate fully in team meetings. John added that they proposed that team members could use the hand gesture feature in the meeting software when they wished to speak. Phil asked if he had to use it since he was in the same room as John who normally led the meetings. John, not knowing if Phil's question was meant sarcastically, answered it directly by saying that the use of the feature was voluntary. John added that others might lead the meeting on occasions and gave the example of Yumiko presenting something when Phil might wish to use the hand signal if he could not get a word in edgeways. Phil smiled at this scenario, recognising that he was not usually backward in coming forward. Yumiko asked if everyone was happy with the proposal, and there was general agreement that it was a good idea.

John turned to the subject of task delegation and said the team had now reached the time when they would have to start to pick up quite a lot of work between meetings both as individuals and in small groups. John added that the reason that it had taken so long for this to happen was because of the need for the team to learn about the OFTEN framework. John also said that in working outside of meetings, he would be available for advice and help on the OFTEN approach. Seeing that Andy had raised his hand on the meeting software, John invited him to speak. Andy thanked John and said he was not sure of John's expectations, but, while he would be available to support most Oe375 team meetings, providing that they did not get any more frequent, he did not have time to do anything outside of meetings. As Andy was speaking, John thought to himself that this was an example of the downside of the hand gesture since he was just about to make the point about the availability of the external team members, although he reminded himself that he did not have to bring people in as soon as they raised their hand digitally. John thanked Andy and said that his expectation of the external team members was as Andy had described and asked Wolfgang and Yumiko to confirm their position. Wolfgang said that, like Andy, he saw his input to the team as being primarily through team meetings. John was pleasantly surprised, however, when Yumiko said that her manager had told her that she could devote whatever time was necessary to Oe375, adding that she would keep her manager appraised of what she was doing and how much time she was devoting to the programme.

John turned to the Oxton team and asked for their comments on task delegation. Phil responded by saying that while he had been allocated to the Oe375 programme full time, he was too long in the tooth to believe that he would not be asked to do additional tasks. He continued by saying that he had always been able to manage his time in the past and he did not expect Oe375 to be any different, adding that he had found plenty to occupy himself with while Oe375 had been ramping up. Sarah said that, as they had discussed with John, Jennifer and she had been told they should work full time on Oe375 but expected to help on other things occasionally. John asked Sam for his reaction, and Sam said that he was not allocated full time to Oe375 but was happy to participate in work outside of meetings and would let the others know if he was experiencing workload difficulties. John thanked the team members for their comments which he said helped to confirm where people stood. He added that while the OFTEN framework dictated numerous tasks to be completed by each of the programme timing milestones, his experience of programme work was that a programme progressed despite of the timing rather than because of it.

John continued by explaining that what he meant by this was that they would have to manage the task–time balance both as individuals and as a team.

John then moved the meeting on to the next agenda of Questioning for Clarification by sharing a slide (Figure 5.2).

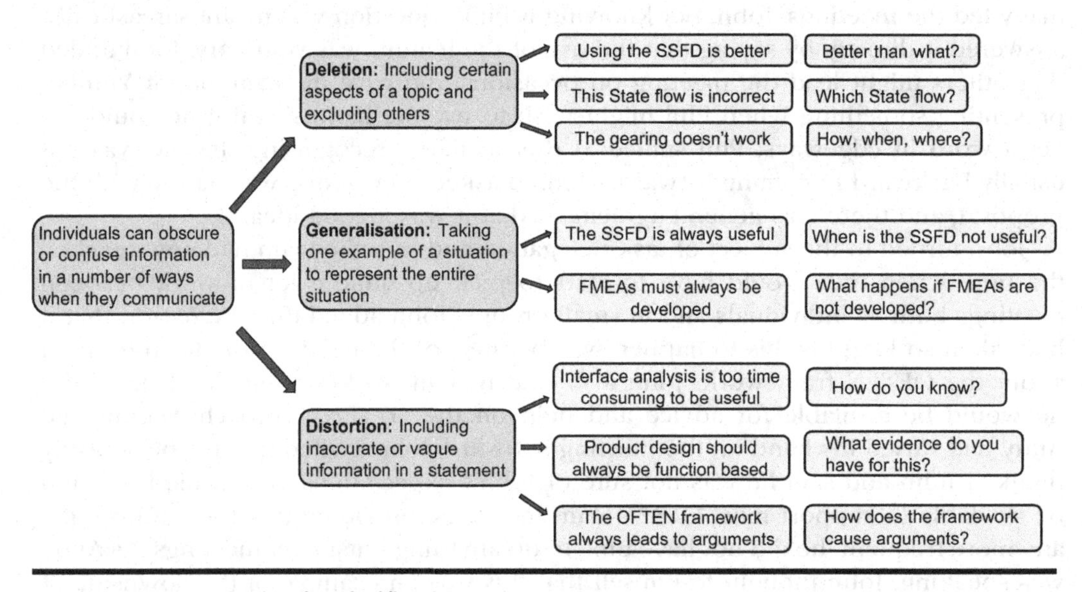

Figure 5.2 Questioning for clarification.

Setting the topic of *Questioning for Clarification* into context, John said that, as team members knew, questioning was an important People Skill in an engineering team, adding that he was sure that team members had needed to ask questions of each other when they were using visualisation in identifying state flows. John noted what he took to be a few nods of agreement by team members and continued by saying that team members had also used another valuable communication skill of *Restatement* in the Warm Up. John went on to explain the content of the slide by saying that there were many ways in which people obscure information when they speak, leaving others requiring clarification and added that sometimes the obscuration was unintentional but on other occasions it was intended. He said that three important ways in which information might be obscured were *Deletion*, *Generalisation* and *Distortion*. John explained that Deletion was the omission of information and gave the example of people often leaving out the comparator when making a comparison as in the statement *Using the SSFD is better* which, he said, begs the question *Better than what?* Jennifer intervened to say that, much as she loved her Mum, she was forever doing this. Jennifer said at least now she could tell her Mum the technical term for what she was doing, adding that this would probably go down like a lead balloon, which caused laughter.

John moved on to Generalisation as the second way in which people confuse information, which he explained was taking one aspect of situation to represent the whole situation. John cited the statement that *The SSFD is always useful* which he said prompts the question *Are there times when the SSFD is not useful?* John continued by saying that in his experience, there are situations in which developing an SSFD does not provide any information over and above that which is already known. John said that an SSFD is not useful for simpler systems where the Physics of the system is well understood but is better applied to more complex systems, although he added the caveat that care has to be taken to ensure that the scope of the analysis is not too broad. John then cited the rule of thumb in functional decomposition of identifying between 4 and 10 states that he had mentioned in the team's previous meeting. Realising that he was getting distracted from the People Skills topic he was discussing, John said that he would pick up this point later and apologised for getting sidetracked.

Returning to the slide, John explained the third way in which people obscure information is through *Distortion* which he said was done when people include inaccurate or vague content in a statement. As an example of this category, he referred to the statement on the slide *Interface analysis is too time consuming to be useful* and added that answering the question *How do you know?* might prompt the need for a cost–benefit analysis. John said that in his experience, this type of analysis would be likely to demonstrate that, although being time consuming, interface analysis can identify safety critical items that might otherwise not be discovered until the product had been launched, thereby avoiding the significant cost associated with a product recall. John stopped himself saying any more and apologised if what was supposed to be a discussion of a key People Skill sounded more like a selling pitch for OFTEN.

John brought himself and the team back to the subject in hand by asking team members if the example of ways in which people obscure information sounded familiar. Phil said that he heard examples of all categories regularly both in and outside of work. Yumiko said she also quite often heard people speak in the ways indicated, adding that she felt that many times people do not realise they are doing this. Sam said that he thought that politicians must have training in the ways of obscuring information shown on the slide, which prompted laughs from some team members. John said he agreed with Sam, thinking back to the incident while sitting in his car yesterday morning. John thanked team members for their input and, in picking up on Yumiko's point, he agreed that obscuring or confusing information is often unintentional. He added, that like other People Skills, the key step in stopping yourself doing this type of thing is by being aware that you are doing it in the first place. John said that there would be plenty of opportunities to practise using this skill as the OFTEN Phase 1 analysis progressed. John noticed the time, and on checking the agenda, he suggested to the team that they move on to the next agenda item.

In introducing the presentation Yumiko was going to lead, John said that although he had stressed the solution neutrality of the SSFD, there is a need at times during the design process to choose between design concepts. He added that this was the case for the Drivetrain as this may have a significant impact on the detail of the design architecture. He reminded the team that in their last meeting they had demonstrated the Drivetrain SSFD was applicable to both a convention chain drive and a belt drive and said that the team needed to consider which concept to use. John continued by saying that Yumiko has expertise in this area and added that she was going to take the team through the advantages and disadvantages of each option, and so saying, handed the meeting over to Yumiko.

Yumiko began by saying that Agano supplied both chain-driven and belt-driven Drivetrains, adding that the company did not manufacture the belts. She explained that belt drives have been around since the 1980s but only began to be popular nearly 30 years later when the carbon fibre belt was introduced. Yumiko then shared a table (Table 5.1) saying that it was not an exhaustive comparison, but it summarised the key points.

Table 5.1 Comparison of Chain- and Belt-Driven Drivetrains

Feature	Chain	Belt
Life / km	6,000	30,000
Maintenance	High	Low
Weight / g	~250	~100
Noise	Silent/Quiet	Quiet
Frame	Conventional	Lubrication
Lubrication	Yes	No
Complexity	High number of parts	No moving parts
Efficiency	Higher at low power input	Lower at low power input
Efficiency loss	Low	Moderate
Gearing	Derailleur, hub	Hub
Corrosion	Low	None
Cost	Low	High
Replacement availability	Very high	Low
Tyre size	Flexible	Restricted
Full suspension	High compatibility	Low compatibility
Lateral flexibility	Some with derailleur	None – straight

Table developed by Y Izumi, Agano

Yumiko said that while she could take the team through the detail of each feature, John had asked her if she would allow people to read the slide for themselves and then practise questioning for clarification, adding that she thought most items were self-explanatory. John smiled to himself and was pleased he had emailed Yumiko about conducting this agenda item in this way. Jennifer asked Yumiko what "Exposable rear triangle gap" meant as a feature of the frame. Yumiko said that most belts are made as one continuous piece which requires that a gap can be exposed in the frame rear triangle when fitting or removing a belt. Yumiko said that this means that belts normally required a compatible frame, although she said that split belts, in which the belt ends are riveted together, are becoming available. Phil asked Yumiko what was meant by tyre size being flexible and restricted, to which Yumiko responded that belts are wider than chains and this places a smaller upper limit on tyre width. Sarah picked up on what Yumiko had written about "Gearing" and said that she would like to check her understanding of what was meant saying that she thought that it meant that a belt drive was incompatible with derailleur gearing, so had to have hub gearing. Yumiko told Sarah that she was right and explained that the incompatibility was because belts do not have the lateral flexibility required for derailleur gears, adding that this accounted for the final feature in the table of the straightness of a belt drive.

John picked up on what had just been discussed between Yumiko and Sarah by saying that this was an important point for the design of the Oe375 eBike and said that since the mid-drive motor was a company requirement that left the choice of a chain-driven drive with derailleur gears, or a chain or belt-driven drive with rear wheel hub gearing. Anxious to move the meeting on, John thanked Yumiko for what he called an interesting and stimulating presentation and additionally thanked the team as a whole for their input. John said that he would discuss the team's findings with Jim Eccleston and get back to them. Noticing that team members in Oxton were starting to get fidgety in their chairs, John said it was time for the meeting break. As Jennifer was getting up from her seat, John walked over to her and told her that he was impressed by her facilitation of the Warm Up saying that he particularly liked the way she had managed the team by giving clear instructions. Jennifer, looking somewhat embarrassed, said that she had rehearsed the Warm Up with Sarah before the meeting and, in thanking John for his feedback, made her way out of the room.

Following the break, John said that he wanted to take the team through the meeting's core technical task of including the Electric Drive in the Drivetrain SSFD that they had developed last time. John said that in the interest of task delegation, he had asked Sarah and Jennifer yesterday if they would develop a draft of the SSFD triads for the Electric Drive, and in looking to Sarah, seeing that she was giving him the thumbs up, he handed the meeting over to her. Sarah said that Jennifer and she had found the development of the Electric Drive SSFD not quite as straightforward as the Drivetrain SSFD and so she would take the team through the progression of their thinking. She then said that like the Drivetrain they started with developing the high-level SSFD and shared a slide (Figure 5.3) on which this was depicted.

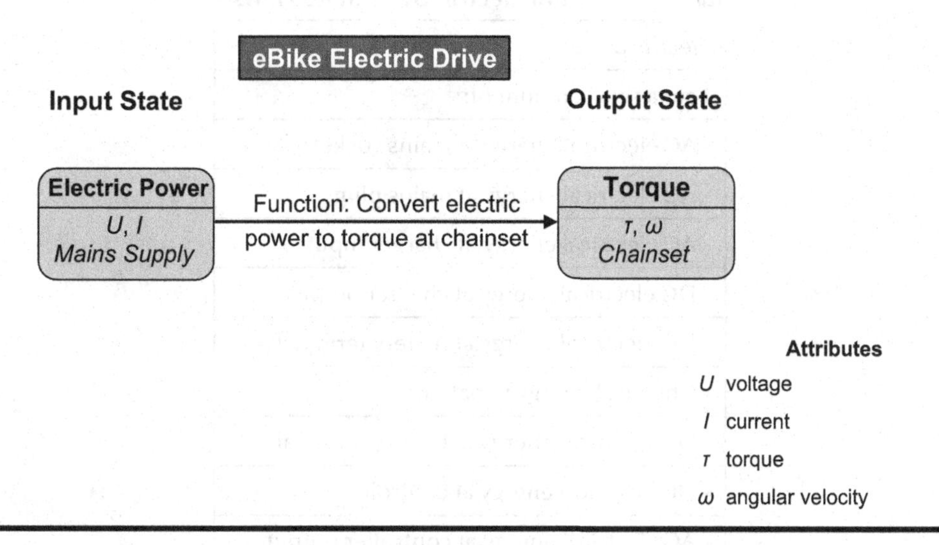

Figure 5.3 High-level SSFD for eBike electric drive.

Sarah explained that the attributes of the input state on the graphic were quantified and described in terms of power rather than energy since this was the more usual way of quantifying electrical flow. Sarah added that to make the input and output states compatible, Jennifer and she had stated the attributes of the output state as torque and angular velocity since the product of torque and angular velocity was the power of a rotating body. Phil said that he thought that the symbol for voltage was capital *V* and not *U*. Sarah responded by saying that both U and V were widely used alternative symbols and added that Jennifer and she had used U here since V had been used as a symbol for volume in the kettle high-level SSFD.

Sarah said that before developing a system-level SSFD for the Electric Drive, Jennifer and she confirmed their high-level mental model of the proposed architecture of Oe375 eBike which Jennifer had sketched on the whiteboard which she shared (Figure 5.4).

Jennifer said that as team members could see it was a standard eBike architecture.

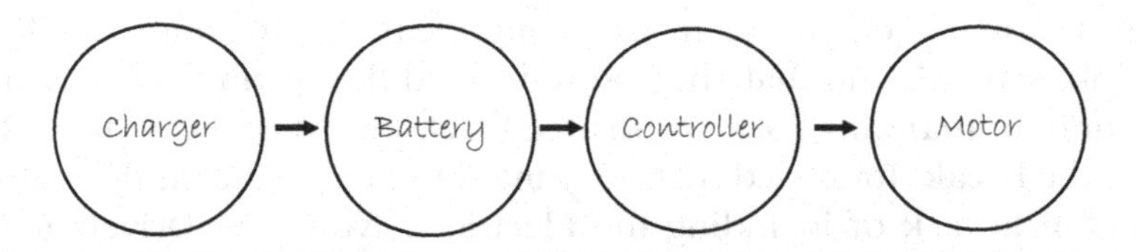

Figure 5.4 High-level architecture of Oe365 eBike electric drive.

Sarah said that Jennifer and she then started to consider the system-level state transitions and said that John had recommended that they thought of these initially in terms of energy transitions. Sarah shared the state flow list that she and Jennifer had developed by thinking that they were a small amount of electrical energy passing through the Electric Drive (Table 5.2).

Table 5.2 List of Electric Drive State Flows

Electric Drive
I am a small amount of:
- AC electrical energy at mains socket
- AC electrical energy at mains plug
- AC electrical energy at charger input
- DC electrical energy at charger output
- DC electrical energy at battery terminals
- Chemical energy in battery
- DC electrical energy at battery terminal
- DC electrical energy at controller
- AC electrical energy at controller output

(*Continued*)

Table 5.2 (Continued)

- AC electrical energy at motor input
- Electrical energy at stator windings
- Torque energy at motor rotor
- Torque energy at chainset
12 state flows

Sarah said that as they had a total of 12 state flows, this contravened the rule of thumb that John spoke about of between 4 and 10 states and so they went to see John to see if he would give them a special dispensation, which caused laughter. Sarah continued by saying that unfortunately John did not grant us one, which prompted more laughter. After the laughs had subsided, Sarah explained that she was joking about asking for a dispensation and went on to say that John told us what we had done was not at all unusual in that we had looked at things in too much depth. Sarah said that John had explained that by visualising what happened inside battery and inside the motor we had effectively started to develop the SSFDs for the battery and motor. Sarah added that John had also told them that keeping an SSFD at the right level of analysis was difficult for most engineers to do, as generally engineers love delving into things. Sarah then told the team that John had told them another helpful rule of thumb in this case which was to ensure that each subsystem only appeared once on the visualised flows list. Sarah then shared the edited list they had developed after speaking with John and paused to let team members read it (Table 5.3).

Table 5.3 Revised List of Electric Drive State Flows

Electric Drive
I am a small amount of:
- AC electrical energy at mains socket
- DC electrical energy at charger output
- DC electrical energy in battery
- AC electrical energy at controller
- Torque energy at motor/chainset
4 state flows

Sarah said that she and Jennifer then developed the State flow triads from the list, and Sarah then shared a slide showing the triads (Figure 5.5).

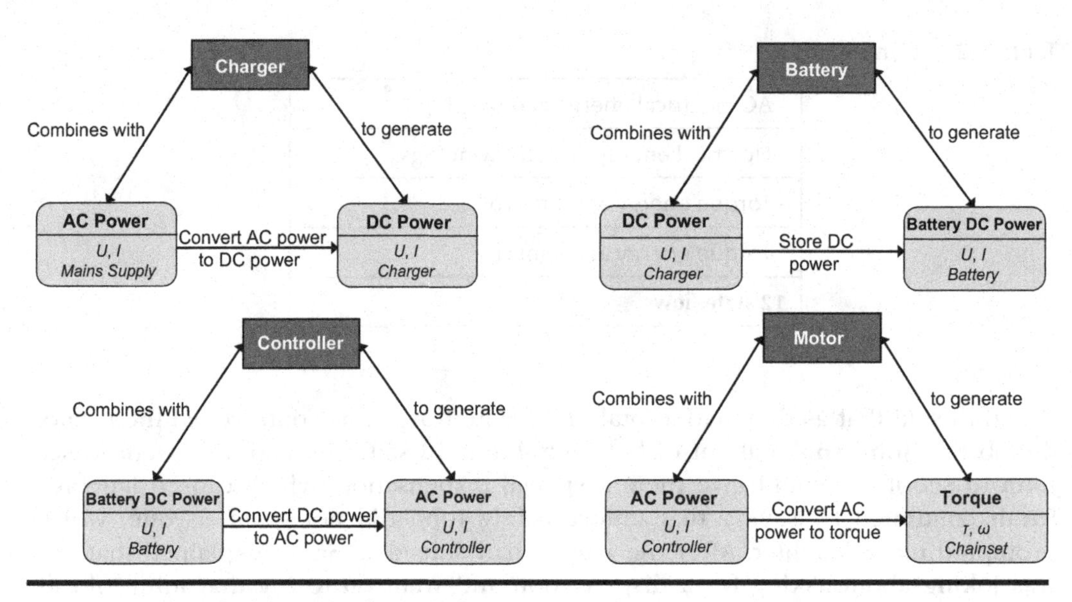

Figure 5.5 Electric drive state flow triads.

Sarah asked team members to note that the state attributes were quantified in terms of power and added that Jennifer and she had been able to do this easily despite the state transition list being expressed in terms of energy. Sarah asked team members for their comments and the resulting consensus was that the triads were sound. Sarah then said that although John had only asked them to develop the Electric Drive triads, it did not take them long to put them into an Electric Drive SSFD and shared a slide showing this (Figure 5.6).

After a pause, during which the team studied the SSFD, Phil said there was something strange about it, explaining that the linear representation of the flow implied that the battery was being charged while the controller and motor were operating and added that this would need, what he described as, a frigging long charger cable. After the laughter had subsided, John said that Phil was right, explaining they could take account of this when they incorporated this flow into the Drivetrain SSFD shortly. John continued by saying that Phil's point did not invalidate the flow that Sarah and Jennifer had depicted on the slide, adding that he was impressed by the work they had done. He said that he particularly liked the way they had described their thought processes during their visualisation which resulted in a good learning experience for the whole team. John started to clap which was picked up by the team. Sam then said that he had a technical question, saying that he thought that Oe375 was to have a brushless DC motor, and if so, why did the electric drive require a controller to convert the DC output of the battery to AC. Sarah said that Sam might recall from his Physics classes at school that a conventional DC motor required a commutator to reverse the current flow through the motor windings on every half-turn of the rotor and this was achieved by having a split copper ring on the motor armature on to which carbon brushes make contact. Sam said that he did remember that, and Sarah continued by saying that in a brushless DC motor the commutation

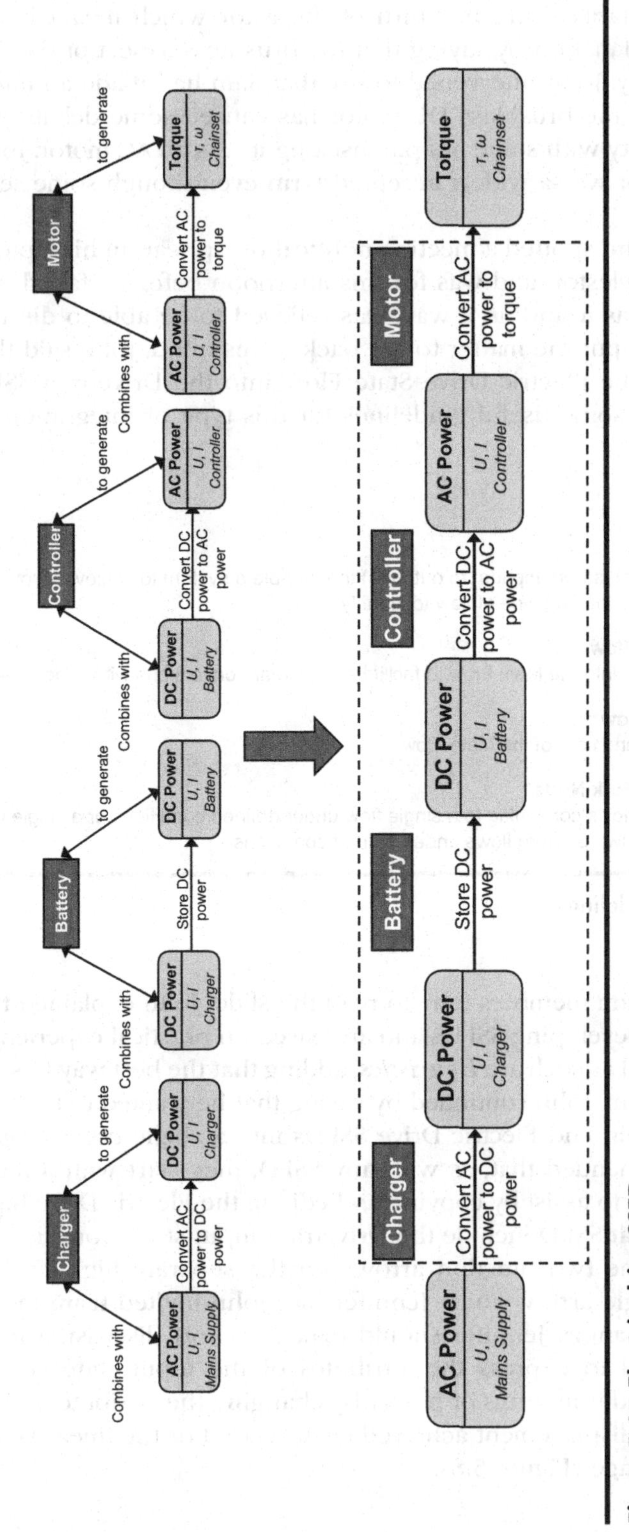

Figure 5.6 Electric drive SSFD.

is achieved electronically in the controller which has a square wave output switching the voltage polarity every half-turn of the rotor which meant it was AC. Sarah concluded her explanation by saying that the brushless aspect of the motor contributes to its reliability. John intervened to say that Sam had made an interesting point and the naming of the brushless DC motor has caused some debate in the broader electrical community with some people insisting it was an AC motor. John added that brushless DC motor was a widely accepted term even though some see it as technically inaccurate.

At this point, John spotted a meeting notification appear on his laptop, seeing that it was from Jim Eccleston and was for this afternoon before it faded. He thought to himself that this was it and, in a way, was relieved to be able to discuss the matter with Jim. Trying to put the matter to the back of his mind, John said that before the team incorporate the Electric Drive State Flow into the Drivetrain SSFD, he would like to show them some useful guidelines for this type of integration and shared a slide (Figure 5.7).

Main Flow
The flow of states from input(s) to output(s) that enable a system to achieve its overall function; this is the first type of flow to identify.

Connecting Flow
Flows directed into the Main Flow to facilitate particular conversions within the main flow.

Branching Flow
Flows that branch out of the Main Flow.

Conditional Fork Node
Two or more flows combining to a single flow under defined conditions or a single flow diverging into two or more flows under defined conditions.

Figure 5.7 SSFD guidelines.

Having given team members time to read the slide, John explained that the guidelines are useful in developing SSFDs and are based on practical experience rather than scientific theory and as such are *Heuristics*, adding that the best way to see their usefulness was to use them. John continued by saying that he wanted to look at the integration of the Drivetrain and Electric Drive SSFDs into a single *eBike propulsion system SSFD*. John recommended that, as with any SSFD, they start with a high-level SSFD. John asked Jennifer to assist by copying and editing the Electric Drive high-level SSFD.

John advised that SSFD include the Drivetrain input state alongside Electric Drive input state, with the two function arrows on the separate high-level SSFDs being combined to a single arrow which Jennifer did. John invited team members to suggest what other changes Jennifer should make. A short discussion followed which identified the need to express the attributes of the input state corresponding to torque due to the rider in terms of power by changing the attribute of linear displacement to the linear displacement achieved each second or the linear velocity. Jennifer duly made the change (Figure 5.8).

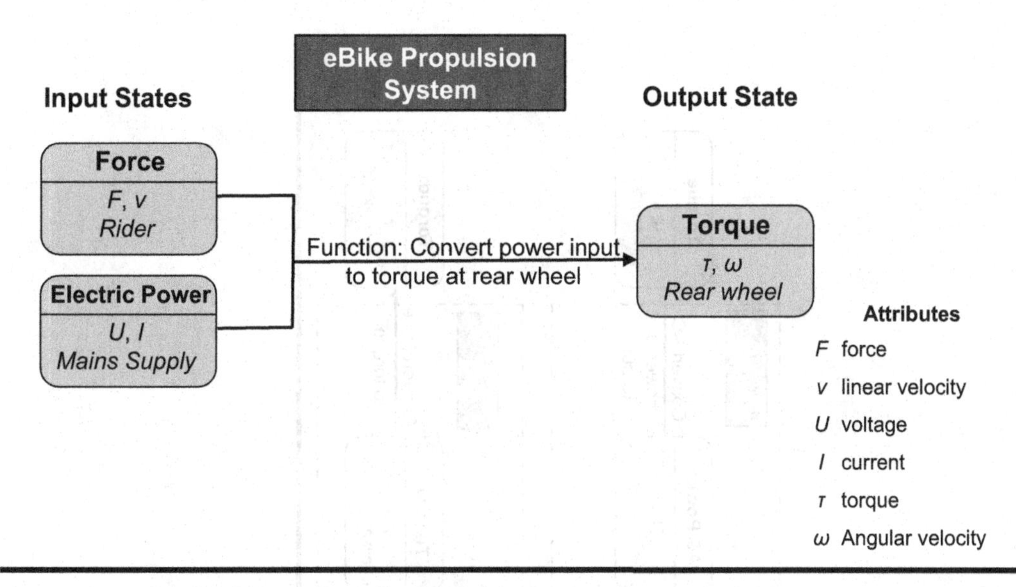

Figure 5.8 High-level SSFD for eBike propulsion system.

Sam then observed that the high-level SSFD showed that both the rider input on the pedals and the electric power input resulted in output torque at the rear wheel, and yet on the Electric Drive SSFD, the output was torque at the chainset. John thanked Sam for what he said was a good point and told Sam that, as he would know, with a Mid-Drive motor both the rider input and the motor assistance drive the eBike rear wheel through the Drivetrain. John added that this might be easier to see once they had developed the propulsion system SSFD and derived the boundary diagram. Sam said that he would wait to see what John meant when the two graphics had been developed.

Returning to the task in hand, John said that having developed the high-level SSFD, they should now develop the detailed SSFD for the propulsion system. Sarah suggested that they start by putting the flows from the Drivetrain and Electric Drive side by side after changing the state flow on the Drivetrain SSFD to power, which Jennifer did on a new slide that she then shared (Figure 5.9).

John asked for comments on the graphic, and Wolfgang said that the output of the Electric Drive SSFD main flow was a state within the Drivetrain SSFD main flow. John, pleased that Wolfgang was adopting the language of the SSFD heuristics, said Wolfgang was right and asked him which heuristic this might correspond to. Wolfgang replied that because the two main flows combined into a single flow, under certain conditions, it was a Conditional Forked Node. In agreeing with Wolfgang, John asked the team under what conditions might the flows combine, and after a pause, Yumiko said that the Electric Drive would integrate with the Drivetrain when the eBike was in the "Pedal bike with power assist" mode shown on the UCD (Figure 4.2). John agreed with Yumiko and asked when this might happen, and she replied that it would happen when the eBike rider switched pedal assist on. John thanked Wolfgang and Yumiko for their input and said that the rider switching pedal assist on and off could be represented on the SSFD as a flow and added that he would come back to this shortly.

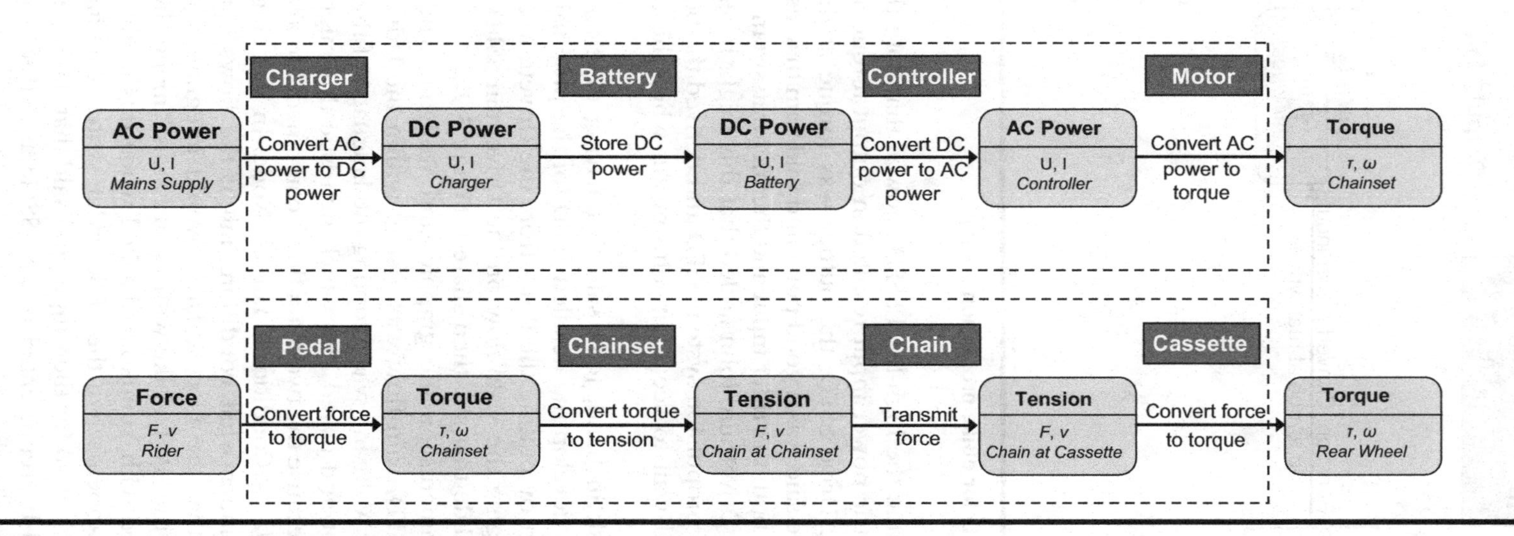

Figure 5.9 Initial draft of eBike propulsion system SSFD.

In picking up Phil's point from earlier, John then asked what type of flow the battery charging represented. After consideration, Phil said that it was a *Connecting Flow* since storing charge in the battery facilitated the main flow from battery DC power to torque at the rear wheel. John agreed with Phil and showed the team how they might represent both a connecting flow and condition fork node by sketching on the whiteboard. Jennifer edited the slide based on John's sketches and shared it with the team, after which John recommended that it be called a draft since it was still not yet finished (Figure 5.10). John asked Jennifer to identify the Connecting Flow and Conditional Forked Node on the SSFD by adding a dashed line around the former and a dotted and dashed line around the latter, which she did.

John said that they needed to include the rider input from the Pedal Assist Selector signal to switch on and off the pedal assist mode as well as the signal generated by the chainset Pedal Assist Sensor which was needed for legal reasons in order that pedal assistance only occurred when the rider was pedalling. John suggested that team members visualise these state flows. John said that, while up to now they had gone through the routine as a team of using visualisation to agree a state flow list and converting this to state flow triads before developing an SSFD, for this relatively simple situation, team members might like to sketch the state flows directly for themselves and agree the modifications to the SSFD. John gave team members 10 minutes to develop the flows during which time he observed team members sketching on pieces of paper. After calling the team members together, it did not take them long to agree the modifications to the SSFD, and at John's prompting, team members also agreed what type of flow to represent (Figure 5.11). John also asked Jennifer to add "Mode" as a metric to the state at the controller which she did.

In explaining the inclusion of the mode metric to the state at the controller, John explained that there were two high-level controller modes corresponding to the state output from the controller of AC power, or no AC power, dependent on the signal inputs from the Pedal Assist Selector and Pedal Assist Sensor. Sam asked what John meant by high-level modes, and John said that in addition to the AC power output being off and on, a typical eBike has various rider-selectable levels of power assist which could be seen as sub-modes of the high level on mode. Wolfgang then said that he was not clear why the connecting flows joined to a State and not a Design Solution, saying that if the PA Selector control signal was controlling the controller should it not go to the controller. John said that the SSFD models the state flow in a system and as such the interest is to depict the effect of the connecting state flow on the main state flow. John continued by saying that the effect of the connecting flow of the control signal was to change the output state of the Controller from AC power to no AC power or vice versa. John emphasised the word state in concluding his explanation. John asked Wolfgang if he was OK with his explanation, and Wolfgang said that he would have to think about it.

John then asked if there were any further comments on the SSFD, and Phil said that he still felt that it was not clear from the diagram that the charging flow was only for limited times, adding that he also thought it was not clear that the control signal to the Controller happened every so often and that the Electric Drive only operated for finite periods. John agreed and said that this was a shortcoming of the methodology, saying that it was not a dynamic method and explained that there were

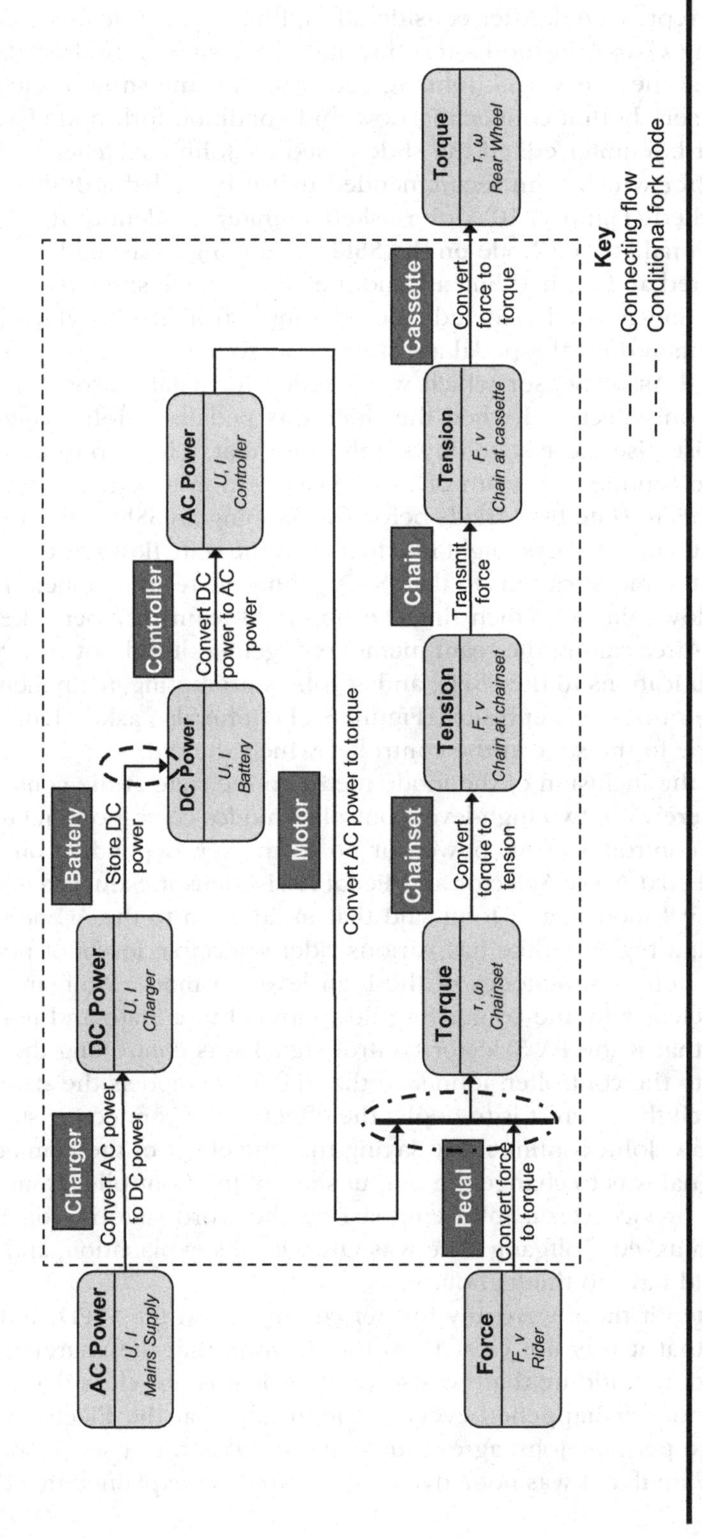

Figure 5.10 Draft eBike propulsion system SSFD.

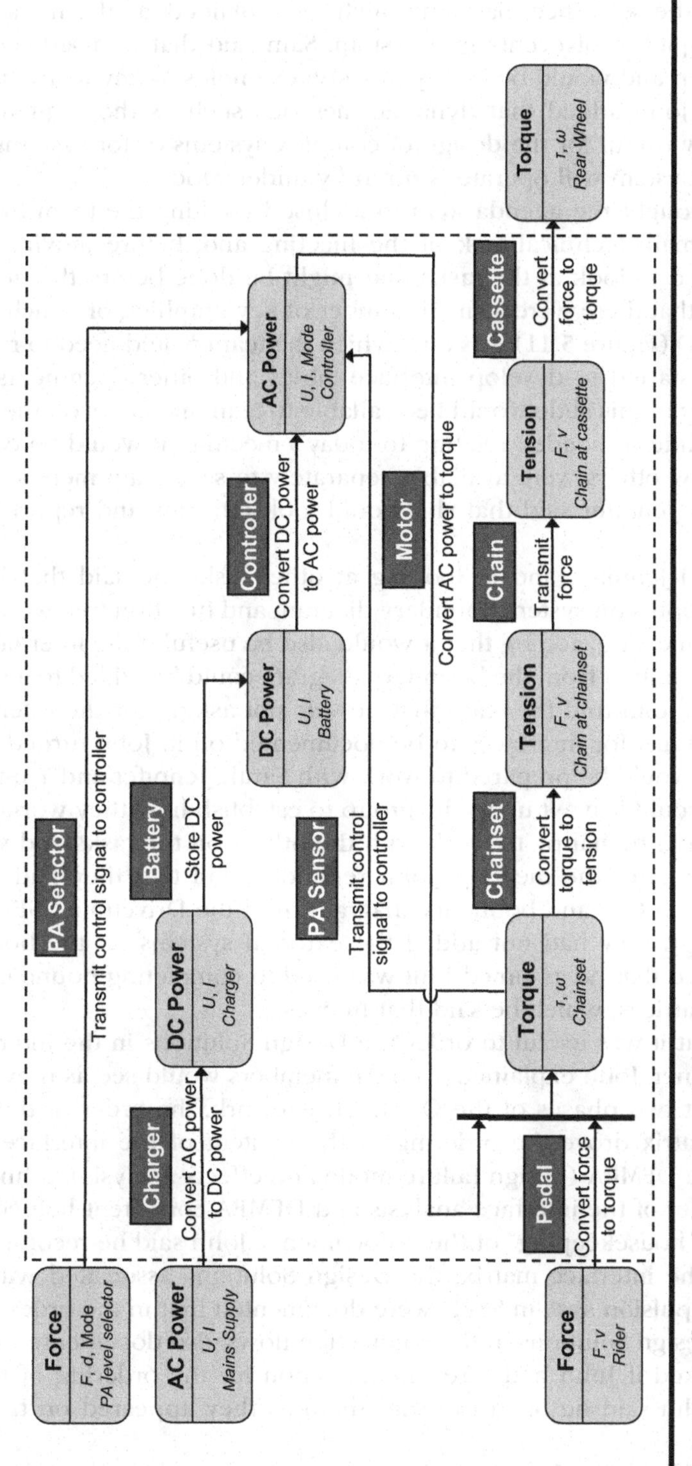

Figure 5.11 eBike propulsion system SSFD.

complementary graphical methods that could be used to analyse the dynamics of a system such as the sequence diagram which he explained, as the name suggested, depicted the sequence of events in a system. Sam said that he used sequence diagrams quite often and would be happy to show examples to any team member who was interested. John added that dynamic methods such as the sequence diagram were particularly useful for the design of complex systems or for systems where the way in which a system will operate is not fully understood.

John then brought the agenda item to a close by telling the team that they had completed the main technical task of the meeting and, before moving into Warm Down, he wanted to look at the tasks that might be done before the next meeting. John explained that there were a small number of key graphics, of which the propulsion system SSFD (Figure 5.11) was one, which the team would need to refer to regularly once they started to develop interface tables and other documents. John said that while a copy of this slide would be available to team members on the ShareDrive in the pack of slides uploaded relating to today's meeting, it would be convenient if this slide and few others were available separately to save team members having to search for them. Jennifer said that she would look into this and report back at the next meeting.

John thanked Jennifer, and in looking at other tasks, he said that it would be helpful if the propulsion system boundary diagram and function tree were developed before the next meeting, adding that it would also be useful if the internal and external interfaces identified from the boundary diagram could be added to a system-level interface matrix. John told the team that he was not asking for the interface matrix to be completed just for interfaces to be documented on it. John turned to Phil and asked him if he would be prepared to work with Sarah, Jennifer and Yumiko on this, saying that he would leave it up to the group to establish how they worked together. Phil said he would be happy to work with the others on the tasks and would be in touch with them after the meeting. John reminded Phil that they had derived the Drivetrain function tree and boundary diagram from the Drivetrain SSFD in the last meeting, although they had not added the external systems on the boundary diagram. John added that he assumed Phil was used to completing boundary diagrams and interface matrices, which he said that he was.

John said that it was useful to order the Design Solutions in the interface matrix in a logical manner. John explained, as team members would see as they progressed through the first two phases of the OFTEN framework, the order of the content of the interface matrix drove the ordering of the content of the interface tables and ultimately of the DFMEA (design failure mode and effects analysis). John added that keeping the order of the interface analysis and DFMEA consistent helped with what John called "the housekeeping" of these documents. John said he recommended that in developing the interface matrix, the Design Solutions associated with the main flows in the propulsion system SSFD were documented first in the order of flow, followed by the Design Solutions in the connecting flows also documented in the order of flow. Phil asked if John had a recommendation for the ordering of the external systems, and John said he liked to order them as they appeared on the boundary diagram.

John found the feedback from the Warm Down encouraging particularly because team members seemed to have enjoyed the meeting with several saying it was good to continue the engineering design work. Phil said that he was looking forward to doing some real engineering, saying that this would be a refreshing change since over the past few years he felt that he was acting more as a project manager than an engineer. John noticed Phil humming to himself as he helped the others set the conference room furniture back to where it was before their meeting at the end of the Warm Down and felt it would be nice to be so happy with the world. Before shutting his laptop down, John accepted the meeting notice from Jim seeing that it was for early that afternoon and had been flagged as urgent.

John waited outside Jim's office with mixed feelings, glad he was going to speak to Jim and apprehensive as to what the meeting would bring. He did not have long to wait as Jim ushered him in and asked him to shut the door behind him. Sitting down at the opposite side of Jim's desk to Jim, John saw the single sheet of paper with the Blade logo on the top on Jim's otherwise tidy desk which he took to be a letter from Blade. Not standing on ceremony, Jim turned the sheet round and pushed it towards John so he could read it and said that it had been emailed to him that morning. He asked John what he made of it. The first thing John noticed about the letter was its brevity and that it was signed by Georgia Lee, the ever-cheerful Aussie lawyer. In reading through the letter, John found that it merely referred to the fact that Blade had received information that material proprietary to Blade was allegedly being shared within Oxton Bikes and politely asked Jim if he would investigate the allegation from the Oxton Bikes perspective. The letter also mentioned that Blade was conducting its own internal investigation. John paused before replying, firstly because he did not wish to appear that he knew about the allegation and secondly because he was genuinely shocked to see the claim in black and white under a Blade letterhead. Taking a deep breath, John said that he assumed that the accusation was directed at him and that he was shocked and surprised by the letter, adding that he refuted the allegation it contained in the strongest possible terms.

Jim responded to John's rebuttal of the allegation by saying that he thought John would say that and, while the letter did not make any mention of who had been sharing the information, he had also concluded that the allegation could only be aimed at John. Jim asked John if he knew Georgia Lee, and John said that he did as she was the lawyer who had been very helpful when the company filed the two patents that resulted from the FTED project. Jim said that he had called her this morning after receiving the letter and, while she was very guarded in what she said, she seemed to him to be a friendly person. He said that Georgia had recognised his name from his pro-cycling days and being a keen sportswoman was interested that he had cycled with Noah White, adding for John's benefit, the well-known Australian professional cyclist. Jim continued by saying that he was finishing his grand tour career at about the same time Noah was starting his. Jim said that following some general cycling chit-chat, Georgia and he returned to the subject in hand, and after some pushing, he had managed to find out from her that the allegation had been made in an anonymous voicemail message. Jim told John, that, although they had not known each other for long, he felt sure that the allegation was not true, adding that this was only reinforced in his mind by the suspicious way in which the allegation had been made. John felt relieved to hear this from Jim, although his relief was tempered by the fact that Jim then went on to say that, despite his reservations about the allegation, he did not feel he could dismiss it out of hand and had decided to conduct an investigation within Oxton. Jim asked John for his thoughts, and John said that he was pleased with Jim's trust in his integrity but found the incident very disturbing.

Jim then told John that, while he was sorry, he would need to take John's laptop to check what material was on it to ensure that his investigation would be seen by Blade as being thorough. Jim then asked John if he had any files on the laptop which might in any way be viewed as incriminating by Blade. John said that whatever Blade proprietary material he had ever possessed was on his Blade laptop and flash drive which he gave back when he left the company and went on to describe the Blade files that he had retained. John was pleased when Jim responded to his description of the files that he had kept by saying that they sounded innocuous and he would not be the first person, nor the last, to have retained that kind of material when they left a company. Jim added that he was going to judge the laptop content in the same light as he would for anyone leaving Oxton, albeit looking at it from the Blade perspective, and that he was not intending to send Blade an inventory. Jim said he would speak to Joe Bowen, the IT manager, explain the situation in general terms and ask him to inspect the laptop himself and to keep his report confidential so as not to spread undue suspicions in Oxton. Jim added that at the same time he would ask IT to lend John a temporary replacement laptop and, looking in his desk drawer, gave John a company flash drive saying that he could copy any files that he might need in the short term to it.

Jim also told John he intended to speak to each of the Oxton members of the Oe375 team individually to ask what type of material had been shared with them. He added that he was not sure of how he would frame this so as to not raise suspicions and asked John if he had any ideas about how he might do it. John thought for a while and then suggested to Jim that he might tell the Oe375 team members that he wanted to reassure himself that the OFTEN framework was really applicable to bicycle design and was discussing it without him (John) being present because that might influence what they would say. John also explained that all the material that he had used with the Oe375 team was on the team ShareDrive and said that he would send Jim the access details. John continued by saying that Jim could confirm with team members that they had not used any other material and also ask them if he had used any automotive examples in the form of anecdotes in describing what he had done at Blade. Jim thought about John's suggestion for a while and then said that he liked it. John also advised Jim to see team members as a group or in pairs so that the meeting would seem less like an interrogation. Jim asked John if he had anything else that he would like to say and, when John said he had nothing to add, Jim concluded the meeting by telling John not to worry and that together they would soon sort out, what he described as, this unfortunate storm in a teacup.

John got back to his desk in a similar mixed frame of mind to that which he had been in before his meeting with Jim. While he was pleased to have spoken to Jim, and apparently gained his support, he wondered why Jim had seen it necessary to check his laptop if he fully believed in his innocence, hoping that it was just as Jim had said, that he was going through the motions. John had just finished copying the files he wanted from the laptop when he found Joe Bowen standing at his desk, laptop in hand. Joe said that Jim Eccleston had asked him to swap over laptops. John asked if he wanted to check the files that he had copied to the flash drive, and Joe replied that it would not be necessary. John reflected on what Joe had just said and thought to himself that does not mean that he trusts me as he can surely recover whatever files had been deleted, not that he had deleted any. At this point, John thought that as an innocent man he was getting paranoid. Joe took John's laptop and, in leaving the replacement, thanked John and left him to his thoughts.

Background Reading and Viewing

Communication Skills. Listening, Restatement	Lambert, B. (2018) *Listen Better: 5 Essential Phrases for Active/ Reflective Listening*, How Communication Works, You Tube, www.youtube.com/watch?v=tgLfz3dh5UE, Accessed 18 December, 2022
Communication Skills: Deletion, Generalisation, Distortion	Drake-Knight, N. (2018) *Deletion Distortion Generalisation*, Continue & Begin Ltd., YouTube, https://www.youtube.com/watch?v=rr9uj3xrCd4&t=2s, Accessed 18 December, 2023
SSFD Heuristics	Yildirim, U., Campean, F. and Williams, H. (2017) Function Modeling Using the System State Flow Diagram, *Artificial Intelligence for Engineering Design, Analysis and Manufacturing*, 31(4), 422–435, https://doi.org/10.1017/S0890060417000294 Pages 422–433

Chapter 6

Crossing Boundaries

John was relieved to get home that evening since although he thought the meeting with the Oe375 team had gone well he did not realise how stressful his subsequent meeting with Jim Eccleston had been until sometime after it was over. He had spent most of the rest of the afternoon trying to prepare for the next Oe375 meeting, but his mind kept returning to his encounter with Jim and what it might lead to. Jane greeted him at the door anxious to hear the details of his day. John had texted her to tell her that the meeting with Jim was happening and texted her again after the meeting, although not being sure himself how well it had gone, he just told her that it had gone OK. Although pleased to see Jane, after returning her greeting, he went straight up to the bedroom after removing his shoes to change before sitting down with her and the cup of tea he knew she would have ready for him. John's routine of changing into casual clothes on returning from the office gave him time to straighten out his thoughts before sitting down with Jane at the kitchen table. Reflecting on their chat afterwards, John thought that it was a good example of questioning for clarification since Jane seemed anxious to know more with every snippet of information about his meeting with Jim that he shared with her. Once Jane seemed satisfied with the amount of detail that she had extracted from him, she sat back while John sipped at his tea, which by now was getting cold. Jane's verdict on the meeting was that it sounded as if Jim believed that the allegations against John were ridiculous and was anxious to convince Blade of this through an objective investigation. She added that Blade probably realised themselves the questionable nature of the accusation they were asking Oxton to investigate.

Jane got up from the table and told John that he had better change into something that had a semblance of smartness as she had booked a table for dinner in the local pub in 45 minutes, adding that he had better get a move on if he wanted a drink first. John wondered why Jane had not told him before he had changed into his casual clothes that they were going out and then remembered that he had not really given her a chance in going straight up to the bedroom on entering the house. John was rather surprised that Jane had booked the meal as they tended not to go out in the evenings in the week, preferring to relax at home after a tiring day's work. Jane was a physiotherapist at the local hospital and they both wondered at times how they would have managed to bring up children as well as holding down full-time jobs as many of their friends did. Not that they would not have liked children, but by now had resigned themselves to just the two of them and their two dogs forming their family. John was pleased to be going out and savoured his pint at the bar, speaking to a couple of acquaintances. He enjoyed his meal even though Jane scolded him for having burger and chips from what she described as an extensive menu. John had thought to himself as he ordered the meal that he deserved burger and chips at this time.

John was in a much better frame of mind on arriving at work the next morning. He had missed his regular early morning walk on the moor with the dogs the day before even if it was only for a day. On looking at his calendar, he was pleased to see no unexpected meetings alongside his regular one-to-one with Jim in the morning and the Oe375 Review/Preview meeting with Sarah and Wolfgang in the afternoon. His meeting with Jim went as if the two had not met the day before with no mention of the matter that they had discussed. As usual, Jim was interested in both the People Skills and the Oxton first-time engineering norm (OFTEN) framework and having asked John to show him the slide on Questioning for Clarification reflected that it accurately depicted some of his conversations both inside Oxton and in his family and social life. Although he was getting to know Jim better each time they met, John was still surprised with his interest in, and eye for, detail. John's interactions with senior managers in Blade was limited but he was aware of the company's widespread use of Executive Summaries which in treating senior managers as extremely busy and capable people provided them with an overview of a situation so saving them the time and effort of having to consider the full detail. In his discussions with Numa McGovern during his time shadowing management colleagues in Oxton, she had told him that Jim did not like to ask anybody in the company to do anything that he could not do himself. John assumed that the level of detail covered in his one-to-ones with Jim was one manifestation of Jim's approach to business.

John's attention was brought back to his current meeting when Jim, in returning to the Warm Up in the last Oe375 meeting, said that he liked the formalisation inherent in the restatement process, adding that he would do better to heed it on occasions. When John showed Jim the slide that Yumiko had developed comparing chain- and belt-driven Drivetrains, Jim said that he liked the objectivity of the design concept comparison. Following a short discussion on the relative merits of the two options, Jim said he would prefer a conventional chain drive with cassette and derailleur gears for Oe375 unless the PD team could put together a compelling argument for a design based on the use of a belt. John responded by saying that he did not believe the case for a belt-driven Drivetrain was sufficiently overwhelming to go against Jim's preference.

Having told Jim how important he thought it was to exploit both tacit and explicit knowledge in engineering teams and taken him through the Tacit and Explicit Knowledge Transfer slide, John explained how team members had developed a list of state flows for the Drivetrain through visualisation. John then took Jim through the development of both the Drivetrain and propulsion system state flow diagrams (SSFDs) introducing the SSFD Heuristics as he went. Jim said that he liked the documented process that tracked the conversion of tacit knowledge in an engineer's head to explicit knowledge expressed as a state flow list, through the SSFD triads, to the

SSFD itself. Jim added that he assumed that the SSFD Heuristics were helpful in structuring the propulsion system SSFD. John was pleased by this response from Jim since it signalled that Jim had understood the cognitive process by which the SSFDs had been developed beginning with tacit knowledge. John was also impressed by the rapid way Jim had understood the process. John said that he had chosen the propulsion system as the starting point for OFTEN since it was a relatively self-contained and interesting system which had the bonus of aligning to Margaret Burrell's suggestion of involving Agano in the initial application of the OFTEN framework. John added that, by being largely a visible mechanical system, the propulsion system was useful when he was asking people to use visualisation as the starting point for design analysis. Jim responded to John's reasoning by saying that it made sense to him and asked him what system he intended to look at next.

John was pleased with Jim's question as this was something that he had been thinking about himself and welcomed the opportunity to discuss it. In showing Jim the eBike Partitioning and Decomposition slide (Figure 4.3), John explained that following on from the SSFD and the derivation of the boundary diagram, the next major step in the OFTEN framework was interface analysis to which Jim retorted that he did remember John's conference paper. John said that it was important to conduct a detailed interface analysis at the system level on all the systems of the eBike. John added that his thinking was to take the propulsion system through the entirety of the OFTEN framework before looking at another system as this would give the Oe375 team members a deeper understanding of the whole framework before looking at other systems. In thinking about what he had just proposed, John realised that design verification could not be conducted at the eBike level until all design work had been finalised. In stating this point to Jim, John modified his proposed strategy to taking the propulsion system through Phases 1 to 3 of the framework before considering other systems.

Jim said that in participating in a grand tour like the Tour de France he would like to maintain a top-down view of the race. Jim explained that while focusing intently on each individual stage, he would like to see each stage in the context of the forthcoming stages and rest days and ultimately in the overall context of getting onto the final podium. John was pleased that his new-found interest in competitive cycling allowed him to know that each day's race in a grand tour was a stage and the tour itself typically comprised of 21 stages with 2 rest days. John was not exactly sure, but he assumed that Jim was backing the strategy of taking the propulsion system through Phase 1 to 3 of the framework and was thinking of the implications of this when Jim said that he had another meeting in a couple of minutes. Jim then concluded their meeting by thanking John for what he said was another informative and interested meeting and told him to keep up the good work.

On returning to his desk, John found that he had a voicemail message from Sarah asking if they might meet 15 minutes earlier for the Oe375 Review/Preview meeting that afternoon. John was intrigued, wondering if it had anything to do with the Blade matter, although he thought that it would be surprising if she had seen Jim already. Letting his mind wander, he concluded that she probably wanted to discuss the Oe375 task she was working on with Phil, Jennifer and Yumiko, although, he thought, if that was the case why was only Sarah wanting to discuss things with him. He decided he would have to wait until this afternoon to find out what she wanted and directed his attention to the discussion he had just had with Jim about the overall strategy for applying the OFTEN framework to Oe375.

Sarah knocked on the open door to John's office a few minutes earlier than the rescheduled time for the Oe375 Review/Preview meeting and asked John if it was OK for her to come in. Getting up from what he was doing, John invited Sarah in, motioning to a seat at the table in one corner of the room where they usually sat for their discussions. Sarah shut the door behind her and sat down at the table with him. Leaving her laptop closed to the side of her, which John also took as an unusual sign, she began by apologising for wanting to start the meeting earlier which John brushed aside and asked her what was on her mind. Sarah began by saying that she was not sure where to begin and continued by saying that she and Jennifer had had a rather strange meeting with Jim Eccleston late yesterday afternoon. John sensed that Sarah was somewhat uncomfortable and in order to not make things more difficult for her he said nothing allowing her the space to sort out how she would frame what she wanted to say. After a pause, Sarah said that at the beginning of the meeting, Jim said that he wanted to learn what Jennifer and I thought about the use of the OFTEN framework on the Oe375 programme. Sarah told John that Jim had said that he (John) knew about Jim's intentions to speak to them and had thought it a good idea. Sarah said that Jim had told them that he was also meeting with Phil and Sam to do the same thing. Sarah continued by saying that at first Jim seemed genuinely interested in what they were doing and seemed to know as much, if not more, about the OFTEN framework than they did. Sarah said that the meeting then drifted into a discussion about what you had told us, what automotive examples you had used and what anecdotes you had told relating to your time with Blade. She added that this was when the meeting got a bit strained as Jim did not seem to believe Jennifer and I when we told him that you had not used any automotive examples or told any anecdotes about your experiences in the automotive industry. Sarah said that Jim effectively asked them the same question again, although wording it differently, leaving Jennifer and her only able to give Jim the same response. Sarah said that since the meeting was called late in the afternoon, she had to leave before the end to pick the kids up from their after-school club, adding that she had learnt from Jennifer this morning that Jim had finished the meeting soon after she had left.

John asked Sarah if Jennifer felt the same way as she did about the meeting, and she confirmed that she had also found it strange. Sarah then said that in talking it over with Jennifer they wondered if they had been doing something wrong or perhaps said something out of turn that had got back to Jim. On hearing this, John thought that Sarah and Jennifer deserved an explanation of why Jim had asked to see them, although he did not wish to open the flood gates to rumour and innuendo. Trying to make it sound as if it were an everyday occurrence and therefore nothing to worry about, he told Sarah that there was a rumour that he had been spilling trade secrets. Sarah asked him what trade secrets were being spilled, and John said that they were to do with his time at Blade. Sarah thought for a minute and then told John that if only Jim had said that in the first place, then Jennifer and she could have reassured him that you had said little about your time at Blade to the Oe375 team and they would not have had to go round the houses to answer what she described as Jim's obscure questions. Sarah added that she realised it was a delicate matter and continued by asking John if he knew how the rumours originated. John paused to think and reminded himself that he had learnt by experience that once you tell one half-truth which sounds a bit fishy you end up by telling half a dozen more half-truths as people question you further and so decided to be frank with Sarah. In answering Sarah's question, John said that what he was going to tell her was in confidence and then explained about the letter that Jim had received from Blade, the anonymous voicemail message and the investigation that Jim was conducting. Sarah said that she was amazed that someone would do such a thing and asked John if he had any suspects in mind to which he replied that he had suspicions about one or two people. Sarah asked John if any of his suspects worked at Oxton, and John replied that this might be the case. Suspecting there might be some friction between Paul Clampin and John, Sarah told John that if Paul was one of his suspects, she felt it very unlikely that he had started the rumour. She continued by telling John that she was now talking to him in confidence and went on to say that, while she knew that Paul manipulated situations to his own advantage, she thought that it was unlikely he would spread untruths about people. Sarah continued by saying Paul was too clever to spread untrue rumours as he would know that there was a possibility of him being found out with potentially ruinous consequences to his career.

John sat back and thought about what Sarah had said, thinking to himself that his chief suspect had been demoted down his short list of one. Sarah asked if she might tell Jennifer about the situation since she was as worried as Sarah was, and John, in agreeing Sarah could share what he had said with Jennifer, asked them both to respect his confidence at least until the matter was fully cleared up. Sarah thanked John for being straight with her and said she would stress the need for confidentiality to Jennifer. John responded by telling Sarah that he appreciated her frankness and of course would also respect the confidentiality of what she had said. Sarah asked John if he would like a coffee, and he said he would very much like one, as much as anything because he thought that the time Sarah took in getting the coffee would allow him to refocus his attention on the Preview/Review meeting, knowing that Wolfgang would sign in at any moment. Sarah had just returned with two cups of coffee when Wolfgang signed in.

After John, Sarah and Wolfgang had exchanged greetings, John welcomed Wolfgang to his first Oe375 PD team Review/Preview meeting. He then outlined how he saw the meeting progressing by stating that he would like to get his and Sarah's perspective of how the last team meeting had gone before agreeing on the Warm Up and draft agenda for the next Oe375 meeting. John reminded Wolfgang that, as he had said in the meeting notice, he did not expect him to continue in the current meeting beyond the discussion on the Warm Up, and Wolfgang confirmed that he intended to leave the meeting at that point. John asked Wolfgang how he thought the last Oe375 meeting had gone, and Wolfgang said that he found it useful to have clarified his working relationship with the team and reiterated that outside of attending the team meetings he had little time available to devote to the programme. John agreed with Wolfgang that it was in everyone's interest to make clear where they stood and, changing tack, he asked Wolfgang what he thought of the inclusion of People Skills in the meetings. Wolfgang said that at first, he had been surprised, and, given his limited availability, felt that he and the team were wasting time discussing, non-engineering matters. He continued by saying that more recently he had seen the benefit of some of the People Skills, particularly visualisation, adding that he felt they would be more helpful to the people working full time on the programme. Wolfgang said that he did not see himself using People Skills in Teutoberg as people might think he was acting a little strangely. John thanked Wolfgang for his feedback and agreed with him that it was difficult for one person to introduce a subtly different way of working in a large organisation. As a result of what Wolfgang had said, John made a mental note to say something to the team in the next meeting about the formalised way in which the People Skills were being introduced. John asked Sarah if she had any further reflections on the last team meeting, and she said that she had nothing to add to what she had said in the Warm Down.

John said that he would like to consider the important People Skill of feedback in the next meeting and proposed that the core of the Warm Up was based on an exercise he called *Mirroring*, which he said was a form of feedback. John explained that sitting or standing opposite each other in pairs for two minutes, one person copies everything the other does both in terms of what they say and any gestures that they make. John added that in effect the observer acts as a mirror reflecting to the other person an image of themselves. John continued by saying that the pair then swap roles and the person who was being mirrored becomes the mirror for two minutes after which the whole team will come back together and discuss their experience. John asked Wolfgang what he thought about facilitating the exercise, and Wolfgang paused before saying that he thought that people might feel a bit awkward and stupid in doing the exercise. John agreed with him that this might happen and said that, if it came out in the team discussion, this would be useful as this would reinforce the fact that giving and receiving feedback between individuals is often not an easy thing to do. John then asked Wolfgang if he was comfortable in facilitating the exercise, and Wolfgang said that since there was a good reason for doing it, he was prepared to manage the exercise. John said to Wolfgang that if he did have any misgivings about the exercise, it would be better if he did not project these onto others by, for example, in introducing the exercise saying that people might feel what he was about to ask them to do was silly. Wolfgang reiterated that he was fine about facilitating the exercise and asked John if he should mention that the exercise was an introduction to the topic of feedback. John responded that he would rather that this came out of the discussion after the exercise, adding that this allowed people to make the connection for themselves.

John then said that he wanted Wolfgang's advice and explained that he had only ever done the exercise before with co-located teams and asked Wolfgang for his opinion as to whether he thought the exercise would work with remote team members. Wolfgang appeared to deliberate for a while before saying that it would be more limited with remote team members because they are only likely to see their partner's head and shoulders. After pausing again, Wolfgang said he thought it would be better if remote team members mirrored each other and Oxton team members paired up together, adding that this would mean that each pair saw a similar image. John said he thought that this was a good idea and told Wolfgang that he suggested that the two of them form a remote pair, leaving Andy and Yumiko to form the other remote pair. John added that he would create the two breakout rooms and assign team members to them. Sarah, who had been listening to the discussion between John and Wolfgang, said that she thought that it sounded an interesting Warm Up. John asked Wolfgang if he wanted to discuss anything else about the Warm Up, and

Wolfgang said that he did not and added that he needed to leave the current meeting, exchanged goodbyes with Sarah and John and signed off.

John and Sarah turned their attention to the rest of the agenda for the next Oe375 team meeting. John asked Sarah how the functional analysis work that Phil was leading on the eBike propulsion system was going, to which Sarah replied that it was going well. John told Sarah about Jim Eccleston's view that the eBike design should be based on a conventional Drivetrain with chain, cassette and derailleur gears, and Sarah said she thought this sensible. John asked if the group would add the derailleur gearing to the propulsion system SSFD, and Sarah said that she would speak to Phil about it, adding that she thought it should not be a problem. Moving on to the rest of the agenda for the Oe375 meeting, John said that the core technical task he would like to introduce in the meeting was interface analysis and, after a minimum of discussion, they agreed on the agenda as Warm Up, Feedback in the Oe375 Team, Report back on Tasks, Interface Analysis and Warm Down.

John then told Sarah that, given the nature of the beginning of their current meeting, he would like to end the meeting with what he called an *appreciation pair* in which each of them would tell the other one thing that they particularly appreciated about them, and to model the process John spoke first. John told Sarah that he really appreciated the positive way in which Sarah was working on the Oe375 programme and he appreciated Sarah's candidness in this meeting. He quickly followed his statement by saying that he had cheated by telling her two things that he appreciated about her. Sarah responded to John by saying that she appreciated his listening skills and frankness at the beginning of the meeting and his enthusiasm for the OFTEN framework, adding that she had gone one better than John and appreciated him on three counts. They concluded the meeting by agreeing that it had been a useful meeting in more ways than one.

After the Review/Preview meeting, John switched his attention to the addition of derailleur gearing to the Drivetrain SSFD and found that it raised a good learning point and wondered if Phil and the rest of the team would realise it too. He did think about dropping Phil a note but then decided not to, reminding himself that learning by discovery was far more powerful than someone solving the problem for you. John decided to have a meal in the canteen for lunch rather than buying a sandwich and eating it at his desk which was what he usually did. Following a relaxing lunch, John directed his attention in the afternoon to preparing for the forthcoming Oe375 meeting.

Unusually, John found himself rushing to the Oe375 meeting next morning having been held up on his way into work by heavy traffic resulting from an accident. He was relieved to see that everyone but Sam had arrived, or signed into the meeting, and that the room furniture had been set out to its normal position. Apologising for his lateness, he gave his excuses, although on looking at his watch he saw that in the end he was only a couple of minutes late. He was reassured when Jennifer said that she had heard about the traffic pile-up on the radio on her way into work since he felt that the heavy traffic excuse was overused by people to the extent that it often lost its credibility. He was pleased that he was not doing the Warm Up as this gave him time to collect his thoughts and asked Wolfgang if he was ready to start, to which Wolfgang responded by saying that they were just about to start anyway.

John assumed that Wolfgang had already reminded the team that he was taking the Warm Up and so he left it to Wolfgang to continue.

Wolfgang began by saying he noticed that the room at Oxton was all set up and it looked as if Sam had not arrived yet which Phil confirmed. Wolfgang said that he wanted the team members to do an exercise in pairs which he would explain in a minute, although given that John was included as part of a pair, Sam's absence meant that they would need to shuffle things round a bit. John suggested to Wolfgang that, as he had done the exercise before, he drop out and that one of the Oxton team members take his place as Wolfgang's partner. Phil volunteered to be Wolfgang's partner, and Wolfgang thanked him and confirmed the other pairings as Andy and Yumiko, and Sarah and Jennifer. John added that he would be the timekeeper if that would help, and Wolfgang thanked him for his offer. Having sorted out the logistics, Wolfgang explained how the exercise was to be conducted. John liked the way in which Wolfgang suggested that the person doing the mirroring act like they were an instant video replay. The three pairings duly completed the exercise with some laughter.

Once the pairs exercise was finished, Wolfgang called the team back together and asked how the team members felt about what they had just experienced. Sarah said that at first it was very off-putting, and she stopped what she was talking about to Jennifer in mid-sentence. Jennifer said that she also found it strange with the strangest bit being the reflection of what she called her nervous movements. Wolfgang asked her what she meant by nervous movements, and Jennifer gave the examples of touching her nose, ear and chin. Yumiko said that she was conscious of Andy's reflection of what she was doing, and this made their conversation more difficult. Wolfgang asked Yumiko if she had learnt anything about herself, and she said what struck her most was Andy's reflection of the way in which her voice was going up and down, adding that she thought these undulations were probably due to the fact she was watching what Andy was doing.

In giving their comments on the exercise, the other team members, including Wolfgang himself, agreed that while it was an unnatural situation, they all noticed their opposite members making physical movements or changing the pitch or flow of what they were saying without them being aware that this was a behaviour that they were exhibiting. In bringing the discussion to a conclusion, Wolfgang asked if team members changed their behaviour towards the end of the exercise in response to seeing and hearing their reflection at the beginning of the exercise and they all agreed that they had. Wolfgang told team members that, while he agreed with them that the exercise that they had just conducted by its nature would have influenced their behaviour, nevertheless they might have learnt something about themselves that they might consider changing in the future. Wolfgang then said that of course the mirror was giving them feedback and, after a pause, moved on by outlining the meeting purpose and agenda. John was impressed by the way in which Wolfgang had conducted what he thought was not an easy Warm Up to manage and, before taking over the meeting facilitation, sent Wolfgang a quick chat note saying how well he thought that he had done.

After thanking Wolfgang, for what he said was an interesting Warm Up, John said that before the next agenda item, he wanted to say a few words about the framing of

People Skills. John continued by saying that some of the People Skills were framed using a particular phraseology and gave the example of the restatement that the team had met in the last meeting which was typically introduced by the phrases "I think what you are saying is …" or "What I heard you say is …". In continuing, John said that many people find that using such phraseology is rather artificial and formalised and is not how they would normally say things. John explained that the People Skills framing is rather like the maxim "Mirror, signal, manoeuvre" which people might well have used when learning to drive adding that while people still go through the routine once they have passed their driving test, most will not go through it formally but will do it without consciously thinking about it. John said that while there is no need to continue to use the standardised phraseology once the People Skills become second nature, in his experience, many people in the teams where he had introduced People Skills continue to do so. He added that he thought this was probably due as much as anything to people wanted to demonstrate to other team members that they were using the People Skills.

John then introduced the agenda item of Feedback in the Teams by sharing a slide (Figure 6.1).

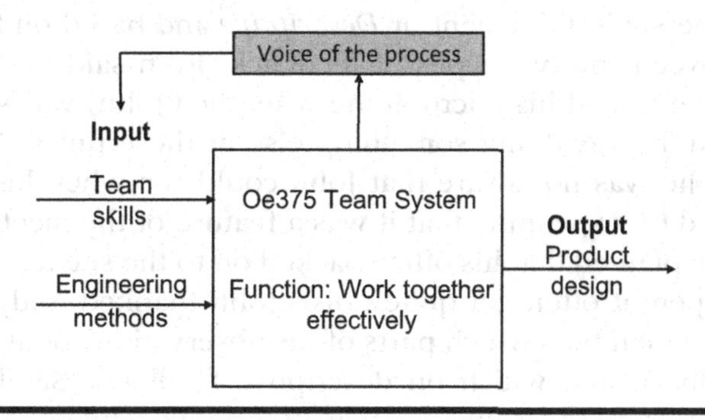

Figure 6.1 The Oe375 team system.

In explaining the slide, John said that as engineers team members would be very familiar with the concept of using feedback to control the performance of a system. John added that the Oe375 team was a system and as such its performance could be controlled using feedback. John continued by saying, as team members also knew, in one feedback strategy the system output is sampled and where the output is not as expected, revisions are made to the input. John said that this feedback strategy is problematic in their case because, firstly measuring output is late in the process, and secondly, knowing which aspect of the team skills and/or engineering methods caused the deviation from what was expected was often not immediately obvious. He went on by saying that a better tactic was to use explicit feedback to control the performance of the team in real time and explained that explicit feedback in team meetings would normally be expressed verbally and could be seen as the "Voice of the Process" as depicted on the slide. At this point, John shared a slide representing the different types of explicit feedback (Figure 6.2).

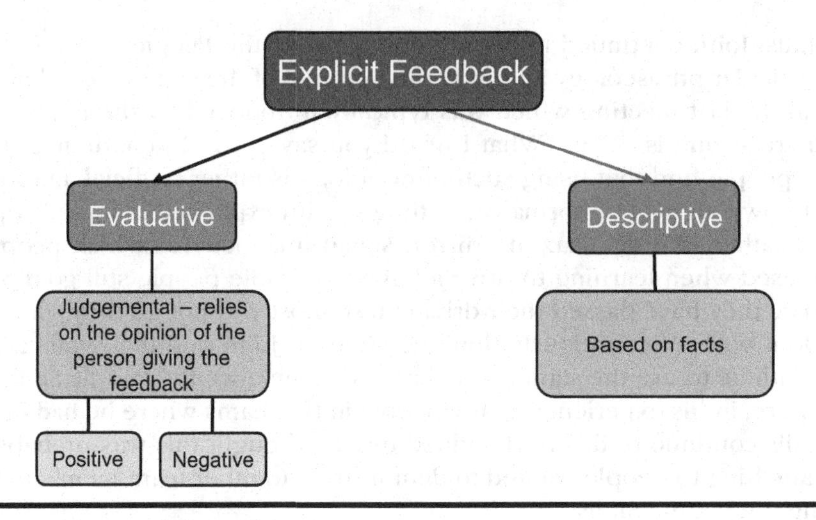

Figure 6.2 Types of explicit feedback in team meetings.

John said that *Explicit Feedback* in team meetings could either be *Evaluative* by being based on a person's judgement or *Descriptive* and based on facts. To illustrate the difference between the two types of feedback, John said to Andy that he had noticed that he often muted his microphone when he (John) was speaking and that he assumed that Andy was doing something else at these times. Andy's immediate response was that he was not aware that John could see when his mic was muted, and John responded by explaining that it was a feature of the meeting software they were using. Andy explained that his office backed on to the site truck yard, and when his window was open, it often got quite noisy. John thanked Andy for the explanation and asked team members which parts of his observation about Andy were based on evaluative feedback and which on descriptive feedback. Sarah said that when John said he saw Andy mute his mic that was descriptive while in saying he assumed Andy was doing other things that was evaluative. John agreed with Sarah, saying that while team members could not dispute that Andy mutes his mic, they could argue whether Andy was doing other things and added that if you cannot dispute a statement, it is descriptive but if you can argue with it, it is evaluative. John then said that his assumption was incorrect on this occasion as Andy had ably demonstrated.

Continuing the discussion of feedback in team meetings, John said that evaluative feedback, which he said was often called judgemental feedback, could either be positive or negative. John said he was going to give an example of both types of evaluative feedback and said to Yumiko that he thought that her comparison of the chain and belt Drivetrains in the last meeting was excellent. John then said to Wolfgang that he thought his management of the Warm Up this morning was appalling and told the team that, in case they did not realise, the feedback to Yumiko was positive and that to Wolfgang was negative, which prompted laughter. John added that his example of feedback to Wolfgang was just that, an example, saying that he really thought that Wolfgang's management of the Warm Up was also excellent. John then asked Yumiko how she felt on receiving his feedback, and she said that she was pleased. Next, John asked Wolfgang how he would have felt if he had actually received the negative feedback as stated. Wolfgang said that he would have been angry as he had put quite a bit of effort into preparing for the Warm Up and he thought that he had done a good job. John explained that Yumiko's and Wolfgang's replies were typical of those receiving positive and negative feedback in that positive feedback can make people feel good while negative feedback tends to provoke a defensiveness or a denial. John continued by asking Yumiko if she could tell him, based on the feedback that he had given her, what it was that John had found excellent about her presentation so that she might ensure that these features were part of any future presentation she might give. Yumiko replied that John had not told her anything about her presentation apart from the fact that it was excellent. On being asked a similar question about the negative feedback he had been given, Wolfgang said he had learnt nothing about why his management of the Warm Up might have been appalling. John explained that neither positive nor negative feedback motivates improvement in team performance, adding that positive feedback tended to reinforce current behaviour arbitrarily and negative feedback was often demotivating.

In continuing the discussion on feedback within the Oe375 team, John said to Jennifer that he noticed that when he was speaking, she often started to type on her laptop, to which Jennifer replied that she was normally taking notes when she typed on her laptop. John explained that his statement to Jennifer by being based on his observations was descriptive and asked Jennifer if she would have felt any differently if he had given her evaluative feedback instead by saying to her that in typing, she was not listening to what he was saying. Jennifer said that while she was a bit annoyed at John telling her that he noticed her typing when he spoke, she knew she had a good explanation. However, in responding to John's judgement that she was not listening, she said that she would have felt a lot angrier since in taking notes while John was speaking, she had to listen more carefully to him than she might otherwise have done. John thanked Jennifer, saying that her response was typical in that she said that she would have been a lot angrier in response to his evaluative feedback than to his descriptive feedback. John continued by explaining that the descriptive statement allowed Jennifer to make the connection between her observed behaviour and what she was doing rather than John, which meant that she would be less likely to feel the need to justify her actions and more likely to adjust her behaviour if she felt that this was required. John apologised to Jennifer for using her as an example, and she responded by saying that this was fine, although adding that she was concentrating so intently on what he was saying she had not taken any notes, which raised smiles. John then summarised the point that he was making by saying that, by being factual, descriptive feedback is less likely to be challenged and hence can be a powerful way of promoting improvement in team performance.

John added that, although team members might not recall, he had said in their first meeting that he would discuss a way in which personal disagreements between team members might be discussed openly and objectively and added that he was referring to the use of descriptive feedback at the time. John then concluded the agenda item by encouraging team members to make any feedback that they gave to others descriptive rather than evaluative.

Before moving on the next agenda item, John observed that Sam was still absent and asked if he had said anything to anyone and, since no one had heard from Sam, John handed the meeting over to Phil for his report back on the tasks team members had taken away from the last meeting. Phil said that before looking at the development of the graphics based on the propulsion system SSFD that the Oxton team and Yumiko had agreed to develop, Jennifer wanted to talk about access to key slides that John had mentioned in the last meeting. Jennifer thanked Phil and reminded team members that John had said that it would be convenient if a few frequently used slides were available in a separate location on the team's ShareDrive. Jennifer explained that she had set up a folder called "Appendix" which she said that paradoxically being the only folder beginning with the letter "A" team members would find at the top of the team's root folder on the ShareDrive. Jennifer added that John had sent her an email suggesting that she put what he called the Roadmap slides (Figures 3.4 and 3.5) in the folder along with the propulsion system SSFD (Figure 5.11). Jennifer explained she had used the code for each slide in the Appendix of *"A dot X (A.X)"* where X was the slide number. Jennifer asked if there were any questions, and Yumiko asked Jennifer why she had used the word paradoxically in relation to the Appendix folder's location. Jennifer replied that an Appendix usually comes at the end of a document rather than at the beginning and, since there were no more comments, she handed the meeting back to Phil.

Phil thanked Jennifer and said that he had asked Sam if he wanted to join Sarah, Jennifer and Yumiko and him in working on the tasks from the last meeting, adding that Sam had said that he was rather busy. Phil reminded Andy, Wolfgang and John that the group had been tasked at the end of the meeting with developing the propulsion system boundary diagram, function tree and interface matrix template. Phil explained that, because the development of the interface matrix template depended on having the completed boundary diagram first, Sarah and Jennifer said that they would develop the boundary diagram and function tree, with the whole team then getting together to develop the interface matrix, adding that this was not quite how things worked out as he was going to show.

Phil reminded Andy and Wolfgang that the boundary diagram and function tree were derived from the corresponding SSFD and shared the Oe375 eBike propulsion system SSFD that the team had developed in their last meeting (Figure 6.3). Phil added that team members could test the accessibility of this slide on the ShareDrive being Figure A.3 in the Appendix folder.

Phil then dissolved the SSFD slide into the Boundary Diagram slide apparently taking away everything from the SSFD slide but the Design Solutions and then showing the flow of energy and information through the system by adding arrows between the inputs, Design Solutions and output. Phil said that the external systems were then added to the graphic to form the boundary diagram (Figure 6.4).

Phil checked that everyone was familiar with the concept of a boundary diagram and referred team members to the coding used in Oxton to denote the flow of energy, material and information through the system. As no one indicated that they were not familiar with the boundary diagram, Phil asked team members if they had any comments on the detail of the propulsion system boundary diagram. Wolfgang asked Phil to remind him what the letters *PA* stood for, and Phil said that it was *Pedal Assistance*. Andy then said that the rear forks were missing as an external system, to which Phil responded that the group had assumed that the rear wheel was attached to the rear triangle of the frame. Andy agreed that this could be the case but if the eBike was to have a full suspension, then the rear triangle would be separated from the frame as a component and act as a fork. Phil said he would add this element to

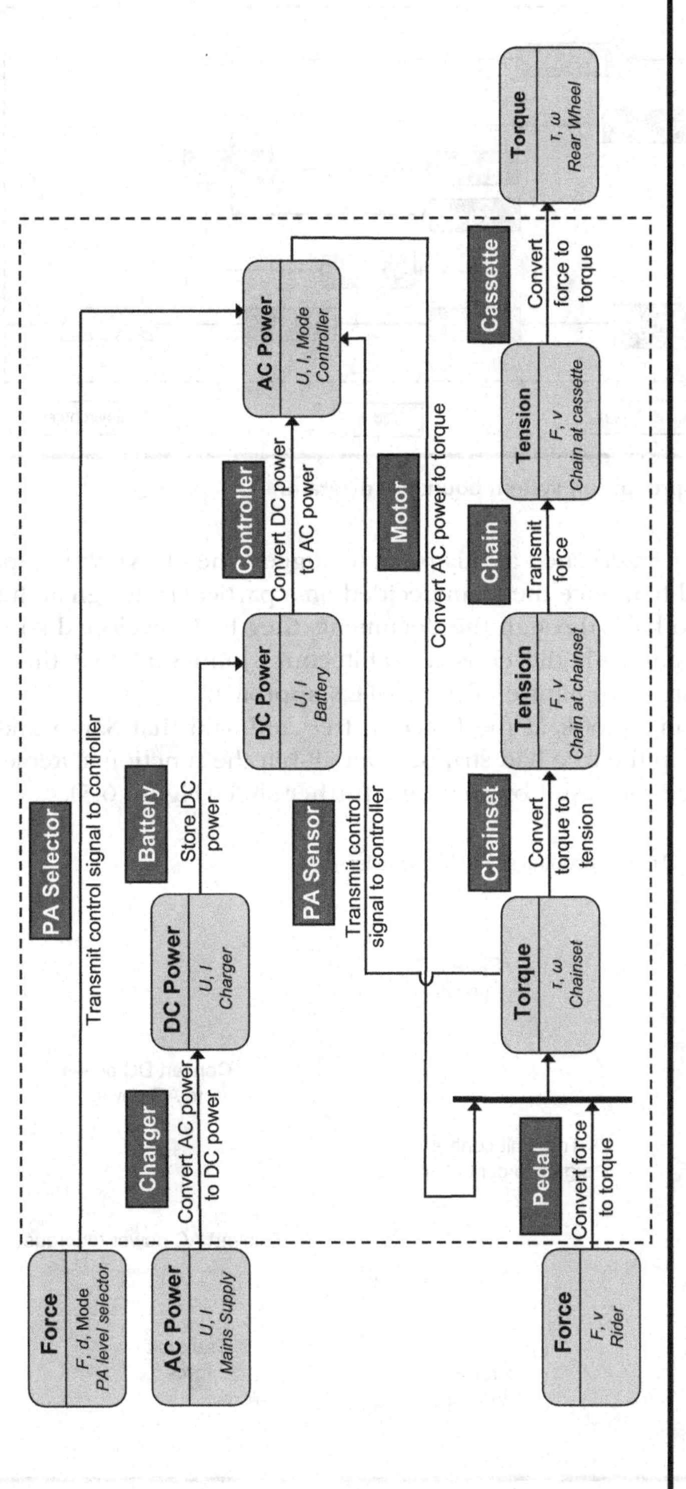

Figure 6.3 eBike propulsion system SSFD (copy of Figure A.3).

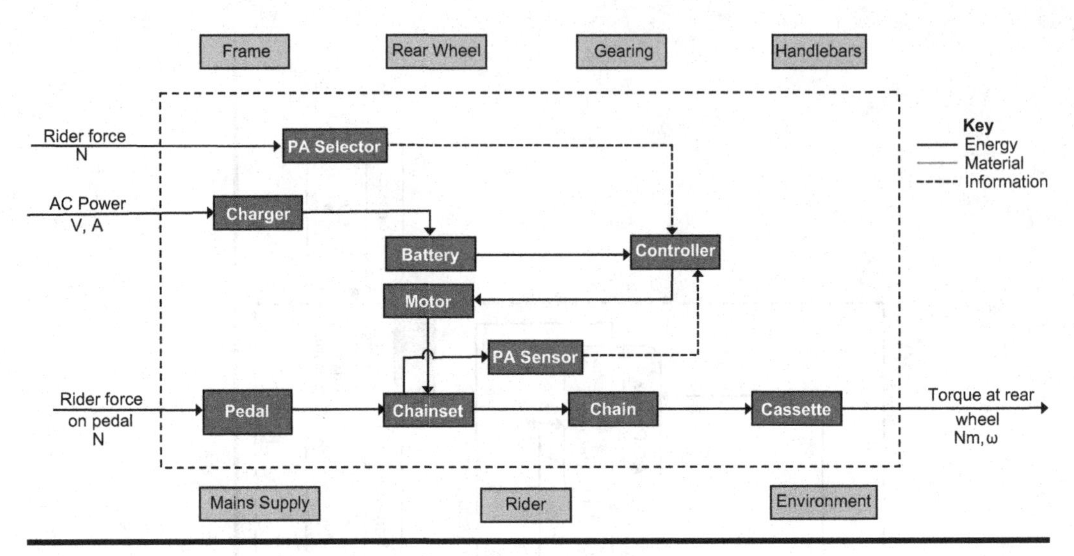

Figure 6.4 eBike propulsion system boundary diagram.

the diagram. John intervened at this point to suggest they leave the graphic as it was for now and said that once the team decided on a particular design architecture, they might have to go back through the documents they had developed so far to ensure they are consistent with the chosen architecture. John said that this was a good example of the iterative nature of the OFTEN approach.

Phil moved on to look at the function tree and said that Sarah and Jennifer in starting to develop the tree had stripped out all but the function statements from the SSFD and showed the result by sharing a further slide (Figure 6.5).

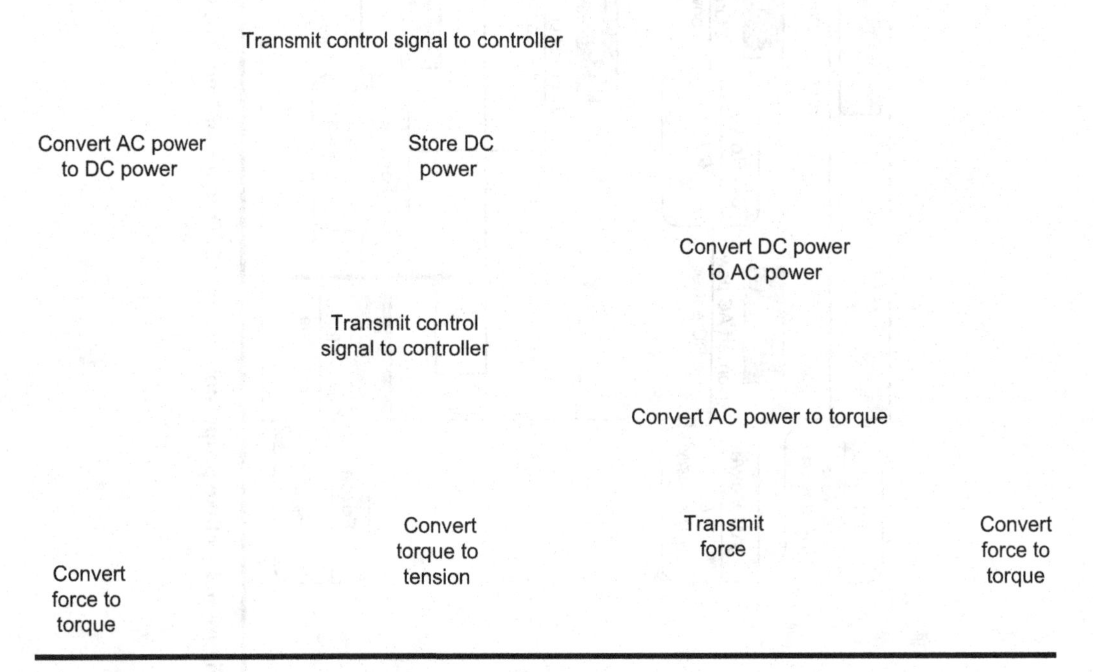

Figure 6.5 Draft propulsion system function tree.

Phil asked Sarah and Jennifer to explain what happened at this point, and Jennifer responded that Sarah and she scratched their heads, which triggered laughter. Sarah added that the problem was that the tree structure was not apparent and there was ambiguity in the function statements, pointing out that there were two *Transmit control signal to controller* statements and a *Convert AC power to DC power* as well as a *Convert DC power to AC power*. Jennifer continued by saying that she had a word with John who advised that the structure of the function tree and more precision of the function statements was inherent in the SSFD. Jennifer added that armed with this information they decided to get together with Phil and Yumiko working on the basis that four heads were better than two. Phil picked up the discussion again and said that Yumiko pointed out that, in the Drivetrain function tree that John showed us, the top function was that stated in the high-level SSFD. Phil added that Yumiko also reminded them that the high-level SSFD for the propulsion system had already been developed. Phil said that on reviewing the SSFD for clues to the function tree structure, it was clear that the SSFD had two main flows with one flowing into the other through a Conditional Forked Node and two connecting flows. Phil explained that the SSFD flows formed the basis of the function tree structure as he would show in a minute. Phil then shared the slide showing the fully developed propulsion system function tree (Figure 6.6).

In explaining the content of the slide, Phil said that the graphic in the bottom right quadrant related the structure of the tree to the flows in the SSFD. Phil said that, as Yumiko had advised, the system function at the top of the tree was that given by the propulsion system high-level SSFD (Figure 5.8). Phil added that the team also moved this system function from where it was at the bottom left of the structure graphic to be in the middle of the tree to give the graphic more of a tree-like appearance. Phil then explained that the "to and from" functional information inherent in the SSFD, which was defined by the States, enabled them to increase the clarity of the function statements. He pointed out the additional information in italics in lighter font and said that this also enabled the function tree to be analysed in isolation from the SSFD. In conclusion, Phil said that the group verified the tree using the *How-Why* test. Phil then asked Andy and Wolfgang if they had any questions or comments.

Andy asked Phil if he would explain the How-Why test that he had mentioned. Phil said that the question "How" of a function should be answered by the functions immediately beneath it in the tree structure and added that this meant immediately to the right in propulsion system function tree. Phil illustrated this through the example of the question "How does the system *Convert controller AC power to torque at chainset?*", saying that this is answered by "*Converting DC power to AC power at controller*" which relies on "*Storing DC power in battery*" and "*Converting AC power to DC power at charger*". Phil continued by saying that equally the "Why" of a function should be answered by the function immediately above it in the tree structure and gave the example "Why does the system *Transmit a control signal from PA selector to controller?*" being answered by the statement "In order to *Convert DC power to AC power at controller*". Phil asked Andy if his explanation was clear, and Andy thanked Phil saying that he understood what he meant. Phil said that in Oxton up to now they had brainstormed functions in developing a function tree and used the How-Why test as the principal way of structuring the tree. He said that this sometimes proved difficult and led to disagreement between team members, adding that he was impressed how the SSFD could be used to determine a function tree structure in an objective manner.

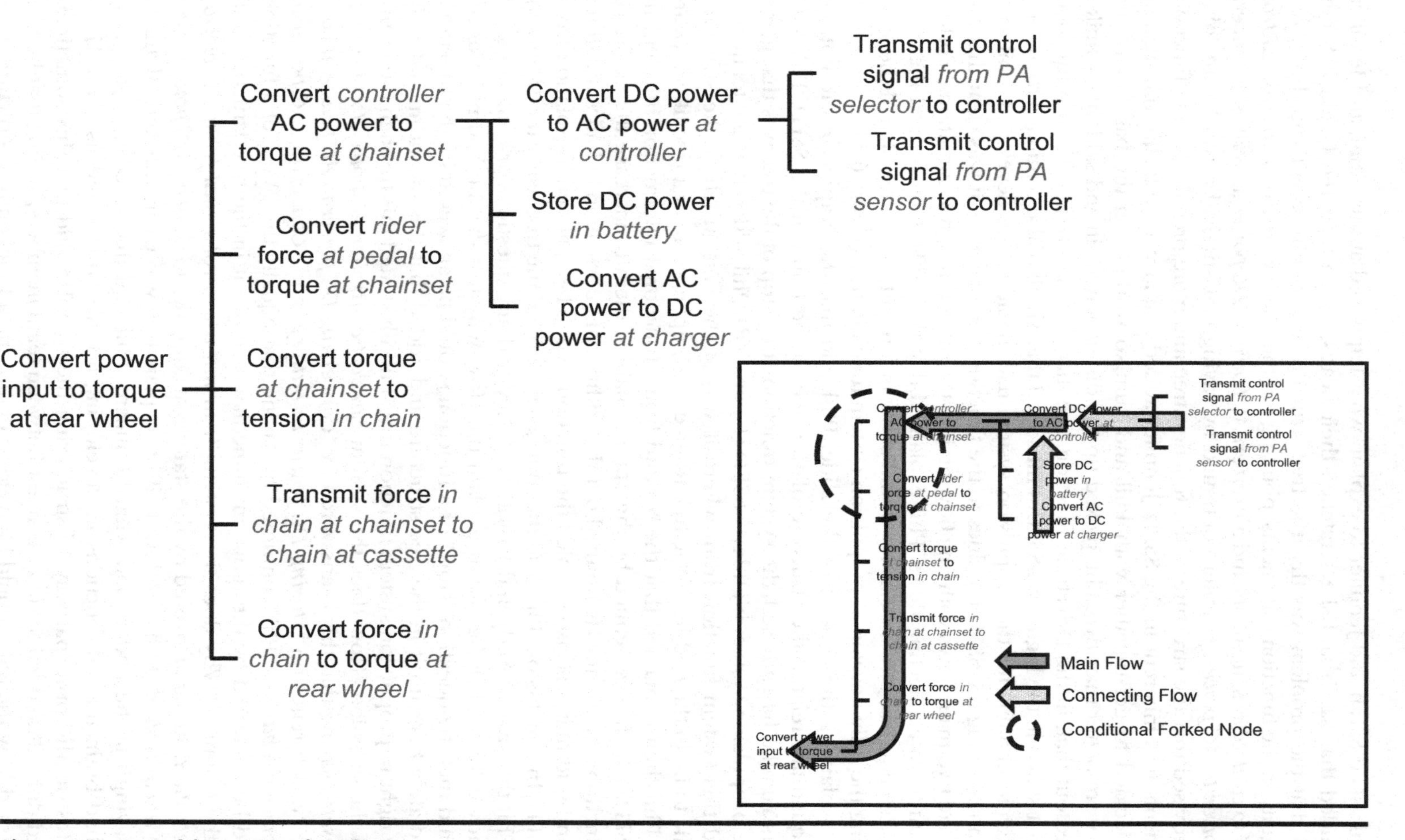

Figure 6.6 Propulsion system function tree.

In looking directly at John, Phil then observed that since the function tree was an inherent feature of the SSFD, he wondered why they needed to develop it at all. John said that Phil had partially answered his own question a few moments ago when he had described how Team B had increased the clarity of the function statements on the function tree by adding the "to and from" information. John said that although the "to and from" information was represented graphically on an SSFD, as team members would see, this information was useful in developing both the interface analysis within Phase 1 of the OFTEN framework and the failure mode and effects analysis (FMEA) in Phase 2. John added that defining functions in precise terms allows these two types of document to stand alone from the SSFD. John referred team members to the OFTEN framework roadmap (Figure A.2) and said that team members could see the information flow from the function tree to the interface analysis and onto the FMEA represented on the roadmap.

At this point, Sam entered the Oxton meeting room and apologised for his late arrival, saying that his girlfriend was ill in bed and that he had waited for her mother to arrive before leaving her. John enquired as to the seriousness of Sam's girlfriend's illness, and Sam said that they rang the Health Service helpline and it seems as if she has the flu so hopefully she would be up and about in a day or so, adding that her mother lived 50 miles away. Sam said that he had texted his manager and asked him to let John know, to which John responded that he had not looked at his emails since the meeting began. John brought Sam up to speed with what they had been doing before he arrived.

Phil then moved on to discuss the propulsion system interface matrix template that the team had developed and shared a worksheet (Figure 6.7).

Oe375 Propulsion System		INTERNAL INTERFACES										EXTERNAL INTERFACES						
		A	B	C	D	E	F	G	H	I	J	E1	E2	E3	E4	E5	E6	E7
		Pedal	Chainset	Chain	Cassette	Charger	Battery	Controller	Motor	PA Selector	PA Sensor	Frame	Rear Wheel	Derailleur Gearing	Handlebars	Mains Supply	Rider	Environment
1	Pedal																	
2	Chainset																	
3	Chain																	
4	Cassette																	
5	Charger																	
6	Battery																	
7	Controller																	
8	Motor																	
9	PA Selector																	
10	PA Sensor																	

Figure 6.7 eBike propulsion system interface matrix template.

Phil said that he assumed, from what people had said previously, that everyone was familiar with how the interface matrix was developed from a boundary diagram. John said that it might be useful to go through the basis of the development of the graphic to help Sam catch up. Phil said that he was happy to do so and, starting with the propulsion system SSFD slide, he quickly explained how the boundary diagram had been derived from it and then in comparing the boundary diagram to the interface matrix explained that the left-hand part of the matrix had the propulsion system Design Solutions listed as both rows and columns with external systems listed as columns on the right side of the matrix. In explaining the format of the matrix, Phil said that each cell represented an interface between the elements listed in the rows and columns, and in using the matrix, they would look at the potential interfaces between all of the elements internal to the propulsion system as well as the potential interfaces between each internal element and each external element. He noted that in performing interface analysis, the interfaces between the external elements themselves were not of interest and said that this was the reason that the external elements were only listed as columns. Sam said that he was a bit confused as to what was a *Design Solution* and what was an *element*. Phil explained that he was using the term element in a general sense to cover both the entities internal to the propulsion system and those external to the system, with the entities internal to the system being the Design Solutions on the propulsion system SSFD. Phil added that on reflection, he might have caused more confusion by introducing the third word entity. Sam thanked Phil and said his third word had sorted out his confusion.

In finishing his explanation of the interface matrix, Phil said that the diagonal row across the internal interface section was greyed out as this represented the interfaces between a Design Solution and itself and the bottom left-hand half of the internal interface cells was greyed out because otherwise, they would be considering each internal interface twice. Sam thanked Phil and said that he had a question and asked why the interface between the Frame and Rear Wheel was not of interest. Phil said that while this was an important interface to consider, it would be analysed as a part of the interface analysis that focused on these parts directly and added that currently they were only considering the propulsion system. Phil asked for comments from the whole team, of which there were none. John intervened to say that he was impressed with the work that Phil, Jennifer, Yumiko and Sarah had done and appreciative of the way in which they had explained their thought processes. John then asked Jennifer if she would upload the slides showing the propulsion system boundary diagram, function tree and interface matrix to complete the Appendix, although recognising that the interface matrix needed to be populated. He then said that the team deserved a break, adding with a smile that this would include Sam, Andy and Wolfgang.

As the others were starting their break, Phil came over to John and apologised for not having had the time to look at the inclusion of derailleur gearing in the Drivetrain SSFD. John said that it did not matter and that he had given the group very short notice. Phil said that Sarah and he had been asked to go out to a bicycle distributor to look at a so-called emergency. John asked Phil if the emergency had been resolved to which he replied that it had. Phil added that the solution was obvious when they investigated the problem, and he wondered why Sarah and he had been bothered in the first case. After Phil had gone to get himself a coffee, John took the opportunity to look at his email inbox and saw that there was an email from Sam's manager passing on Sam's apologies for being late. John then checked that the files for the propulsion system boundary diagram and interface matrix, that Phil had just spoken about, had been uploaded to the team ShareDrive and was pleased to find that they were there.

In restarting the meeting after the break, John said that, as Phil had described, the boundary diagram was derived from the SSFD which could be compared to the more conventional approach of basing the diagram on the hardware associated with the existing product. John explained that, as they had seen, the use of the SSFD had enabled them to be fully conscious of any design choices they made rather than unthinkingly adopting design concepts used in the existing product. John added that the function tree had also been derived from the SSFD. John said that he would like the team to note that he had spoken of deriving the boundary diagram and function tree from the SSFD rather than them being extracted from the SSFD. John explained that what he meant by this distinction was that it was not simply a case of lifting the boxes representing the Design Solution and the flow arrows from the SSFD to form the boundary diagram since the flow arrows represent functions in the SSFD and the flow of energy, material or information in the boundary diagram. John added that Sarah had very nicely illustrated what happens when an attempt is made to extract a function tree from an SSFD by lifting out all the function statements, so leaving a number of somewhat vague statements and little apparent structure. John continued by saying that, as the group had also nicely illustrated, the clue to the precision of the functional statements and the function tree structure was implicit in the SSFD. John compared the derivation of the function tree from an SSFD with the conventional approach of developing a function tree from what he termed scratch through brainstorming. John said that as Phil had pointed out it was sometimes difficult to determine the structure of the tree by this method and added that in his experience brainstorming also often led to functions being identified over a range of system levels rather than at the single system level being analysed.

John said that the propulsion system SSFD, boundary diagram and function tree are key graphics which they would need to refer to quite frequently and added that they could be found in the Appendix file. In getting back to interface analysis, John said that he assumed that, in using the interface matrix before, team members had used the standard convention of identifying exchanges of energy, material and information as well as what he called the special interface of physical. Sarah agreed that this was how they used the interface matrix in Oxton, and John asked her if she could give examples of each type of interface and explain what was meant by a physical interface. Sarah gave *force, heat, electric current* and *vibration* as examples of exchanges of energy, and *fluid, powder, mud* and *dust* as examples of exchanges of material. She continued by saying that in her experience information exchange usually occurred as an *electrical* or *optical signal*, adding that an electrical signal exchange could be both wired and wireless. In coming to physical as an exchange, Sarah said that in her experience, this was a bit more problematic and cited the situation where some people would wrongly categorise exchanges like a force between mating surfaces as a physical exchange arguing that for the force to be exchanged the surfaces must be in physical contact. Sarah explained that the way in which she thought about a *physical interface* was in terms of the relative geometry of the two surfaces and added that this would include interfaces where contact should not take place. Sarah paused at this point, and Andy took the opportunity to say that he had a couple of questions. Andy asked Sarah what she meant by an optical information exchange, and she said that she was thinking mainly of fibre optics. In asking his second question, Andy said that he assumed, from what Sarah had said, that if two surfaces had to be in contact to transmit a force between them that this would be both a physical and an energy exchange, and Sarah agreed that this was the case.

John thanked Sarah for what he called her clear explanation and said that he wanted to look at interface analysis next. He then asked Wolfgang, Andy and Yumiko if they had dual monitors and, while Wolfgang said he had two monitors, both Andy and Yumiko said that they only had a single monitor. John said that this being the case, he would cut and paste images onto a single-screen display as and when required, adding that this would be easy for him to do as most of the time he would be using a spreadsheet. John continued by saying that he recommended that they start the Oe375 propulsion system interface analysis by looking at the internal interfaces, and after a cut and paste, he shared a worksheet which showed the internal interface section of the propulsion system interface matrix with the boundary diagram alongside (Figure 6.8).

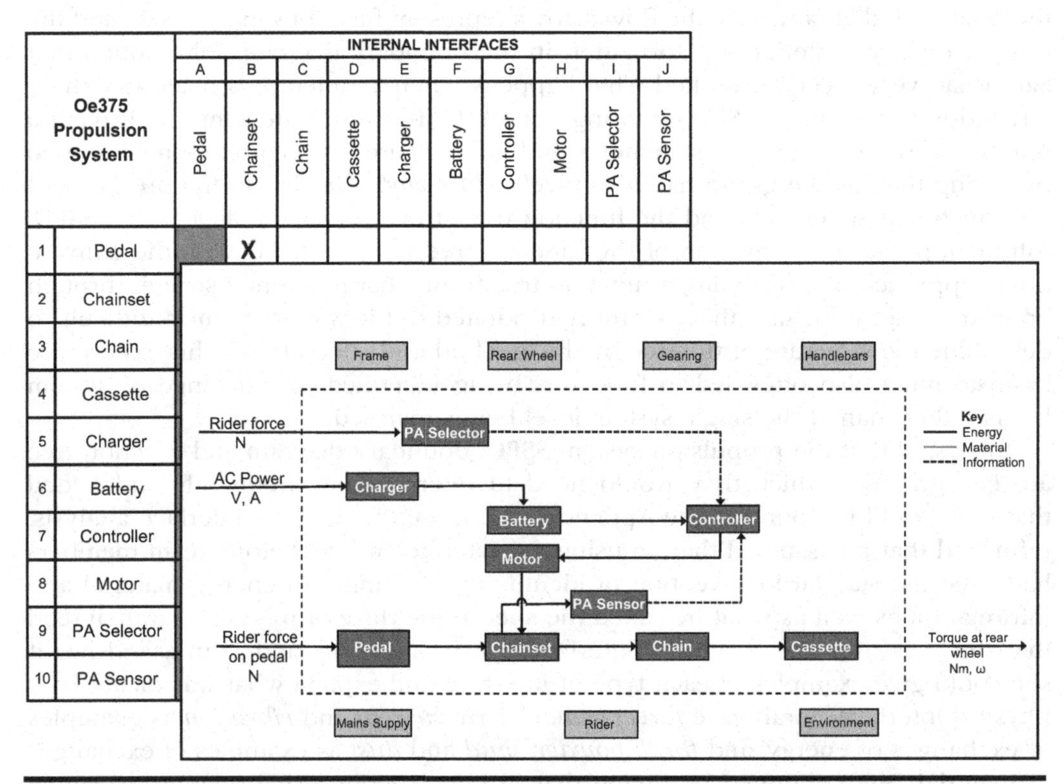

Figure 6.8 eBike propulsion system internal interface matrix and boundary diagram.

John said that they needed to consider the interface represented by each cell of the matrix in turn and identify what exchanges, if any, occurred at that interface. He added that if there was a minimum of one exchange at an interface, they would put a cross in the cell corresponding to that interface. He said that he had put a cross in the first cell corresponding to the interface between the pedal and the chainset as it was clear from the boundary diagram that an exchange occurred at this interface. Andy said that rather than put a cross in a cell in HJK they indicated the type of exchange, to which John replied that his recommendation was to put that information in an interface table for the reason he would explain very shortly. John then shared a blank interface table for the propulsion system (Table 6.1).

Table 6.1 eBike Propulsion System Blank Internal Interface Table

Propulsion System Internal Interface Table										
Interface Ref	Interface		Type	Description of Exchange	Metric	From	To	Impact	Interface Function / Requirement	Affected System Function

John said that his preference was to populate the interface matrix before considering the interface table but on this occasion, he would look at them side by side for a single interface. John said that he had uploaded templates for the interface matrix and interface table to the team ShareDrive. John explained that the format of the external interface table was identical to that of the internal table, adding that it was convenient to keep the documentation of the analysis of internal and external interfaces separate. John said that he would take the team through the analysis of the first interface on the interface matrix and hence interface table and explain the table column headings as he went (Figure 6.9).

Propulsion System Internal Interface Matrix

Oe375 Propulsion System		INTERNAL INTERFACES									
		A	B	C	D	E	F	G	H	I	J
		Pedal	Chainset	Chain	Cassette	Charger	Battery	Controller	Motor	PA Selector	PA Sensor
1	Pedal		X								

Propulsion System Internal Interface Table

Ref	Interface		Type	Description of Exchange	Metric	From	To	Impact	Interface Function / Function Required	Affected System Function
1-B	Pedal	Chainset								

Figure 6.9 Interface code and name transferred from matrix to table.

John said that there was clearly at least one interface between the pedal and chainset and this was indicated by the "X" in the appropriate cell of the interface matrix. He then began populating the interface table with the interface reference which he said was the coding by row and column specified on the interface matrix and typed *1-B* into the left-hand cell of the table. John continued by saying that the interface identification was also copied directly from matrix to table and typed *Pedal* and *Chainset* into the appropriate cells on the table (Figure 6.9). John said that what he had just done was the easy bit, adding that the next step was to document further details of any exchanges at the interface. In doing this, John recommended that the first exchange that the team documented was what he called the *main exchange* being that directly associated with the system achieving its overall function. John added that he found it useful to review the system SSFD and boundary diagram (Figures A.3 and A.4) and focus on the interface being analysed and shared the relevant section of these graphics along with a section of the interface table for the pedal–chainset interface (Figure 6.10). John asked the team to describe the exchange shown, and Wolfgang said that the rider force on the pedal was transferred to the chainset as torque.

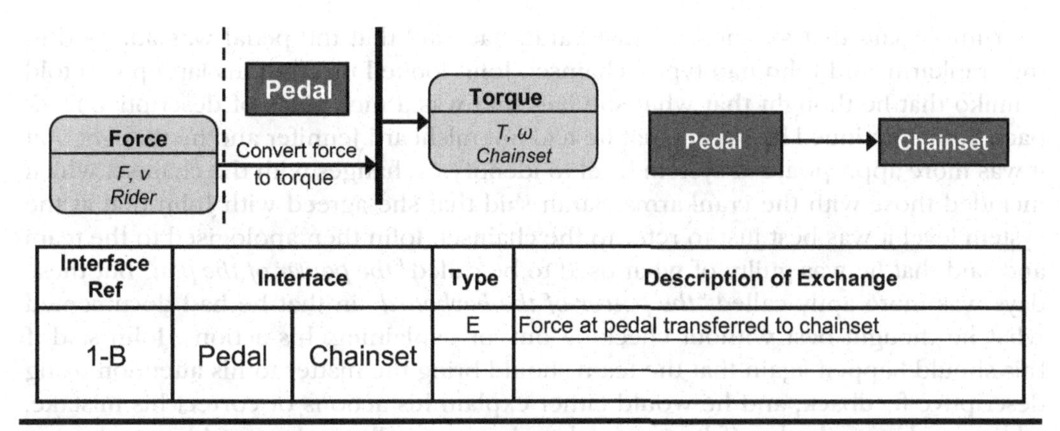

Interface Ref	Interface		Type	Description of Exchange
1-B	Pedal	Chainset	E	Force at pedal transferred to chainset

Figure 6.10 Sections of propulsion system SSFD, boundary diagram and internal interface table.

John documented the description in the interface table based on what Wolfgang had said, zooming in on the columns that he was completing. Wolfgang asked John why he had not documented the exchange exactly as he had described it and had missed out three words "Rider" and "as torque". John asked Wolfgang if he would hold his question and added that he would address it shortly. John then said that, as the team could see, they also need to indicate what type of exchange it was using the code *E, M, I, P* for *Energy, Material, Information* or *Physical* and typed E into the *Type* cell in the table. John explained that the extract of the propulsion system SSFD he was sharing was an extract based on State transitions quantified in terms of power while the extract from the boundary diagram was quantified in this case in terms of energy flow. John added that he reconciled this difference in his mind by thinking of the SSFD State transition as the transition of energy that took place each second.

John said that next all other exchanges that took place at the interface needed to be identified, and explained that in doing this he liked to visualise the exchanges in the same way that team members had done for state flow and added that this meant that if team members followed suit, they would be exploiting their tacit knowledge. John added that in visualising the exchanges, he liked to mentally go through the types of exchange. John asked what other exchanges took place at the interface, and Sarah said that the pedal would need to be attached to the crankarm and that this would need energy in the form of force to which Sam added that the pedal would need to be in physical contact with the crankarm. As Sarah and Sam spoke, John documented what they were saying in the table. As John was typing, Phil said that he thought they should add the caveat that the pedal should be attached so that it could rotate freely, and John added this requirement (Figure 6.11).

Interface Ref	Interface		Type	Description of Exchange
1-B	Pedal	Chainset	E	Force at pedal transferred to chainset
			E	Pedal attached to chainset so pedal rotates freely
			P	Pedal located on chainset so pedal rotates freely

Figure 6.11 Details of exchanges at the pedal/chainset interface.

Yumiko said that she noticed that Sarah had said that the pedal was attached to the crankarm, and John had typed chainset. John looked up from his laptop and told Yumiko that he thought that what she had said was a nice piece of descriptive feedback. John continued by saying that he had not misheard Jennifer and he thought that it was more appropriate at system level to identify exchanges with the chainset which included those with the crankarms. Sarah said that she agreed with John that at the system level it was best just to refer to the chainset. John then apologised to the team and said that he was guilty of what used to be called "*the power of the pen*" but these days was more aptly called "*the power of the keyboard*" in that he had documented what he thought best without checking out, or explaining, his actions. John said if this should happen again that the team should bring the matter to his attention using descriptive feedback, and he would either explain his actions or correct his mistake, adding with a smile that if he received evaluative feedback, he would not take any notice. Immediately, Phil retorted by asking John if he would stick to his new rule even though his feedback to Yumiko had been evaluative. John smiled in response and said that he was impressed with Phil's ability to distinguish types of feedback.

Feeling that things were beginning to go away from the task in hand, John brought team members' attention to the fact that he had separated the E and P exchanges associated with locating and attaching the pedal to the crankshaft in the way Sarah had described earlier. John then asked the team to notice that two "E" energy exchanges had been documented in the interface table and reminded the team of the comment that Andy had made a short while ago about noting the type of exchange in the interface matrix. John checked with Andy that this was done in HJK using the E, M, I and P coding, and Andy said that this was how they did it. John said that while this was accepted practice, it was not normal to indicate multiple exchanges of the same type in the matrix due to the limitation of space, and again Andy concurred that HJK recorded each type of exchange only once on the matrix. John said that since multiple exchanges of the same type at an interface were common, he felt that the interface table was a better place to record this information and asked Andy for his view. Andy said that he had to agree with John.

John continued the documentation of the detail of the pedal/chainset interface by saying that further details of each exchange were recorded in the table, and that these included the units in which the exchanges would be measured and the direction in which the exchange took place. After a minimum of discussion, John included the team's agreement of this detail in the interface table (Figure 6.12) and went on to explain what was meant by the next column headed *Impact*.

Interface Ref	Interface		Type	Description of Exchange	Metric	From	To	Impact
1-B	Pedal	Chainset	E	Force at pedal transferred to chainset	Nm, s^{-1}	Pedal	Chainset	2
			E	Pedal attached to chainset so pedal rotates freely	Nm	Pedal	Chainset	2
			P	Pedal located on chainset so pedal rotates freely	mm	Both	Ways	2

Figure 6.12 Further detail of exchanges at the pedal/chainset interface.

John explained the term Impact by saying that the exchanges at the interface that the team were considering would influence the system function, adding that the three exchanges that they had identified so far all had to occur if the propulsion system high-level function of *Convert power input to torque at the rear wheels* was to be achieved. John continued by saying that where an exchange is necessary for system functionality, it is coded as a 2 with the coding scale extending from +2 to −2, adding that −2 is an exchange that must be prevented if the system function is to be achieved. John continued his explanation of the impact scale by saying that the middle of the scale 0 denotes that the exchange does not affect system functionality, adding there were also +1 and −1 codes which he said that he would explain as the team came across them. John then asked team members if they agreed that the impact of the three exchanges that they had identified so far should be coded as 2, and after quick and unanimous agreement, John typed 2 against each exchange in the Impact column (Figure 6.12).

John said that the next thing that the team needed to establish was the functions that were required to ensure that each of the exchanges that they had described occurs where the exchange is beneficial to system operation, or that it does not occur if the exchange is detrimental to system operation. John said that starting at the top row of the Interface Function/Requirements column, the team should document what he said was the *Main Function* which was the function that was required to ensure that the main exchange takes place. John reminded team members that the description of the main exchange was also documented in the top row of the interface table extract they were considering. John shared the relevant section of the interface table extract (Figure 6.13) and added that to make this more legible he said that he had hidden some of the columns that they had already completed. John reminded the team that the main exchange was the first exchange that they had listed in the Description of Exchange column. John suggested that based on this description of the exchange, the interface function should be *Convert force at pedal to torque at chainset* and documented this in the interface table extract.

Interface		Type	Description of Exchange	Interface Function / Requirement
Pedal	Chainset	E	Force at pedal transferred to chainset	Convert force at pedal to torque at chainset
		E	Pedal attached to chainset so pedal rotates freely	
		P	Pedal located on chainset so pedal rotates freely	

Figure 6.13 Main function required to facilitate main exchange.

John told Wolfgang that he would now answer his earlier question about not describing the main exchange using the exact words that he had spoken. John asked Wolfgang if he could remember what he had said, and Wolfgang said his description of the exchange was "*The rider force on the pedal was transferred to the chainset as torque*". John explained that there were two separate aspects to what Wolfgang had said that he would like to consider. In addressing Wolfgang's inclusion of the word "Rider", John referred the team to the propulsion system interface matrix (Figure 6.7) and pointed out that the *Pedal/Chainset* and *Rider/Pedal* were two separate interfaces. John said that he preferred not to use the word rider in describing an exchange that took place at the pedal/chainset. John added that he saw this as more than semantics since while the rider exerted a force on the pedal, the rider did not directly influence the transfer of force from the pedal to the chainset as this was determined by the engineering of this interface.

In turning to the other two words "as torque", which Wolfgang had used, John said that his preference was to describe the exchange initially in more everyday terms in the Description of Exchange column, recognising that what was recorded in the interface table was meant to be read and understood by others. John added that the statement of the interface function could then use more precise engineering language so that together both columns gave complementary rather than duplicative information. Wolfgang thanked John for his explanation and said that it made sense to him.

After some discussion, the team then agreed the wording of the two other interface functions, and John documented them in the interface table (Figure 6.14).

Interface		Type	Description of Exchange	Interface Function / Requirement
Pedal	Chainset	E	Force at pedal transferred to chainset	Convert force at pedal to torque at chainset
		E	Pedal attached to chainset so pedal rotates freely	Fix pedal to chainset
		P	Pedal located on chainset so pedal rotates freely	Locate pedal on chainset

Figure 6.14 Interface functions required to facilitate exchanges.

In referring team members to the OFTEN framework roadmap (Figure A.2), John said that, as team members would see in the next Oe375 meeting, the information gained in conducting interface analysis in Phase 1 of the framework feeds directly into the development of FMEAs in Phase 2 of the framework. John added that a key component of this information flow was what is called an *Affected System Function* which was documented in the last column on the interface table. John explained that an Affected System Function is the system function or functions that were directly affected by the exchanges that occurred at the interface being analysed. John explained that for an internal interface each of the functions generated by the Design Solutions that comprise the interface could potentially be an Affected System Function and as such would be documented on the corresponding SSFD and stated more precisely on the function tree. John said that the concept of the Affected System Function was probably easier to understand by considering an example, and he shared a worksheet showing extracts taken from the propulsion system boundary diagram, SSFD and function tree (Figures A.3, A.4 and A.5) alongside the interface table extract that the team were currently analysing (Figure 6.15).

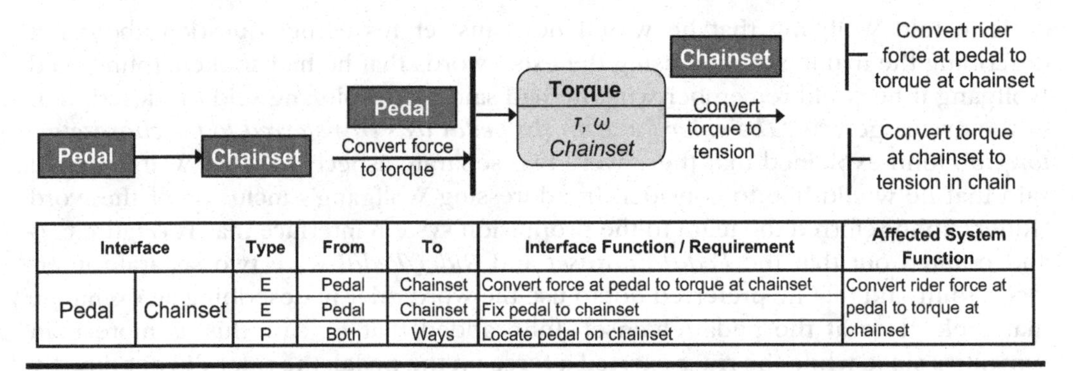

Interface		Type	From	To	Interface Function / Requirement	Affected System Function
Pedal	Chainset	E	Pedal	Chainset	Convert force at pedal to torque at chainset	Convert rider force at pedal to torque at chainset
		E	Pedal	Chainset	Fix pedal to chainset	
		P	Both	Ways	Locate pedal on chainset	

Figure 6.15 Affected system function.

John recommended that the team look at each of the two Design Solutions that formed the interface in turn, starting in this case with the pedal. John directed team members' attention to the interface functions/requirements listed in the interface table extract. John said that team members should answer the question "Is the ability of the Pedal to generate the system function *Convert rider force at pedal to torque at chainset* affected by any of the Interface Functions failing to be fully realised?" After some discussion, team members agreed that all exchanges listed in the interface table extract affected the ability of pedal to generate its function. John documented this function as an Affected System Function in the interface table extract (Figure 6.15).

Turning to the second Design Solution, the chainset, John said that the exchange associated with the Interface Function *Locate pedal on chainset* was seen to be a two-way exchange and so would be directed to both the pedal and chainset. John said that the question in terms of the chainset was "Is the ability of the chainset to generate the system function *Convert torque at chainset to tension in chain* affected by the Interface Function *Locate pedal on chainset* failing to be fully realised?" John added that in identifying Affected System Functions, a good dose of engineering logic combined with common sense was required, and he asked the team members if, for example, having a loose pedal would affect the ability of the chainset to achieve its function. After some debate, the team agreed that while the output torque from the chainset might be more variable and reduced if the pedal became loose, having a loose pedal would not of itself affect the ability of the chainset to achieve its function. John said that for the pedal/chainset interface, there was only one Affected System Function.

John said that they had now completed the analysis of the pedal/chainset interface and added that he would like to run through one more interface before the end of the meeting. In taking Phil's positive response to his request as representing the team, John said that he would jump forward to the chain/cassette interface and explained by jumping forward he was implying that normally other interfaces would be considered before the chain/cassette interface. John said that his recommendation was that both the internal and external interface tables were developed by taking interfaces in turn by going across the rows of the interface matrix from left to right starting at the top row and then working down the rows. John added that this not only helped to structure the interface tables in a logical manner but as team members would see later helped to structure a Design FMEA.

On reviewing the propulsion system interface matrix (Figure A.6), the team established that the chain/cassette interface was coded 3-D, and from the boundary diagram and SSFD, they identified that the main exchange at the interface was the transmission of force from the chain to torque at the cassette and gave the exchange the units torque and speed. Andy said that for the force to be transmitted to the cassette, the chain needed to be in physical contact with the cassette. After further discussion, the team also identified two material exchanges at the chainset/chain interface, and John documented a description of these in the table along with their units of measure (Figure 6.16).

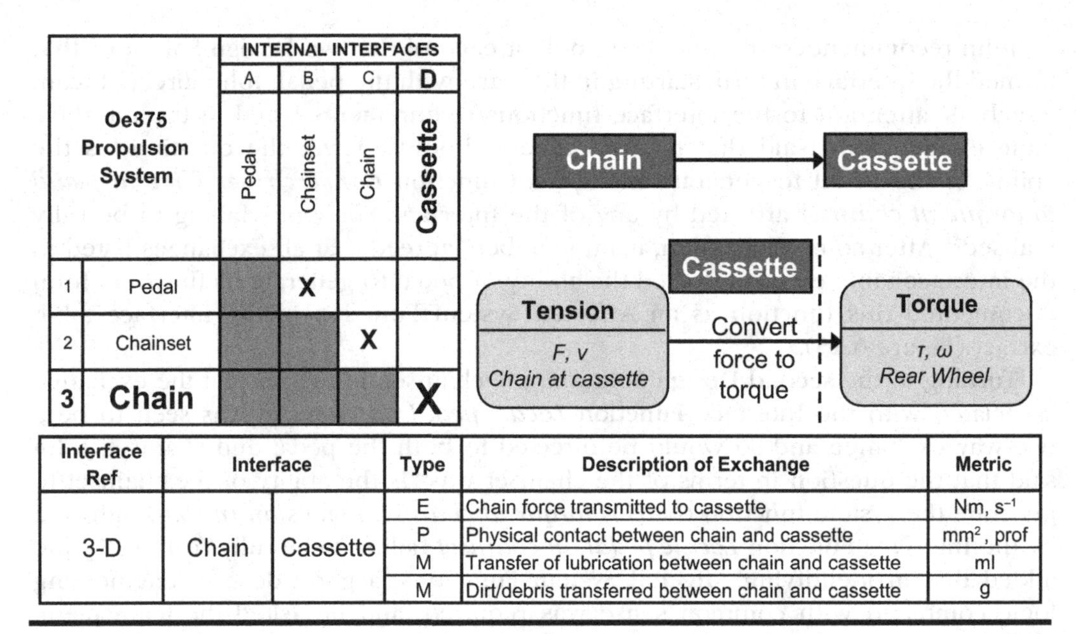

Interface Ref	Interface		Type	Description of Exchange	Metric
3-D	Chain	Cassette	E	Chain force transmitted to cassette	Nm, s^{-1}
			P	Physical contact between chain and cassette	mm^2, prof
			M	Transfer of lubrication between chain and cassette	ml
			M	Dirt/debris transferred between chain and cassette	g

Figure 6.16 Chain/cassette interface details with extracts from interface matrix, boundary diagram and SSFD.

Andy asked John what the metric "prof" was against the physical exchange, and John replied that it was the profile of the chainset teeth. The team identified the direction of the exchanges, and John typed their conclusions into the table (Figure 6.17). When it came to the Impact column, John documented the first two items as 2 and assumed that discussion of these ratings was not required. In looking at the two exchanges of material, John explained that while the transfer of lubrication between the chain and cassette was beneficial to the smooth operation of the eBike Drivetrain, it was not essential, adding that a dry chain and cassette would still work but would result in a reduction in the performance and life of the Drivetrain. Accordingly, John rated the impact as 1 and documented this in the table. Turning to the transfer of dirt/debris between the chainset and chain, John explained that this transfer was inevitable if the chain and cassette were open to the environment and while being detrimental to the performance of the Drivetrain, within given limits, it would not prevent operation of the Drivetrain and rated the impact as −1 (Figure 6.17).

Interface		Type	Description of Exchange	Metric	From	To	Impact
Chain	Cassette	E	Chain force transmitted to cassette	Nm, ms^{-1}	Chain	Cassette	2
		P	Physical contact between chain and cassette	mm^2, prof	Both	Ways	2
		M	Transfer of lubrication between chain and cassette	ml	Both	Ways	1
		M	Dirt/debris transferred between chain and cassette	g	Both	Ways	−1

Figure 6.17 Further detail of exchanges at the chain/cassette interface.

The team identified the chain/cassette main function and also readily identified the functions associated with the physical contact between chain and cassette and the transfer of lubrication, and John documented these in the interface table extract (Figure 6.18).

Interface		Type	Description of Exchange	Interface Function / Requirement
Chain	Cassette	E	Chain force transmitted to cassette	Transmit chain tension to torque at cassette
		P	Physical contact between chain and cassette	Engage contact between chain and cassette
		M	Transfer of lubrication between chain and cassette	Pass lubrication between chain and cassette
		M	Dirt/debris transferred between chain and cassette	

Figure 6.18 Chain/cassette interface functions.

Discussion then ensued in the team led primarily by Phil as to whether the torque generated at the cassette by the chain moving over it should not be specified as the rotational force acting on the rear wheel rather than the cassette. John said that once the propulsion system interface matrix was completed, it would show the internal chain/cassette and external cassette/rear wheel as separate interfaces. John said that the torque acting on the cassette due to the chain should be considered in analysing the chain/cassette interface with the torque transmitted by the cassette to the wheel being considered separately at the cassette/rear wheel interface. John added that in his experience engineers often found the discipline of containing thinking to specific interfaces in performing interface analysis difficult at first.

When the team reviewed the exchange associated with the transfer of dirt/debris between chain and cassette, John explained that this was better dealt with by defining a requirement. John continued by saying the dirt and debris was an issue to be addressed in Phase 3 of the OFTEN framework and that for now the team should identify a suitable requirement which, after a short discussion, he documented in the table (Figure 6.19).

Interface		Type	Description of Exchange	Interface Function / Requirement
Chain	Cassette	E	Chain force transmitted to cassette	Convert chain tension to torque at cassette
		P	Physical contact between chain and cassette	Engage contact between chain and cassette
		M	Transfer of lubrication between chain and cassette	Pass lubrication between chain and cassette
		M	Dirt/debris transferred between chain and cassette	Operate with specified level of dirt/debris

Figure 6.19 Interface functions and requirements.

John said that the last thing that they had to do in analysing the chain/cassette interface was to identify the Affected System Function. John shared the interface table extract along with the relevant extracts from the propulsion system boundary diagram, SSFD and function tree (Figure 6.20).

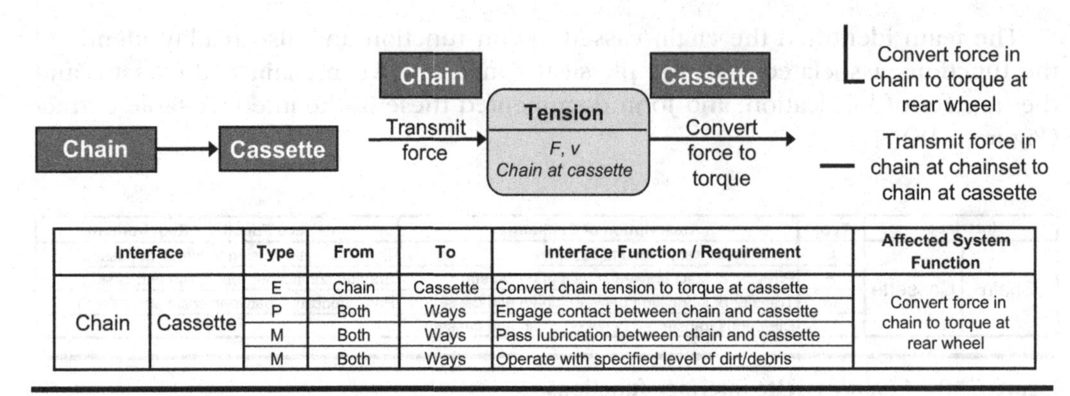

Interface		Type	From	To	Interface Function / Requirement	Affected System Function
Chain	Cassette	E	Chain	Cassette	Convert chain tension to torque at cassette	Convert force in chain to torque at rear wheel
		P	Both	Ways	Engage contact between chain and cassette	
		M	Both	Ways	Pass lubrication between chain and cassette	
		M	Both	Ways	Operate with specified level of dirt/debris	

Figure 6.20 Affected System Function associated with the chain.

John suggested that the team consider the function of the cassette first and asked that if failing to fully realise any of the listed interface functions/requirements affected the ability of the cassette to generate the function *Convert force in chain to torque at rear wheel*. The team agreed that all of the interface functions/requirements affected the performance of the cassette, and John documented the function of the cassette as an Affected System Function.

In turning to the chain, John asked team members to consider if failing to fully realise any of the listed interface functions/requirements would affect the ability of the chain to generate the function *Transmit force in chain at chainset to chain at cassette*. After a brief discussion, it was agreed that the three interface functions/requirements associated with exchanges which were directed as "Both Ways" affected the ability of the chain to function. John recommended that where there were two Affected System Functions for an interface, as was the case here, the details associated with the second Affected System Function should be duplicated in the interface table as this was helpful when it came to developing the design failure mode and effects analysis (DFMEA). John then updated the table extract accordingly (Figure 6.21).

Interface		Type	From	To	Interface Function / Requirement	Affected System Function
Chain	Cassette	E	Chain	Cassette	Convert chain tension to torque at cassette	Convert force in chain to torque at rear wheel
		P	Both	Ways	Engage contact between chain and cassette	
		M	Both	Ways	Pass lubrication between chain and cassette	
		M	Both	Ways	Operate with specified level of dirt/debris	
		M	Both	Ways	Engage contact between chain and cassette	Transmit force in chain at chainset to chain at cassette
		M	Both	Ways	Pass lubrication between chain and cassette	
		M	Both	Ways	Operate with specified level of dirt/debris	

Figure 6.21 Affected System Function associated with the cassette.

John thanked the team for their efforts, and before beginning the Warm Down, recognising that the team had covered a lot of ground in the meeting, John asked the team members how they were feeling. Phil said that while he was OK he had lost sight of why they were doing what they were doing. On being asked by John to say more, Phil said that he understood the importance of considering interfaces during product design but what they had done, although logical, seemed to him more like filling in a table with a lot of detail for the sake of it. John thanked Phil for his feedback and said that his perception was not unusual and described it as not being able to see the wood for the trees by being so deep into the engineering detail that you tend to lose sight of the higher-level reason for what you are doing. John said that when that happened to him, he found it useful to take a breather and look at the OFTEN framework roadmap in the Appendix (Figure A.2) to see where he was. John asked if anyone else felt the same as Phil and several team members said that they knew from what John had said that the interface table was useful in developing the DFMEA but were not really clear where the development of the interface table was leading. John asked team members to recall that the OFTEN framework was based on the approach of Failure Mode Avoidance, the essence of which was the identification of failure modes early in the design process in order to develop design-based countermeasures.

John then asked Phil, based on his experience, roughly what percentage of the design failures that had been discovered in previous Oxton programmes were related to interfaces between parts on a bike or eBike. Phil said he could not state a precise figure but that it was a lot, saying that almost all of the fixes that he had been involved with recently immediately before, or during, product launches were to fix problems that had occurred either at interfaces between parts or interfaces within parts. John thanked Phil and said that this was the reason why they were analysing interfaces in detail. John added that in the next two phases of the OFTEN framework, they will use their interface analysis to conduct detailed failure mode identification and, having identified the failure modes, establish corrective actions to stop them occurring. John said that interface analysis was painstaking, but in his experience very much worth it in the end.

John then asked if, what he termed "the group", defining this as Sarah, Phil, Jennifer and Yumiko, would complete the propulsion system internal and external interface analysis to present to the other team members at the next meeting, and Phil said they would, adding that they realised they still had to add the derailleur gearing to the propulsion system SSFD. In thanking Phil, John thought to himself that it was just as well this last item had not been considered today as they were quickly approaching the meeting close time. He realised that they only had time for a restricted Warm Down, something that he hated doing given how important he thought effective meeting closure to be. He asked the team members to restrict their closing comments to one statement. The resulting statements were generally positive, although John felt that restricting team members to a single statement meant that their feedback was not very useful. Since their meeting had overrun by a couple of minutes, John was pleased by the highly efficient way in which the Oxton members returned the room furniture to its normal position, and consequently, he did not feel quite as embarrassed as he might have done in apologising to those waiting for the next meeting as he left the room.

John caught up with Sam in the corridor and asked him if he had heard anything about his partner. Sam said that Emily's Mum had texted him to say that all was OK, and that Emily was sleeping. Sam added that in the flap this morning he had forgotten to spray the budgies and added that he could not ask Emily's Mum to do it since she was not enamoured with Emily and him breeding birds. John asked Sam why he would spray the budgies, and Sam replied that this was to encourage them to breed as in the wild they breed during the rainy season. By this time, John had reached his office, and Sam and he spent quite some time outside John's office door in what John found was a fascinating conversation about the ins and outs of breeding budgerigars until Sam said he had to dash to another meeting. John told Sam he hoped that Emily would soon be well, and the budgies felt amorous despite missing their shower. Sam thanked John and walked on towards the office where he worked. On sitting down at his desk, John was intrigued with what Sam had told him but quickly directed his attention back to the Oe375 meeting that had just finished. In reflecting on the meeting, John was pleased with what had been accomplished despite the meeting having a rushed beginning and end from his perspective. He was surprised how little he had thought about the Blade allegation during the meeting, realising that he was fully engaged in helping the team achieve the meeting purpose. On rebooting his laptop after plugging in his docking station, he opened his email, and his eye was immediately drawn to two interesting-looking new items in his inbox.

The first email that had caught John's eye was from Kevin Bailey at Blade. Kevin and John had joined Blade as engineering apprentices on the same day and, like John, Kevin had progressed into a management role and was now in HR with responsibilities that included working with the Colleges and Universities that offered the vocational educational courses that Blade apprentices took. John noticed that the email had been forwarded, with the original email having his Blade email address with a sent date of over a week ago. John initially wondered how Kevin knew his new email address at Oxton but then thought Kevin had probably done what he would have done, which is to look up the Oxton CEO email address on a website designed for this purpose and having seen the Oxton email address format take an educated guess. Reading down the email, John saw it concerned the retirement of Mike Holmes the Manager of the Blade Apprentice School. Mike, who was a trainer at the Apprentice School when John and Kevin joined Blade, was assigned as a mentor for their intake and so had a lot of contact with them during their first year in the company. John remembered Mike as being very supportive and never forgot the advice that he gave to all "his" apprentices that they should not see the vocational courses they were required to follow as a hurdle to overcome but as stepping stones to personal development and advancement. The email explained that after 37 successful years with the company, Mike had elected to retire and invited the email recipients to a retirement celebration at a hotel local to Blade that was frequently used for such occasions. John saw that the celebration was taking place the week after next and that anyone who felt inclined to go was invited to pay a sum of money into a bank account set up for the purpose of wishing Mike a goodbye and a long and happy retirement. The email concluded by asking recipients to indicate whether they would be attending the event. John did not need to think twice about going since he always felt that Mike was one of the few people that he had met during his time with Blade that had a significant influence on the direction his career took with the company. John responded to Kevin immediately and made a mental note to pay into the retirement fund when he got home that evening and looked forward to the opportunity of meeting with some of his former colleagues again.

The second email that John had particularly noticed was from Mo Lakhani, the Oxton Quality Manager asking John if he was available for a chat, and, if he was, requesting John to kindly set up a one-hour meeting at his earliest convenience. John was intrigued by the note and wondered why a chat would last for an hour, require a formal meeting and apparently be urgent. A thought crossed his mind that it might signal more trouble for him but, given the favourable impression that he had formed of Mo while shadowing him, he doubted it. He put the wording of the meeting request down to Mo's friendly manner and, in thinking about when he had shadowed Mo, he remembered him referencing numerous "chats" that he had had with people from the cleaners to Jim Eccleston. On checking his and Mo's calendar, he noticed that both had an open hour the next day and sent Mo a meeting notice in which he said that he was looking forward to their chat, thinking that he would soon find out what was on Mo's mind.

Background References

Mirroring Evaluative and Descriptive Feedback	Syer, J. and Connolly, C. (1996) *How Teamwork Works: The Dynamics of Effective Teamwork*, McGraw-Hill Pages 205–210
Boundary Diagram	Campean, I., Henshall, E. and Rutter, B. (2013) Systems Engineering Excellence Through Design: An Integrated Approach Based on Failure Mode Avoidance, *SAE International Journal of Materials and Manufacturing*, 6, 389–401, https://doi.org/10.4271/2013-01-0595 Pages 389–397

Chapter 7

Analysing Failure

While driving to work, John thought that it would be a good idea to review some of the Oxton failure mode and effects analyses (FMEAs) before his meeting with Mo Lakhani since, once the Oe375 propulsion system interface analysis was complete, John intended to move on to Phase 2 of the Oxton first-time engineering norm (OFTEN) framework for the propulsion system which would focus on the development of a design FMEA (DFMEA). Once he had settled down in his office, John called Sarah for the access details for the Oxton FMEA repository. Jennifer answered Sarah's phone telling him that Sarah was not in yet and asking if she could give Sarah a message. John explained what he wanted, and Jennifer replied that she could give him access adding that he would only need to review one eBike document since they were all very similar with only the minimum of administrative changes between programmes. John thanked Jennifer for her help, and almost as soon as he put the phone down, an email from her appeared in his Inbox with the FMEA repository access details and the DFMEA that she advised him to look at.

His immediate reaction on opening the file of the DFMEA that Jennifer had recommended in her email was that it was more than 200 pages in length. Although disappointed, he was not surprised at the size of the document and took this as an indication that once updated, it was probably rarely used between programmes and thought to himself that he certainly had no intention of wading through the whole document. Based on what had been discussed with the Oe375 Product Design (PD) team, John knew that the document would be based on the identification of hardware failures rather than function failure modes. His initial pessimistic reaction to the document was confirmed when he read through the first few pages to see it contained causes documented as effects and vice-versa and had a mix of system levels from the largest parts such as the frame, down to the smallest components such as a clamp bolt, without any discernible structural logic to the separation between levels. The FMEA seemed to him to be the result of the brainstorming of effects and potential failure mechanisms with little thought given to the system level that was being analysed. John found the structure of the FMEA form itself interesting as this appeared to him to reinforce what the Oxton Oe375 PD team members had told him about the practice in Oxton in which failure modes were often identified late in

DOI: 10.4324/9781003286066-7

the programme through the road and off-road testing of bikes and eBikes. Curious to see if his assumption was correct, John looked elsewhere on the FMEA site and was pleased to find an Oxton FMEA Instruction Manual dated some 15 years earlier on the cover page, the year when Oxton Bikes was formed. John assumed that the PDF file had been scanned from a printed manual with the tell-tale blackened edges to pages. On looking at what he took to be the first inside page, he was interested to see that the copyright was attributed to an automotive company and dated some 18 years earlier than the date on the cover.

In glancing through the Instruction Manual, John's eye was drawn to a short paragraph headed "Team Based" at the bottom of the third page and found it interesting that the team process that was described advised what was called "the responsible engineer" to talk to other areas when completing the FMEA. John thought to himself that this sounded more like an information-gathering exercise than a team-based process with the clear implication that the development of the document was primarily down to one person. John recalled that Sarah had told him that DFMEAs in Oxton were completed by one person and, while he was surprised to hear this at the time, he now realised that Sarah's explanation reflected accepted practice in the company. In looking through the summary of the Oxton DFMEA process as described in the Instruction Manual, John initially saw alignment with how he envisaged the OFTEN approach to DFMEA apart from the focus on part failures rather than functional failure. The first step in the Oxton DFMEA process was to identify the part being analysed along with its function. Having identified the part and its function, failure mechanisms were identified along with the potential effects and potential causes of failure. John found the next step in the Oxton DFMEA process of Design Verification interesting since this was a clear deviation from how he saw the OFTEN process. The manual explained that in this step all of the current Design Verification tests that will be used to prevent or detect failures should be listed. While it was clear to John how a verification test might detect a failure, he could not see how such a test might prevent failure unless the same test was used as an integral part of the design process. John's vision of a verification test was that it was performed to confirm that any design actions taken as a countermeasure to a failure mode successfully prevented the failure mode occurring.

In placing emphasis on the use of Design Verification testing in detecting failures, the Instruction Manual gave the examples of rig and lab testing and mathematically based investigations in addition to vehicle testing. The use of the term vehicle indicated to John that the document had not been customised for use in Oxton from its original form in the automotive company in which it had been developed. The mention of rig tests and mathematical investigations alongside vehicle testing left John wondering why Oxton appeared to place such a reliance on the testing of bicycles and eBikes off and on-road in their Design Verification and made a note to himself to discuss this with Mo.

Out of interest, John looked at the Process FMEA section of the manual and saw that what was called *Current Controls* replaced Design Verification in the DFMEA, with the Current Controls in the Process FMEA section of the manual being described in terms of the prevention or detection of failure modes. The FMEA process that John had used in Blade included Current Controls in both Design and Process FMEAs. In considering the relative merits of the prevention of failure modes and the detection of failure modes, John was clearly of the opinion that prevention was the superior strategy and remembered this being clearly stated in the Blade FMEA manual. John's experience was that the rigour and detail of interface analysis conducted within the first-time engineering design (FTED) framework was such that failure modes were identified sufficiently early in the design process to allow prevention measures to be implemented where required. John's reading of the Oxton FMEA Manual served to reinforce his intent to move forward with the strategy of placing emphasis on preventing failure modes.

John had chosen Mo's office for their meeting and knocked on his open door. Mo got up from behind his desk smiling and walked over to John, shook his hand and thanked him for setting up the meeting. Mo's office was spacious with its walls adorned with the type of poster John associated more with a shop floor quality noticeboard than a Quality Manager's office. He thought the poster blazing "Quality is Job 1" was apt since Job 1, being the first saleable product of a new model off a production line, was close to the point in the product development process at which Oxton appeared to achieve product quality. John could not fail to notice that the year planner that covered most of one wall was densely packed with handwritten notes up to and significantly beyond the current date. Mo beckoned for John to sit down in front of his desk and returned to his own spacious desk seat.

Mo began their discussion by saying that he had requested the meeting with John after speaking to Phil Eastman, explaining that he had asked Phil along with Sarah Jones if they would kindly pay a visit to a bicycle distributor. Mo said that the distributor had experienced several problems with the new Oxton off-road bike and told Mo that on this occasion they would not be fobbed off by anyone but Oxton's most experienced people. John said that Phil had mentioned the visit to him. Mo continued by saying that the visit was rather a waste of time as it turned out that oversized disc brake pads had been installed which had seized in the calliper and were catching on the rotor which could easily had been discovered by the distributor. Mo added that Phil had said he did not know how they had been fitted as it was a job to remove them, but that once he had replaced them everything was fine. John remembered from his time in shadowing him that Mo seemed to love telling "War Stories", that is tales of his exploits in fixing problems, and even though this was someone else's story, Mo had told him more detail about the incident than Phil had. In getting back to the reason why he had asked to meet with John, Mo said that both Sarah and Phil had been enthusing about the work that they had been doing with John. Mo added that Sarah had told him that the Oe375 team had been using what she called the OFTEN framework and had explained that this helped to identify problems early in the product design process so that launch issues can be eliminated. Mo told John that he would like to find out more about OFTEN and wondered whether it could help with the kind of problems they were currently having with the new off-road bike. John responded by telling Mo that he would be pleased to tell him about the OFTEN framework, adding that since it was not a problem-solving method it would be unlikely to help a great deal as a find-and-fix tool.

If Mo was disappointed to hear John's opinion that OFTEN was unlikely to be helpful to him with solving his current problems, he did not show it, and in smiling, he said that he would appreciate it if John would outline the OFTEN process to him.

John was pleased that he had copied across his Harrogate Conference presentation to his temporary laptop and, having explained its automotive context, used this as the basis of a relatively quick summary of OFTEN. Mo seemed interested in what John was saying and asked a couple of insightful questions, including the role of Noise in Design Verification testing and the difficulty in moving people away from downstream find-and-fix to upstream failure prevention.

Having addressed Mo's interest in OFTEN, John told him that he would like to discuss the way in which Oxton developed their FMEAs. Mo asked John what was on his mind. John said he had been reading the Oxton FMEA Instruction Manual and reviewed a DFMEA that Jennifer Coulter had recommended and briefly summarised the areas that he wished to cover. As John was speaking, Mo got up and picking up a dry marker listed the topics that John mentioned on what John thought was a remarkably clean whiteboard on the wall behind where he was sitting (Figure 7.1).

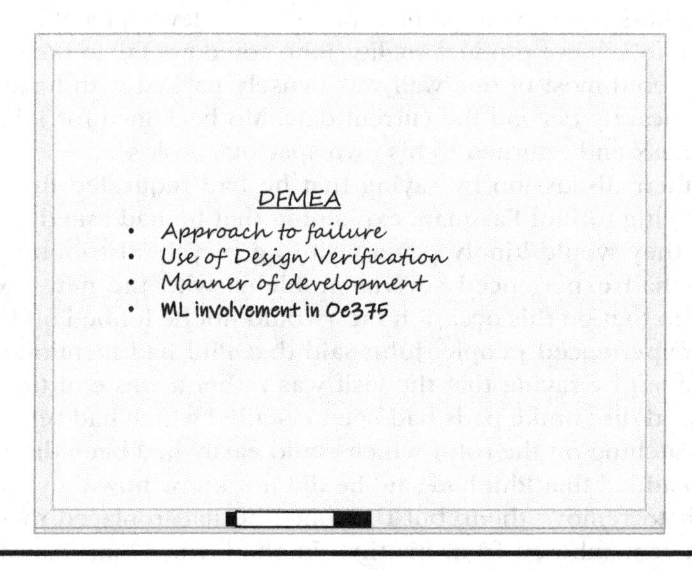

Figure 7.1 Topics for discussion.

Mo then moved a chair from a corner of his office and, in sitting next to John, suggested John turn his chair to be able to see both the whiteboard and him. Once they were seated, again Mo invited John to speak about the first item on the list. John said that while he realised Oxton used the hardware-based approach to DFMEA in identifying failures, he felt the functional approach had several advantages, not the least of which was that it was consistent with the OFTEN framework. In response to John's comment, Mo said that he had used both approaches to DFMEA and that he personally had no equity in the approach used in Oxton. Mo then confirmed what Sarah had said that the primary reason Oxton developed DFMEAs was to satisfy the requirements of the Quality Audit and added that from this standpoint it did not matter a great deal which approach to DFMEA was used. Mo said that he would be interested to see how John intended to use DFMEA within the OFTEN framework and John, standing up and picking up the whiteboard pen, added a fourth line item to their list entitled "ML involvement in Oe375", suggesting that they discuss this after addressing the other three items.

John asked Mo about the history of Oxton's FMEA Manual, and Mo said that he was not sure of its provenance but assumed that it was a direct copy of a manual published about 30 years ago by the automotive OEM that owned the copyright and as such it was way out of date. John asked why it had not been updated by Oxton, and Mo said this was because it was never felt necessary to do so since DFMEAs were only developed to satisfy the Quality Audit with each new DFMEA being piggy-backed off the previous document. John felt that it was not going to be helpful to discuss the merits of this strategy with Mo at this time and moved the discussion forward by confirming his intention to use the function-based failure approach in developing DFMEAs for Oe375. Mo said he was fine with John doing this.

John thought his ideas on the use of Current Controls in a DFMEA might meet with some resistance from Mo but was anxious to broach the subject. John said that he assumed that Mo would know that in the current widely used DFMEA form, the Design Verification column in the Oxton DFMEA had been replaced by two columns; Current Controls Prevention and Current Controls Detection. Mo said that he was aware of this from his previous job. John said that, in line with current FMEA practice, he was going to put the emphasis of effort in the Oe375 programme on the prevention of failure. John said that he was intending to revise the Oxton DFMEA form to include the Current Controls Prevention and Detection as a replacement for Design Verification. John said that most of the time the same test was used for detection and verification with the difference being the time in the programme at which the test was conducted. Mo said that he thought John's idea seemed sound and added that he was interested to see the DFMEA process in action on Oe375.

John was pleased to get Mo's support for what he intended to do about Current Controls, although he had expected some more discussion. John cast his mind back to some of the discussions regarding potential changes to a quality procedure that he had been a part of in the Global Quality Forum in Blade. He recalled that such discussions seemed to go back and forth for some time before a decision was made. In coming back to the present meeting, John reflected that Mo's approach was the more helpful.

As they had covered the second item on their agenda, John went on to consider the third item by telling Mo that he understood that DFMEAs were largely developed by one person in Oxton. Mo agreed that this was the case and pre-empted John's next point by saying that if John was taking a different approach to DFMEA, he assumed that this would include it being more of a team-based effort. John was pleased that Mo seemed to be on the same wavelength as him and confirmed Mo's assumption. John said that he was also going to encourage the Oe375 PD team to develop a series of smaller interrelated DFMEAs for each of the bicycle major systems rather than one large overall FMEA for the whole bicycle. Mo said that he thought this was a good idea too.

In discussing the last item on their list, John said that he would be pleased to have Mo's direct involvement in the Oe375, although he would not wish the programme team to become top heavy in management-level members. Mo said that he appreciated the point that John was making but would like to give whatever support he could, bearing in mind his current workload of what he described as firefighting, and John took to mean fixing problems with current Oxton products. After further discussion, the pair agreed that John would keep Mo abreast of the progress of the DFMEA and the subsequent development of countermeasures to potential failure modes, and Mo would then become directly involved in whatever Design Verification testing that the Oe375 team felt should be done. John and Mo then agreed that it had been a useful meeting. Mo said that while his intent had been to see if he could enlist John's help in supporting the current problem-solving effort, as it turned out, he saw his input to the Oe375 programme as an investment which, if things worked out as John envisaged, should reduce some of his firefighting workload in the future.

On returning to his office and sitting down at his desk, John felt pleased with the way in which the meeting with Mo had gone and was appreciative of Mo's positive attitude, although he wondered if he might offer some resistance when it came to the detail of the Design Verification. John decided that his best course of action was to let the OFTEN framework speak for itself by keeping Mo in the loop as Phases 2 and 3 progressed.

On opening his email, John saw that there was a note from Sarah on behalf of what he thought of as the Group of 4. Sarah explained that they had completed the Oe375 propulsion system interface matrix and nearly completed internal interface table and would like to complete the external interface table before the next Oe375 team meeting, although this would necessitate the meeting being postponed by a day. Sarah continued by saying that she had been in touch with all the other team members and the proposed postponement was alright with them, provided the meeting started an hour later and added that this was also alright with Yumiko as a one-off. John checked his own calendar and finding that he could also accommodate the revised meeting timing with a minor shuffle sent off the revised meeting notice followed by the one word reply to Sarah's note of "Done". It was only after he had sent out the revised meeting notice that he wondered if Sarah had checked the availability of the conference room and, on looking at the room booking calendar, was relieved to see that Sarah had provisionally booked this for the new timing. He confirmed Sarah's conference room booking and cancelled the one for the previous day letting Sarah know what he had done, adding that he assumed that the Review/Preview meeting would go ahead as scheduled. In thinking about the request to postpone the next team meeting, John was pleased that the Group of 4 was motivated to finish the whole of the task they had taken away from the meeting and looked forward to seeing the results of their endeavours.

It was only as he was getting ready for the Review/Preview meeting that John realised that he had not asked anyone to do the Warm Up for the next Oe375 PD team meeting and decided that it would be a change if he did it himself this time. On Sarah's arrival, she exchanged greetings with John and shut the door behind her. Having sat down, she asked John if he had heard anything about, what she called, the situation, and John replied that he had not heard anything, adding that he would let her know when he did. In changing the subject, Sarah asked John if she might ask for his advice on an aspect of interface analysis before they got into the Review/Preview meeting. John responded by saying he would be happy to help in any way he could and asked Sarah what was on her mind. Sarah asked John how Affected System Functions should be identified for external interfaces to the propulsion system, and John said that they were identified using the same logic used in identifying an Affected System Function for internal interfaces. John said that there was one essential difference, however, in that the function associated with the external Element is not specified on the propulsion system SSFD. John cited the example of the motor/frame interface and said that the function associated with the frame for this interface was not identified on the propulsion system SSFD. John explained that in this situation, he recommended that only the Affected System Function generated by the Design Solution that was internal to the propulsion system be documented in the external interface table, which for the motor/frame interface would be the motor function. John said that the Affected System Function associated with the frame for the motor/frame interface would be identified in treating the motor as the external Design Element interface during the interface analysis for the frame. John said that once both sets of interface analysis had been completed, it was useful to compare them to see if any additional information that was relevant to either system had been identified.

Sarah thanked John for his help, and in returning to the Review/Preview meeting, John asked Sarah for her reflections on the previous Oe375 PD team meeting, having apologised for the hurried nature of the Warm Down in that meeting. Sarah said that she thought the meeting had gone well to the extent that the group had become quite enthused about the interface analysis, and as John knew from the postponement of the next team meeting by a day, they were determined to finish both the internal and external interface analysis. John said that he was both surprised and delighted at this development, explaining he was surprised since, because of the volume of work involved, interface analysis is sometimes seen as a chore, and pleased because the Group of 4 was keen to complete the analysis. In response to John's comments, Sarah said that she enjoyed the fact that the interface analysis they were doing relied on their engineering knowledge. Sarah then said that she was not over enamoured with the term "Group of 4", saying it sounded to her more like a terrorist group and added that she would ask her fellow group members to decide on a name for themselves. John said that he thought this was a good idea. As a final reflection on the previous Oe375 meeting, Sarah said she found the topic of feedback in team meetings interesting and that she thought that descriptive feedback was a useful tool.

In thinking of the next Oe375 meeting, John said that he had forgotten to identify anyone to do the Warm Up and so while he was happy to do it at this meeting, he was thinking about drawing up a rota for future meetings. Sarah said that she thought that a rota was a good idea and said that she would ask Jennifer to draw up a form and put it on the team ShareDrive for people to sign up to if John agreed. John said that would be very useful, adding that he would tell people about the rota at the meeting. John explained that he would design a Warm Up that he could link to the meeting People Skills item of Framing Information that he intended to discuss. John then told Sarah that he had recently discussed with Jim Eccleston the next steps that the Oe375 team should take as a result of which he had decided that the team should take the Oe375 propulsion system into the next two phases of the OFTEN framework before taking the other eBike systems through Phase 1 of the framework. John said that his thinking behind this strategy was that it would give the team members a better understanding of the first three phases of the framework which he said he felt was important. Sarah asked why John was not intending to take the propulsion system through Phase 4 as well before looking at the other system and then quickly answered her own question by saying that in order to conduct design verification at the eBike level, the design analysis for the other systems needed to have been completed. Sarah added that she thought John's strategy was a good idea.

In looking at the technical content of the meeting, John said that he was going to introduce the use of DFMEA as a part of Phase 2 of the OFTEN framework and get enough done so the Group of 4 could take some tasks away. Sarah said that the group had not yet had a chance to update the propulsion system SSFD with derailleur gearing. John said that this was not a problem, adding that there would shortly be another closely related task for the group and explained that the two tasks could easily be amalgamated. Sarah said that she would pass this message on, adding that the group's interface analysis needed to be discussed in the Oe375 meeting. John suggested that given the limited time available, they should conduct an audit of the interface analysis as if it was a design review. He added that he proposed that he chose a couple of internal interfaces and one external interface to be reviewed. Sarah said that she thought that John's proposal would work well and asked which interfaces he would wish to look at. John said he wanted to look at two internal interfaces next to each other in the state flow through the system and chose the controller/motor and motor/chainset. In addition, John said he would look at an external interface with the motor and chose the environment/motor interface.

John and Sarah then agreed on the Oe375 PD team meeting agenda as Warm Up, framing information, propulsion system interface analysis and OFTEN framework Phase 2. Sarah said that the programme work was getting quite intense and the more easy-going early team meetings, which team members felt like training, were becoming a distant memory. In agreeing with Sarah, John said that this build-up of intensity was his experience of programme work generally.

John made sure that he arrived in plenty of time for the Oe375 meeting and was surprised to see Phil already in the conference room and the furniture set out to the team's requirements. After greeting Phil, he asked him if the group had completed the propulsion system interface analysis, and Phil said that they had completed their task by close of business yesterday. John asked if Yumiko had been able to contribute to the task, and Phil said that Sarah had emailed her and suggested a couple of interfaces that she might look at during her day if she had time. Phil continued by saying that Yumiko had sent across her draft analysis which the rest of the group had reviewed and included with a couple of minor edits.

After setting up his laptop and preparing the meeting software whiteboard for later, John filled the remaining time before the other team members arrived in discussion with Phil about his horticultural exploits on his two allotments. John was surprised to hear of what he considered the exotic fruit that Phil grew in his polytunnel which included nectarines and grapes and asked Phil if such crops were susceptible to the vagaries of the British climate. Phil said that in addition to the polytunnel, he used fleecing to protect from frost and added, that even with this, the succession of late frosts this year had had a detrimental effect on crop yield. John noticed that all team members had arrived or signed in while he had been deep in conversation with Phil, and so after greeting everyone, he started the Warm Up by saying that he wanted team members to consider verbal communication within their team and, walking over to his laptop, shared a slide (Figure 7.2).

Figure 7.2 Verbal communication within teams.

In explaining the slide, John said that there were three key elements to verbal communication with teams and identified these as: the words that were spoken, what he called the dance, and the music. John said that the dance related to body movement and gave eye contact, gestures, limb movement and facial expression as examples. John added that the music concerned things like the speed, tone, volume and flow of what was said. John then asked team members individually to put these three elements into the order of their importance in the context of the Oe375 team and gave them a couple of minutes to do this. He then requested team members to pair up and compare their finding with another over the meeting Chat facility, inviting Jennifer to pair with Andy, Sarah with Wolfgang and Phil with Yumiko and saying that he would pair with Sam, making it clear to Sam that they should also communicate via Chat.

As John expected, Sam and he agreed that while the dance was less relevant to a team with members communicating electronically, it was difficult to separate the words from the music. Sam and John's conclusion was replicated by the other pairs when John asked the team members to come together and discuss their findings. John felt that Wolfgang summarised the team consensus nicely when he said that while the words themselves were clearly important, if they were verstümmelt, they might become meaningless. After using the German word, Wolfgang looked up the English translation which he said was "garbled". The team members also agreed that in face-to-face contact, the dance could also be very important in giving clues as to how a speaker felt about what they were saying, and that this loss of communication through body language could make communication with and between members in remote locations in the Oe375 team more difficult. John thanked team members for their inputs and said that he would pick up their conclusions shortly in looking at Framing Information. In concluding the Warm Up, John then went through the meeting purpose and agenda.

On moving to the first agenda item after Warm Up, John said that he would like to explore verbal communication in teams further by looking at the element of "The Words" in more detail and reminded team members that they had already considered the communication skills of Listening and Questioning for Clarification in previous meetings. John added that he was going to consider the ways in which words might be framed in verbal communication between team members. John then shared a slide (Figure 7.3).

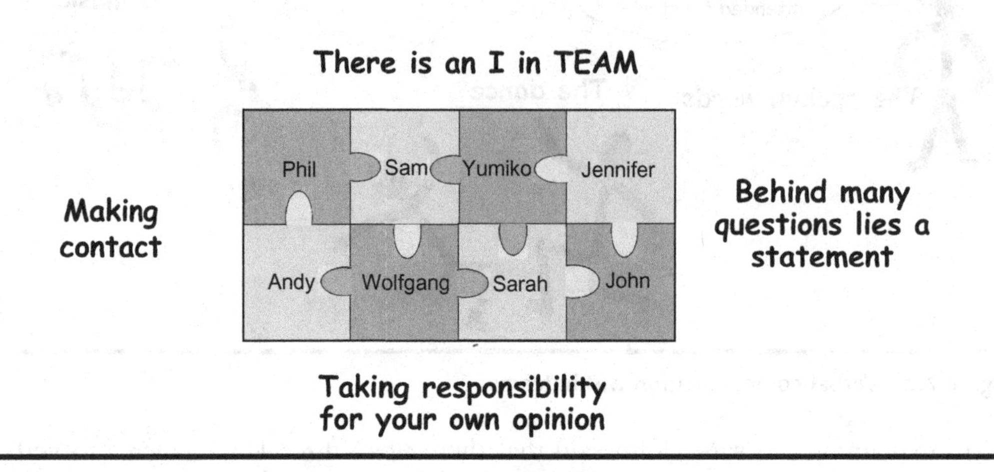

Figure 7.3 Framing information: the spoken word.

John said that he assumed that team members might have come across the expression "*There is no I in Team*" and asked if anyone could explain the thinking behind the expression. Jennifer responded by saying that she understood it to mean that good teamwork involved everyone working together and that an individual working alone could not achieve the same results as a team. John thanked Jennifer and said that he had a similar understanding to her and found it interesting that she had used the word individual. John then asked what would happen if all the individual members who made up the Oe375 PD team depicted on the slide left the team with no one new joining. Phil said that if that happened, then obviously there would not be a team. In agreeing with Phil, John said that all teams were composed of individuals or "*Is*" and added the team is a system and just like with an eBike if you take all the component parts that make up the system away, the system as an entity does not exist. John explained that this was why he believed that the expressions "*There is no I in Team*" and "*There is an I in Team*" were equally valid.

Sam asked where the expression "*There is no I in Team*" originated, and John responded that he believed it was first used in sports and subsequently has become widely used in business. John then asked rhetorically what the fact that the team is composed of individuals had to do with the way in which team members might speak to each other. In answering his own question, John said that people in a team will often speak on behalf of other team members and cited the example of a team member saying, "*We don't need to develop an SSFD for this system*". John said that the word "*We*" made the sentence appear to be definitive and in doing so disguised an opinion as fact. Yumiko asked John what he meant by the word definitive. John said that definitive meant that something was final and not open for discussion. John said that although he had said that an SSFD did not need to be developed for a particular system, other individuals in the team might think that developing an SSFD was worthwhile. Phil said that he had been a member of several engineering teams where the team leader deliberately used the word "*We*" in pretending that a decision based on his or her opinion had the support of the whole team. In supporting what Phil had said, Sarah said that she had heard fellow team members refer to the word "*We*" as "*Teamy*", implying that it apparently helped the team process because it included others in what was being said, when the reality was often the opposite. John thanked Phil and Sarah for their input and explained that fundamentally individual team members needed to take responsibility for their own opinions as this aided the clarity of verbal communication within the Oe375 team by clearly distinguishing opinion from fact. John added that taking responsibility for an opinion could easily be done by using the word "*I*" rather than "*We*" such as in the "*I believe that an SSFD does not need to be developed for this system*".

John continued by saying that people frequently use another way of disguising their opinion by asking a question and quoted the example of someone asking, "*Is it necessary to develop an SSFD for this system?*" John explained that the problem with this approach was that the people to whom the question was directed would probably not know if the questioner asked the question because they were genuinely unsure or as a surrogate for stating their opinion that developing an SSFD for the system was not necessary. Jennifer said that she was very familiar with this technique, although had to admit that she used it herself both within and outside of work. In agreeing with Jennifer, Andy said that he heard this approach every day both in work and his private life and also admitted to being, what he termed "guilty", of using questioning to hide an opinion. John said that when he heard someone whom he thought was expressing their opinion as a question, depending on how frustrated he felt, he would ask the questioner "*Is there a statement behind your question?*" John added that just as with the use of "*I*" rather than "*We*", taking responsibility for an opinion by making a statement rather than asking a question can significantly improve the clarity of spoken communication in a team.

Turning to the last subject indicated on the slide, John said that in using the spoken word it is important that the speaker and the listener or listeners make what he called "contact". John explained that contact is made when the listeners are fully engaged in what the speaker is saying, and the speaker is fully engaged in speaking to the listeners. John said that contact between speaker and listener can be enhanced by taking a few common-sense actions, and he shared the whiteboard that he had prepared earlier (Figure 7.4).

Speaking Guidelines

- Look at the person you are talking to
- Talk to the person, not about them
- Use the name of the person you are talking to occasionally

Figure 7.4 Speaking guidelines.

Sarah asked John what he meant by talking to a person rather than talking about them. John explained that if he was talking about Sarah to the team, he might say something like "*I find Sarah's help in preparing for the Oe375 team meetings invaluable*", while if he was talking to Sarah, he would look at her and say "*Sarah, I find your help in preparing for the Oe375 team meetings invaluable*". John asked Sarah how she felt about the two options that he had just offered, and Sarah replied that she had felt that John was more sincere when speaking to her. John thanked Sarah and said that he would like to explore how the Oe375 team members felt they were currently doing in their PD team and Group of 4 meetings against the three guidelines.

John asked Andy, Wolfgang and Yumiko for their thoughts, and Andy said that in communicating visually through a laptop and webcam, it was almost impossible to look simultaneously into the webcam and at the screen so that meant it was not possible to look at other team members eye to eye and so at best the first guideline could be is "*Look in the general direction of the webcam when talking*". This caused smiles on team members' faces, and John in agreeing with Andy said that this was a shortcoming in achieving contact with remote members of the team and in his opinion served to make the other two guidelines even more important. Yumiko said that particularly in the meetings in which she had worked with Sarah, Jennifer and Phil she was not always clear if a question or comment was directed to her because of the difficulty that Andy had explained of not being able to look individuals directly in the eye. Yumiko added that it would help if other team members would mention her name when asking her a question or directing a comment to her. Following on from Yumiko, Wolfgang said that he felt that the Oe375 team meetings went well and were far better than some of the other virtual meetings that he attended.

John thanked Andy, Yumiko and Wolfgang for their feedback and asked the Oxton team members for their comments. Phil said that although he thought the guidelines were followed well as a natural part of the Oxton team members working together, he was sure team members would try to help the remote members overcome the difficulties that they had mentioned. Jennifer said that she agreed with Phil and reflected that, while the guidelines were common sense, it had been useful to make them explicit as the points made by Andy and Yumiko had shown. John thanked Phil and Jennifer for their comments and, on checking the agenda, handed the meeting over to Sarah to present the work that had been done by the Group of 4 on the propulsion system interface analysis.

Sarah said that before looking at the interface analysis, Phil wanted to say a few words. Phil thanked Sarah and said that he would like to speak about the name of the group that had been conducting the propulsion system interface analysis. Phil said that as just illustrated, John had started to refer to the grouping of Sarah, Jennifer, Yumiko and himself as the "Group of 4" and that a number of the so-called group members were not over enamoured with this title. Phil continued by saying with a smile on his face that after long deliberations and bitter arguments, the group had decided they wanted to be known in the future as Team B, adding they initially thought of themselves as the A-Team but felt that would not be fair on Sam, Wolfgang and Andy. Phil then said that, with this very important business concluded, he would return the meeting to Sarah.

Sarah explained that she had agreed with John that Team B would take the Oe375 team through two internal and one external interface as examples of the interface analysis that Team B had conducted on the propulsion system. Sarah said that before she looked at the interface table extracts for the three interfaces that had been selected for review, she would share the completed propulsion system interface matrix and said that this included the interfaces that they intended to review (Figure 7.5).

	Oe375 Propulsion System	INTERNAL INTERFACES										EXTERNAL INTERFACES						
		A	B	C	D	E	F	G	H		I	E1	E2	E3	E4	E5	E6	E7
		Pedal	Chainset	Chain	Cassette	Charger	Battery	Controller	Motor	PA Selector	PA Sensor	Frame	Rear Wheel	Derailleur Gearing	Handlebars	Mains Supply	Rider	Environment
1	Pedal		X														X	X
2	Chainset			X					X		X	X					X	X
3	Chain				X							X	X	X			X	X
4	Cassette												X				X	X
5	Charger						X	X								X	X	X
6	Battery							X	X			X					X	X
7	Controller								X	X	X	X					X	X
8	Motor											X					X	X
9	PA Selector														X		X	X
10	PA Sensor											X					X	X

Figure 7.5 Propulsion system interface matrix.

Sarah asked if there were any questions or comments, and Jennifer said that she had uploaded a copy of the interface matrix to the Appendix with the file name A.6. Yumiko then asked if there was any difference in the way in which interfaces were viewed on the internal interface matrix giving the "Cassette to Chain" or "Chain to Cassette" as an example. Sarah looked to John who in assuming that she would like him to answer Yumiko's question said that it was the same interface either way, although the nature of the internal interface matrix was such that an interface with a particular design solution might appear in a row or a column of the matrix. John cited the fact that the chainset had three row-wise internal interfaces and one column-wise internal interface and added that it was important to view design solutions both in terms of their rows and columns in the internal matrix. John reminded team members that the interface table indicated the direction of flow of each exchange. Yumiko thanked John for his explanation. Sam then said he did not think that the rider interfaced directly with several design solutions and gave the example of the motor. Sarah said that the rider's main interaction with some of the design solutions such as the motor would be during cleaning of the eBike. Sarah said that the team thought that this was an important interface since if the rider cleaned the eBike with a high-pressure lance, this might cause water ingress. Sam followed up by asking how the rider could be prevented from using a high-pressure lance to clean the eBike. Phil responded to Sam's question by saying that of course if a rider wished to use a high-pressure lance, then that was up to the rider, Phil added that by identifying this interface the Oe375 team should be able to ensure that appropriate guidance for cleaning the eBike would be included in the User Manual.

Anxious to move on, Sarah thanked Sam for his intervention and said that the three interfaces that John had selected to review were 7-H controller/motor, 2-H motor/chainset and 8-E7 motor/environment. Sarah said that Jennifer, Yumiko and she would share the presentation and said that she was going to discuss the analysis for the first interface being the 7-H motor/controller interface. Sarah shared the initial columns of the internal interface table extract for the interface along with the corresponding extract from the boundary diagram and SSFD (Figure 7.6).

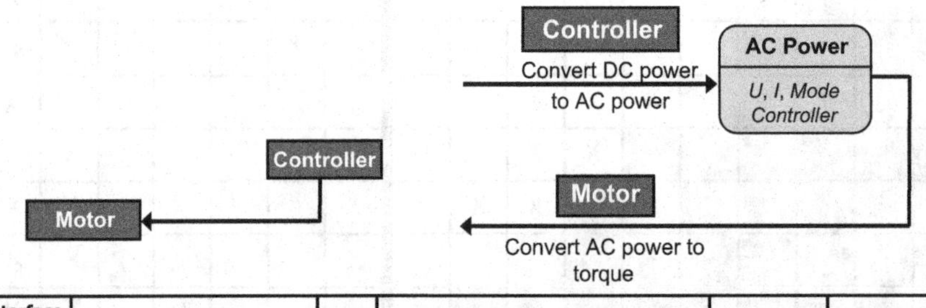

Interface Ref	Interface		Type	Description of Exchange	Metric	From	To	Impact
7-H	Controller	Motor	E	AC power transfer	V,A,Hz	Motor	Controller	2
			E	Electromagnetic interference	dB	Both	Ways	–2
			I	Motor speed and position	ω, θ	Motor	Controller	2
			E	Heat transfer	J	Motor	Controller	–1

Figure 7.6 Exchanges at motor/controller interface with extracts from SSFD and boundary diagram.

Sarah said that Team B had identified *AC power transfer* as the main exchange along with the two other energy exchanges and one information exchange documented on the interface table extract. Sarah asked if there were any questions or comments, and Sam asked why the motor speed and position exchange was needed. Sarah replied that commutation was done electronically in a brushless motor through the controller with the controller requiring to know the rotor position in the motor in order to switch the current direction at the appropriate position. Sam thanked Sarah, and since there were no more questions, Sarah shared the remaining columns of the interface table along with extracts from the propulsion system SSFD, boundary diagram and function tree (Figure 7.7).

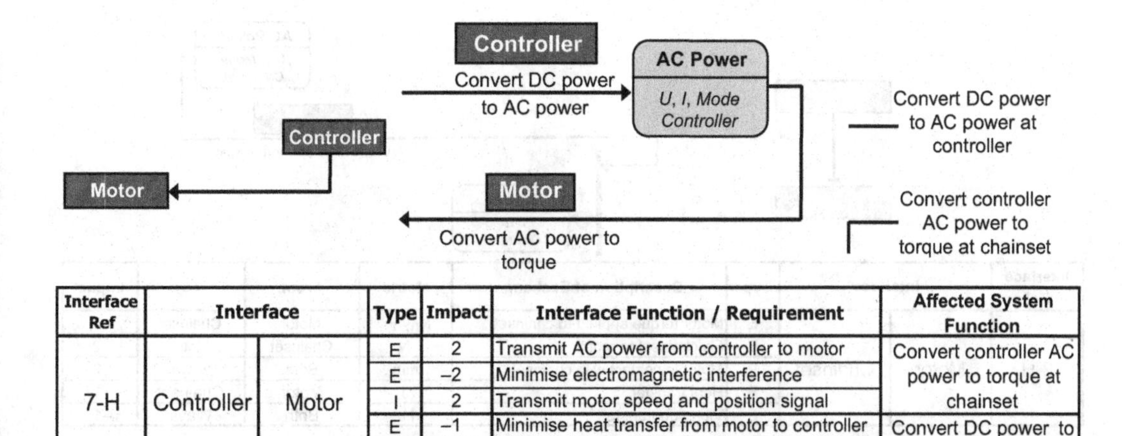

Interface Ref	Interface		Type	Impact	Interface Function / Requirement	Affected System Function
7-H	Controller	Motor	E	2	Transmit AC power from controller to motor	Convert controller AC power to torque at chainset
			E	–2	Minimise electromagnetic interference	
			I	2	Transmit motor speed and position signal	
			E	–1	Minimise heat transfer from motor to controller	Convert DC power to AC power at controller
			E	–2	Minimise electromagnetic interference	

Figure 7.7 Interface function/requirement, and Affected System Functions for motor/controller interface.

Sarah asked the non-Team B members to note that there were two Affected System Functions at the motor/controller interface with three of the four interface exchanges affecting the motor function and the other exchange affecting the controller function. Sarah added that one exchange affected both Affected System Functions. Sarah asked if there were any further questions or comments, and Andy asked why the information flow of the motor position and speed to the controller was not shown on the SSFD and boundary diagram. Jennifer explained that Team B had discussed this very point and concluded that other control flows such as that sending information on the state of charge of the battery to the controller were not shown on either the SSFD or the boundary diagram. Jennifer said that Team B members decided if they included all the control flows on the SSFD, the model might start to resemble a bird's nest and was too much detail for a system-level graphic. Jennifer added that Team B felt that this information was of sufficient importance to include in the interface table. John was pleased with the way in which Jennifer had addressed Andy's question and felt that he had nothing to add. Sarah thanked Andy for the question and said that Jennifer was presenting the next interface and handed the meeting over to her.

Jennifer thanked Sarah and said that she would look at the analysis of the motor/chainset interface 2-H and shared the relevant extract from the propulsion system internal interface table along with extracts from the propulsion system SSFD and boundary diagram (Figure 7.8).

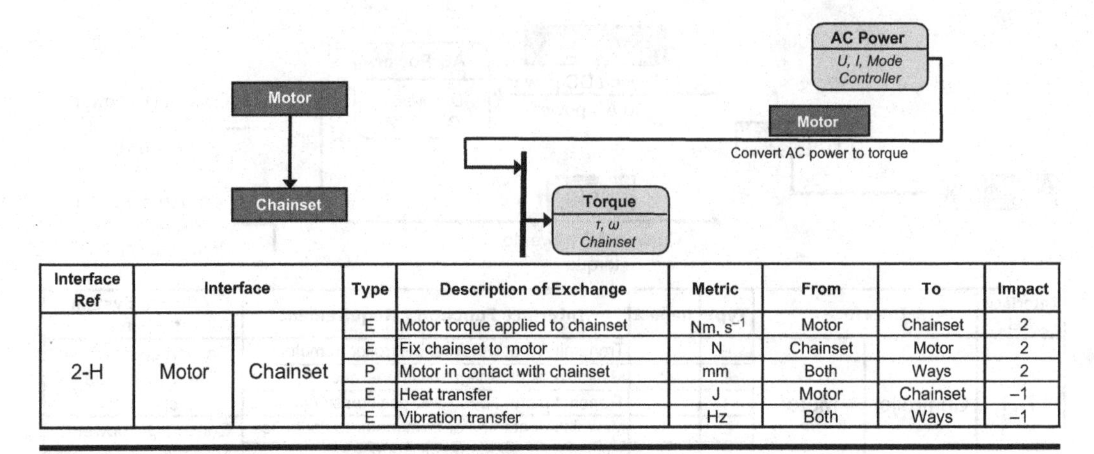

Interface Ref	Interface		Type	Description of Exchange	Metric	From	To	Impact
2-H	Motor	Chainset	E	Motor torque applied to chainset	Nm, s^{-1}	Motor	Chainset	2
			E	Fix chainset to motor	N	Chainset	Motor	2
			P	Motor in contact with chainset	mm	Both	Ways	2
			E	Heat transfer	J	Motor	Chainset	–1
			E	Vibration transfer	Hz	Both	Ways	–1

Figure 7.8 Exchanges at the chainset/motor interface.

Jennifer said that Team B had identified the main exchange from the boundary diagram as the *Motor torque applied to the chainset*. Jennifer added that the team had also identified three other energy exchanges and one physical exchange as documented on the internal interface table extract and asked Wolfgang, Andy and Sam for their comments. Wolfgang said that he noticed that the vibration was transmitted both from the motor to the chainset and the chainset to the motor, adding that in his experience eBike motors were quiet and did not tend to exhibit much if any vibration. Jennifer asked Yumiko if she would answer Wolfgang's point. Yumiko said that Wolfgang's observation about the low levels of noise and vibration emitted by an eBike motor were correct and added that this was because of careful design. Yumiko explained that vibration in a brushless direct current motor (BLDC) arose both from mechanical sources and fluctuations in the electromagnetic forces within the motor. Yumiko said that she did not think it appropriate to go into detail at this time but there were factors in the motor design that could be manipulated to minimise both sources of vibration. Yumiko concluded by saying that she felt that the vibration and noise were important exchanges that needed to be specified in the interface table. After Wolfgang had thanked Yumiko for her explanation, Andy asked Jennifer why audible noise was not identified on the interface table extract. John was pleased when Jennifer answered Andy's point, albeit it seemed to him somewhat sarcastically, by saying that motor noise would be heard by the rider and not the chainset and so was not an exchange at the interface that they were currently reviewing.

Jennifer thanked Wolfgang and Andy for their questions and showed the remainder of the chainset/motor interface table details (Figure 7.9), saying that Team B members identified the Affected System Functions by reviewing the propulsion system SSFD and function tree.

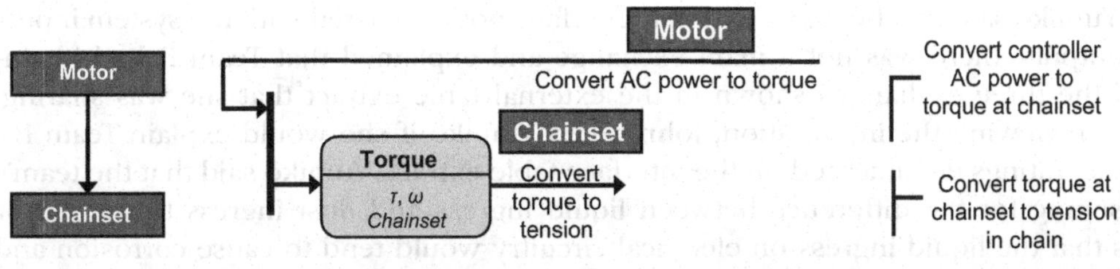

Interface Ref	Interface		Type	Interface Function / Requirement	Affected System Function
2-H	Motor	Chainset	E	Transmit torque from motor to chainset	Convert controller AC power to torque at chainset
			E	Clamp chainset to motor	
			P	Couple motor physically to chainset	
			E	Conduct heat away from motor via chainset	
			E	Operate within specified level of vibration	
			P	Couple motor physically to chainset	Convert torque at chainset to tension in chain
			E	Operate within specified level of vibration	

Figure 7.9 Interface functions, requirement, and Affected System Functions for chainset/motor interface.

Jennifer asked for questions and comments, and Andy said that since the team had decided that a loose pedal would not affect the function of the chainset in the last Oe375 team meeting why had Team B decided that a loose coupling between the motor and chainset would affect the chainset function. Jennifer said that Team B had discussed this point and had decided that the difference was that the pedal was not directly coupled to a chainring while the motor was. Andy said that he might beg to differ on this point, adding that he did not want to disrupt Team B's presentation. Jennifer thanked Andy for his frankness and said that Yumiko was going to discuss the external interfaces with the motor next and handed the meeting over to her.

Yumiko thanked Sarah and in referring team members to the propulsion system interface matrix (Figure A.6) said that there were three external interfaces with the motor and identified these as the *Motor/Frame*, *Motor/Rider* and *Motor/Environment*. Having explained that she would consider one of these interfaces in detail and then consider a common feature of all the interfaces, Yumiko shared what she called the first graphic (Figure 7.10).

Motor

Environment

Interface Ref	Interface		Type	Description of Exchange	Metric	From	To	Impact
8-E7	Motor	Environment	E	Heat flow	J	Motor	Environment	–1
			M	Liquid ingress to motor	ml	Environment	Motor	–2
			M	Dust ingress to motor	mg	Environment	Motor	–1

Figure 7.10 Exchanges at the motor/environment interface.

Yumiko said that being an external interface not associated with the system inputs or outputs, there was not a main exchange and explained that Team B had identified the three exchanges shown in the external table extract that she was sharing. After reviewing the information, John asked Yumiko if she would explain Team B's impact ratings documented on the interface table extract. Yumiko said that the team's reasoning for the difference between liquid ingress and dust ingress to the motor was that the liquid ingress on electrical circuitry would tend to cause corrosion and the team felt that this had the potential of leading to catastrophic failure while dust ingress was likely to have less effect on motor performance. Turning to the exchange of heat between the motor and environment, Yumiko said that an increasing temperature within a BLDC motor would reduce its efficiency by reducing the electromagnetic interaction between the stator and rotor due to both a decrease in current flow in the stator windings and a decrease in magnetic flux density in the permanent magnet of the rotor. Yumiko added that while the motor would continue to operate as it got hot, its performance would reduce if sufficient heat was not dissipated from it and said that this reasoning led Team B to the impact rating of −1. John said out loud, as if he were speaking to himself, "that would teach him to question the integrity of the Team B analysis", which produced laughter. Yumiko asked Sam, Andy and Wolfgang if they had any comments, and when they did not respond, she shared the remaining part of the interface extract (Figure 7.11).

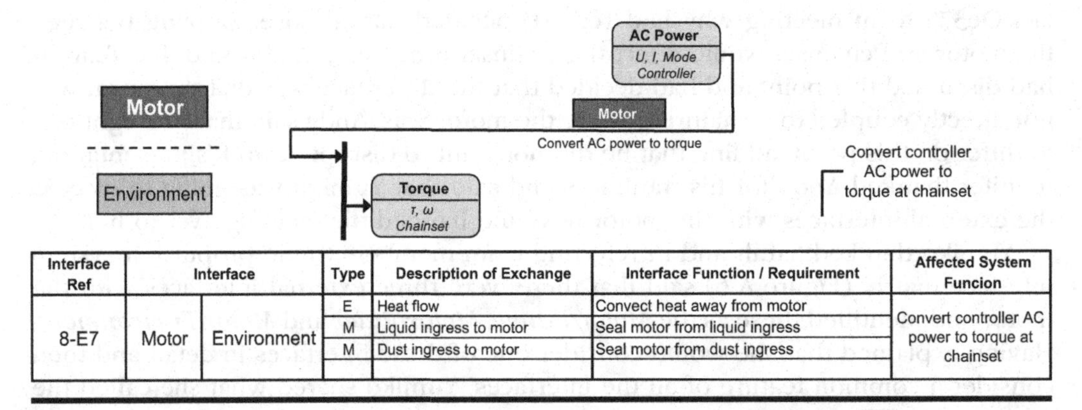

Interface Ref	Interface		Type	Description of Exchange	Interface Function / Requirement	Affected System Funcion
8-E7	Motor	Environment	E	Heat flow	Convect heat away from motor	Convert controller AC power to torque at chainset
			M	Liquid ingress to motor	Seal motor from liquid ingress	
			M	Dust ingress to motor	Seal motor from dust ingress	

Figure 7.11 Interface functions, requirement and Affected System Function for motor/environment interface.

Yumiko said that in identifying the Affected System Function, Team B considered the interface function of *Convect heat away from motor*. Yumiko explained that any deterioration in achieving this was likely to cause the motor to overheat which would affect its performance and hence its ability to achieve its function of *Convert controller AC power to torque at chainset*.

Yumiko then shared a graphic showing an extract from the propulsion system external interface table for each of the three external interfaces with the motor (Figure 7.12).

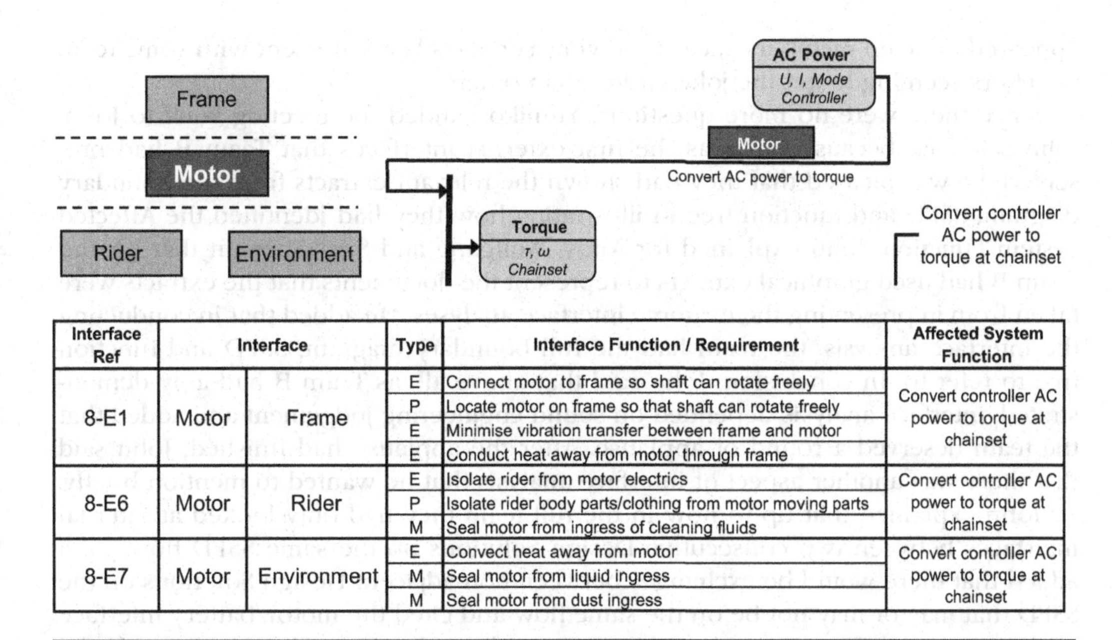

Interface Ref	Interface		Type	Interface Function / Requirement	Affected System Function
8-E1	Motor	Frame	E	Connect motor to frame so shaft can rotate freely	Convert controller AC power to torque at chainset
			P	Locate motor on frame so that shaft can rotate freely	
			E	Minimise vibration transfer between frame and motor	
			E	Conduct heat away from motor through frame	
8-E6	Motor	Rider	E	Isolate rider from motor electrics	Convert controller AC power to torque at chainset
			P	Isolate rider body parts/clothing from motor moving parts	
			M	Seal motor from ingress of cleaning fluids	
8-E7	Motor	Environment	E	Convect heat away from motor	Convert controller AC power to torque at chainset
			M	Seal motor from liquid ingress	
			M	Seal motor from dust ingress	

Figure 7.12 Interface functions, requirement, and Affected System Functions for the external motor interfaces.

In directing team members' attention to the Affected System Function column of interface table extract that she was sharing, Yumiko said that she wanted to verify that the *Motor/Frame* and *Motor/Rider* interfaces shared the same Affected System Function as the *Motor/Environment* interface. By way of example, Yumiko selected the Interface Function "*Connect motor to frame so shaft can rotate freely*" at the motor/frame interface and said that if this function was not fully met, it was likely that the motor would not be able to operate properly. Yumiko said that equally if the interface function "*Isolate rider body parts/clothing from motor moving parts*" at the motor/rider interface was not fully met, the operation of the motor would be impaired. Yumiko asked Wolfgang, Andy and Sam if they agreed that the Affected System Function stated on the propulsion system interface table extract was common to all the three external interfaces with the motor, and Andy said he agreed that it was. Wolfgang and Sam both concurred with what Andy had said.

Yumiko asked if there were any questions or comments, and Sam asked why Team B had documented only one Affected System Function for the external interfaces with the motor. Yumiko referred Sam to the propulsion system boundary diagram (Figure A.4) and explained that where an external system was a part of the eBike such as the frame and rear wheel, the Affected System Functions associated with these systems would be identified in conducting interface analysis for each system and treating the motor as an external system. In smiling, Yumiko added that she doubted Team B would have time to conduct interface analysis on the systems which were not a part of the eBike such as environment and rider. John was impressed by Yumiko's explanation and thought that it was more precise and succinct than the explanation that he had given Sarah in the last Review/Preview meeting when she had asked him the same question that Sam had asked. John also noted the smiles that appeared on team members' faces following Yumiko's last statement with some team members seeming to get the joke earlier than others.

Since there were no more questions, Yumiko handed the meeting back to John. John said that because this was the first external interfaces that Team B had presented, he was pleased that they had shown the relevant extracts from the boundary diagram, SSFD and function tree in illustrating how they had identified the Affected System Function. John explained for Andy, Wolfgang and Sam's benefit that he and Team B had used graphical extracts to represent the documents that the extracts were taken from in presenting the example interface analyses. He added that in conducting the interface analysis, the team had the full boundary diagram, SSFD and function tree to refer to. In conclusion, John said that above all, as Team B had ably demonstrated, interface analysis depended on sound engineering judgement and added that the team deserved a round of applause. After the applause had finished, John said that there was another aspect of interface analysis that he wanted to mention briefly.

John explained that up to now in the full team they had only looked at internal interfaces between two consecutive Design Solutions on the same SSFD flow. John added that there would be exchanges between non-adjacent Design Solutions on the SSFD that may or may not be on the same flow and cited the motor/battery interface as one such interface. John shared an extract from the internal interface table for this interface along with the corresponding extracts of the propulsion system SSFD, boundary diagram, function tree and interface matrix (Figure 7.13).

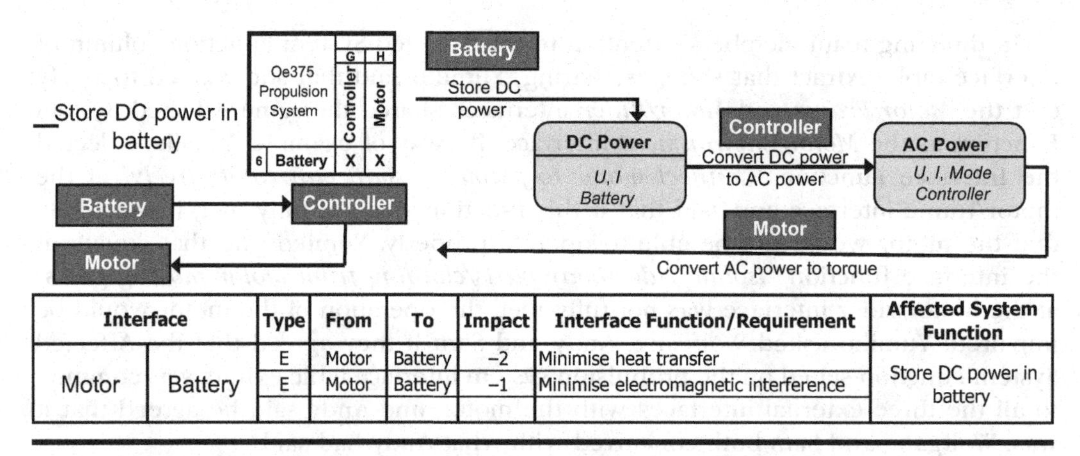

Interface		Type	From	To	Impact	Interface Function/Requirement	Affected System Function
Motor	Battery	E	Motor	Battery	−2	Minimise heat transfer	Store DC power in battery
		E	Motor	Battery	−1	Minimise electromagnetic interference	

Figure 7.13 Interface functions/requirements, and Affected System Functions for the motor/battery interface.

John explained that there was not an exchange on the interface table extract that corresponded to a main function since, as shown in the boundary diagram extract, the principal flow of power between the battery and motor was through the controller and not directly between the motor and battery. John added that nevertheless the direct exchanges associated with the interface requirements documented on the interface table extract and indicated on the interface matrix extract would occur. John said that as these exchanges could have a detrimental effect on battery performance, the Affected System Function was the function of the battery of *Store DC power in battery*. John asked if anyone had a question or comment, and Andy said that the battery would also generate heat and electromagnetic interference that could affect the motor function. John said that Andy had raised a good point and while he had assumed such exchanges between the battery and motor were negligible, it was worthy of further investigation. Since no one else spoke, John decided that it was time for a break.

Following the break, John said that he was going to move on to Phase 2 of the framework and reminded team members that this was entitled *Failure Analysis* by sharing a slide (Figure 7.14).

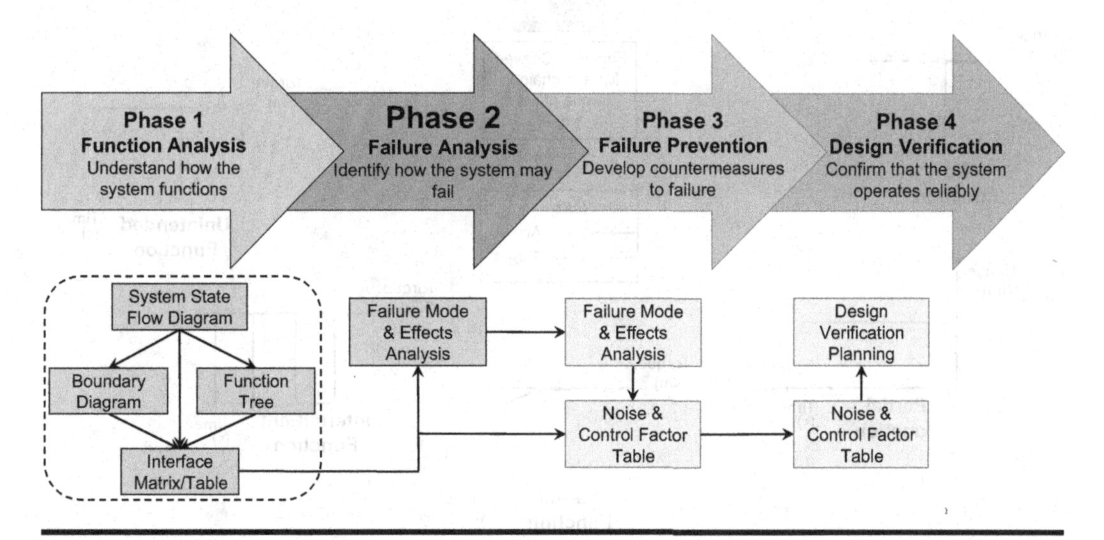

Figure 7.14 OFTEN framework – Phase 2.

John said that having understood how a system functions in Phase 1 of the framework, the objective of Phase 2 was to identify how a system might fail to function by using the information gained from the function and interface analyses in Phase 1 to develop a Design FMEA. John explained that while the other eBike systems would need to be taken through Phase 1 of the framework, taking the Oe375 propulsion system through what he called a slice of Phases 1 to 3 of the framework first would allow team members to gain a much richer mental model of the framework than the one he had depicted on the slide. John added that this should help them during the analysis of other eBike systems. John explained that the design verification at the systems level of the eBike was best conducted when the design analysis of all of the systems of the eBike had matured and so he was proposing not to move into Phase 4 of the framework until most of the eBike systems had been taken through the first three phases. Since no one commented of John's proposal, John brought team members attention back to the slide he was sharing by saying that as the graphic implied the integrity of the analysis conducted in Phase 2 of the framework depended on the integrity of the analysis conducted in Phase 1.

John said that he had recently looked through one of Oxton latest DFMEAs that Jennifer had recommended that he review and had also read the Oxton FMEA manual. John explained that his review of both documents confirmed what team members had already told him that the Oxton approach to FMEA was based on identifying hardware failures and listed cracked, distorted, corroded and waterlogged as example failure mechanisms that he had seen in the DFMEA that he had examined. John continued by saying that the team would use their interface analysis as an input to the development of the propulsion system DFMEA and to do this they were going to use an alternative, widely used, way of developing a DFMEA based on what are called *Function Failure Modes*. John added that he had spoken to Mo Lakhani who was supportive of what the Oe375 team would be doing. John then shared a slide (Figure 7.15).

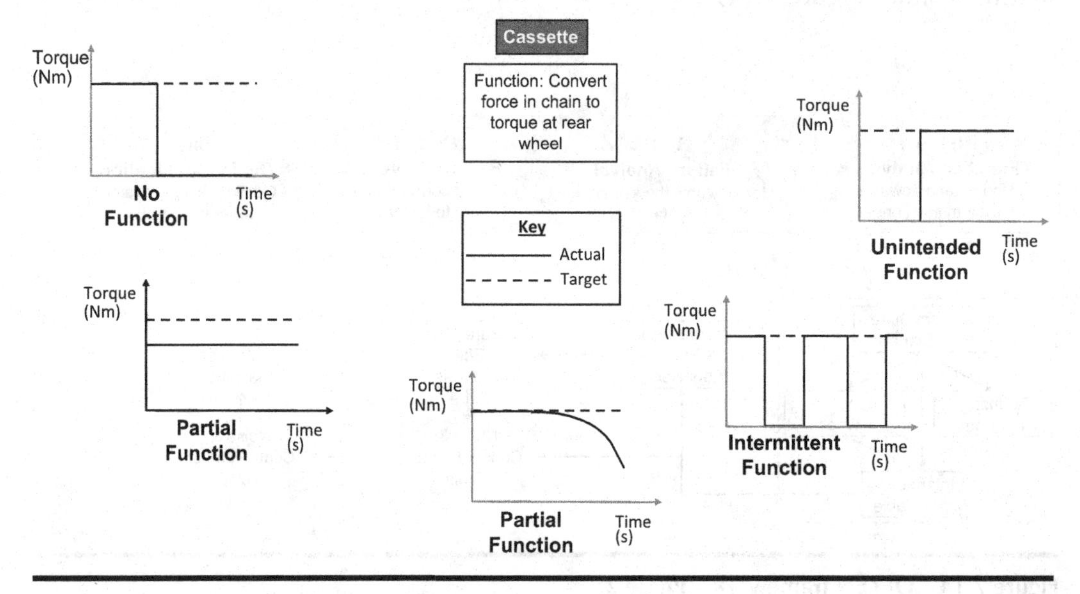

Figure 7.15 Types of function failure mode.

John said that he wanted to consider the concept of a function failure mode and, in referring to the slide, he said he would use the function of the cassette of *Convert force in chain to torque at rear wheel* as a graphical example. John explained that there were four types of function failure mode and listed these as *No Function*, *Partial Function*, *Intermittent Function* and *Unintended Function*. John added that each type was depicted graphically on the slide with two variants of Partial Function shown. John said that the partial function failure mode shown at the centre bottom of the slide was usually said to be a *degraded function* and added that, while this graphic showed a progressively decreasing output, if the output progressively increased over time or if the output was at a constant level above its target value, it was also considered to be a partial failure mode. John asked if there were any questions or comments, and Jennifer asked if John would explain what unintended function was. In response, John gave the example of the electric motor spontaneously providing assistance when the eBike was in the "Pedal bike without power assist" mode.

John then asked the team in what units the failure modes and their corresponding functions would be measured. Sarah said that the graphics implied that a function and its failure mode types were measured in the same units, which in this case was Nm being torque. John thanked Sarah and said that this was an important point and stressed that function failure modes were always measured in the same units as their corresponding function. John said that he wanted to look next at what underpinned the FMEA process and shared another slide (Figure 7.16).

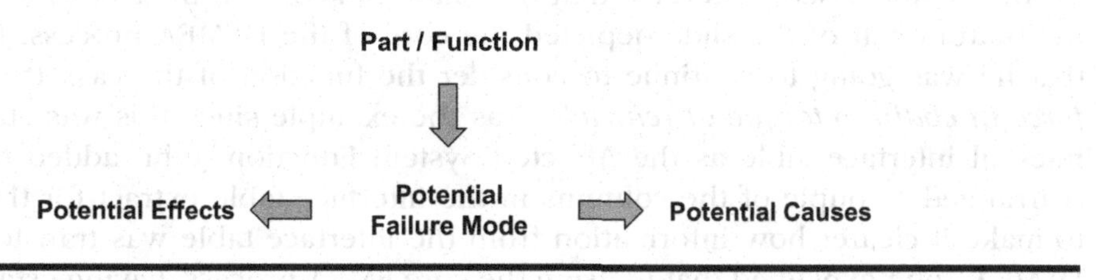

Figure 7.16 Core FMEA process.

John explained that the slide depicted the core of the FMEA process which was probably familiar to team members since it was essentially the same for both the hardware approach to FMEA which started by identifying the part to be analysed, or the functional approach to FMEA which started with the function to be analysed. John continued by saying that because they would be using the functional approach, having identified the function to be analysed, the next step in the process was to identify potential functional failure modes followed by establishing the potential effects and potential causes of the failure modes. John drew team members' attention to his repeated use of the word potential which he said was particularly relevant to the OFTEN approach to DFMEA. He added that the intent of using DFMEA within the OFTEN framework was to identity potential failures during the design of Oe375 and to take appropriate design actions to prevent the failure occurring in customer usage. John said that they would look at the development of preventative design actions when they considered Phase 3 of the OFTEN framework.

John said that he was now going to look at how the initial content of the propulsion system DFMEA could be developed from the information established during interface analysis and shared another slide (Figure 7.17).

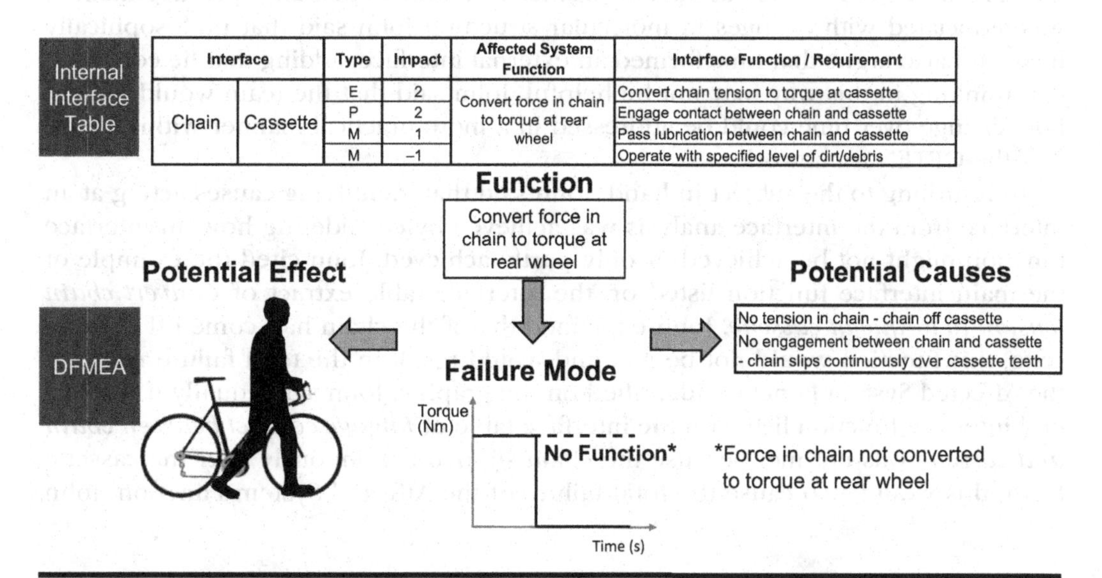

Figure 7.17 Pictorial depiction of initial development of propulsion system DFMEA.

John apologised for the slide being rather busy and asked team members to note that the top half of the slide contained an extract from the internal interface table for the chain/cassette interface that the team looked at in their last meeting, while the bottom half of the slide depicted the core of the DFMEA process. John added that he was going to continue to consider the function of the cassette of *Convert force in chain to torque at rear wheel* as the example since this was stated on the internal interface table as the Affected System Function. John added that he had rearranged a couple of the columns in the interface table extract for this interface to make it clearer how information from the interface table was transferred to the DFMEA. John explained that in using the core FMEA process, having established the *Function* to be analysed, the next step was to identify the *Function Failure Mode* to be analysed and explained that he was going to consider the total failure mode of *Force in chain not converted to torque at rear wheel*. John said that a *Potential Effect* of this failure mode, as depicted by the picture on the slide, would be that *Rider was unable to ride the eBike*. John added that in a system-level DFMEA, the effect was considered from the customer perspective as the effect on the customer and/or system operation.

In moving on to the final step of the FMEA core process, John reminded team members that this was to identify *Potential Causes* of the failure mode. John said that because the propulsion system DFMEA was at the system level of analysis, all potential failure mode causes that the Oe375 team would identify in the DFMEA would occur at an interface since a cause that occurred within a component part should be identified in a component-level DFMEA for that part. John added that in fact he found it difficult to envisage any failure mode that did not take place at an interface, although he admitted with a smile on his face he might have to think on a molecular scale on occasions. Andy cited the example of a failure mode associated with material degrading over time without any external influence as a failure mode that did not occur at an interface. John thanked Andy for his example and added that this was a good illustration of the need to think at a molecular level to justify the assertion that all failure modes occur at interfaces since he assumed that any such degradation was associated with changes in molecular structure. John said that philosophically he would also argue that time formed an external interface, adding that he conceded that thinking in this way may not be helpful. John said that the team would look at how change over time could be addressed in a more practical manner within Phase 3 of the OFTEN framework.

In returning to the subject in hand, John said that identifying causes acting at an interface from the interface analysis was achieved by considering how an interface function might not be achieved or only partly achieved. John cited the example of the main interface function listed on the interface table extract of *Convert chain tension to torque at cassette*. John explained that if the chain had come off the cassette, this function would not be met and would result in the total failure mode of the Affected System Function identified on the graphic. John said equally if the second interface function listed on the interface table of *Engage contact between chain and cassette* was not met because the chain slipped continuously over the cassette teeth, this would also cause the total failure of the Affected System Function. John drew team members' attention to the two causes listed on the slide that he had just described and added that determining causes by considering the reasons why interface functions or interface requirements might not be met required some thought and was best conducted within a team rather than by one person acting alone. John summed up what he had just said by saying that the maxim that he used was "*The failure modes of interface functions or interface requirements were causes of the failure modes of the Affected System Function associated with the interface*", adding that this statement needs to be considered with a clear head, which caused a few smiles.

John asked team members if they had any comments or questions. Yumiko asked why causes associated with the interface functions or requirements with an impact of 1 and –1 shown on the interface table extract were not included in the DFMEA. John reminded Yumiko that an impact of 1 was related to an enhancement of performance, with an impact of –1 was related to a decrease of performance of the system both of which would be likely to correspond to a system partial failure mode and not the total failure mode that they were currently looking at. John said that he would return to this point in a minute, and Yumiko thanked him for his explanation.

John said that the slide he was still sharing (Figure 7.17) depicted DFMEA pictorially and added that, as team members well knew, FMEAs were normally documented in tabular form. John explained that he hoped that the pictorial representation would illustrate that FMEA development was an exercise in thinking rather than exercise in form-filling. John added that continuing to document an entire DFMEA pictorially would be extremely cumbersome and said that team members might be relieved to learn that he was going to revert to the standard FMEA format and shared a slide depicting this transition (Figure 7.18).

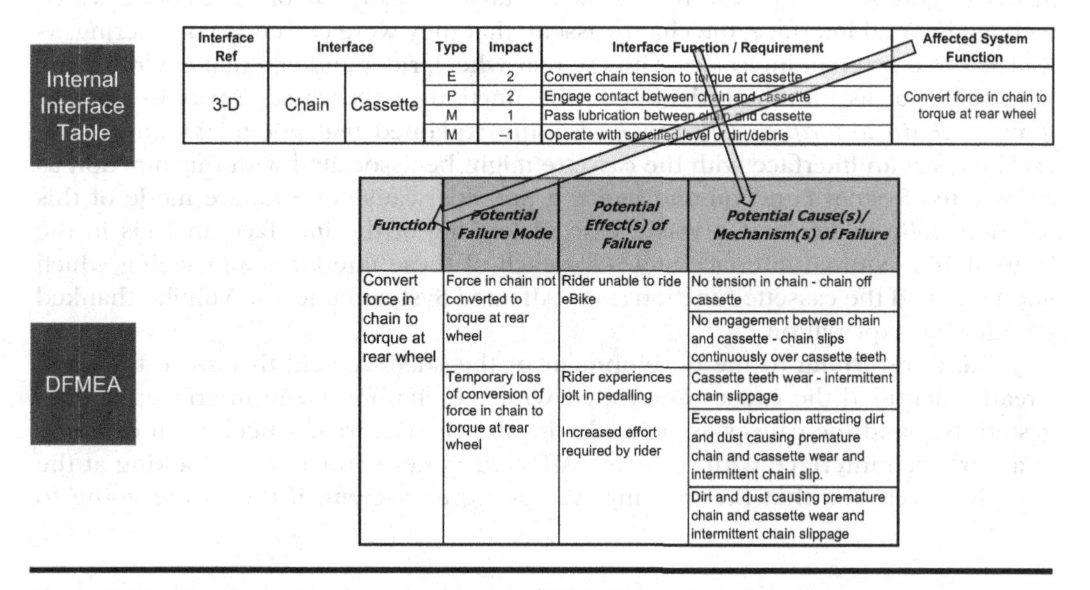

Interface Ref	Interface		Type	Impact	Interface Function / Requirement	Affected System Function
3-D	Chain	Cassette	E	2	Convert chain tension to torque at cassette	Convert force in chain to torque at rear wheel
			P	2	Engage contact between chain and cassette	
			M	1	Pass lubrication between chain and cassette	
			M	–1	Operate with specified level of dirt/debris	

Function	Potential Failure Mode	Potential Effect(s) of Failure	Potential Cause(s)/ Mechanism(s) of Failure
Convert force in chain to torque at rear wheel	Force in chain not converted to torque at rear wheel	Rider unable to ride eBike	No tension in chain - chain off cassette
			No engagement between chain and cassette - chain slips continuously over cassette teeth
	Temporary loss of conversion of force in chain to torque at rear wheel	Rider experiences jolt in pedalling	Cassette teeth wear - intermittent chain slippage
		Increased effort required by rider	Excess lubrication attracting dirt and dust causing premature chain and cassette wear and intermittent chain slip.
			Dirt and dust causing premature chain and cassette wear and intermittent chain slippage

Figure 7.18 Development of tabular format of DFMEA from interface table.

In explaining the slide, John said he had reverted to the standard column ordering in the interface table extract and that the DFMEA included those columns that corresponded to the core FMEA process. John said that the Oxton team members would recognise the DFMEA tabular format shown on the slide as being very similar to the standard Oxton form, except that while the first column on the Oxton form identified the *part*, the first column on the OFTEN DFMEA documented *function*.

John explained that he had included both the *total failure mode* and the *intermittent failure mode* of the function that they were considering in the DFMEA extract since, as he had just said, some of the interface exchanges would not be associated with total failure. John said that the extract now included causes associated with the two interface function/requirements in the interface table that Yumiko had mentioned. John said that the extract of the DFMEA on the slide had yet to include the standard indices used for risk assessment, which he said that they would look at in their next meeting.

John asked if there were any questions and comments, and Wolfgang said that there were other causes of the failure mode of "*Temporary loss of conversion of force in chain to torque at rear wheel*" such as the chain slipping intermittently on the chainwheel or the cassette slipping intermittently on the rear wheel. Wolfgang asked if these causes should be included in the DFMEA extract. John agreed that these potential causes should be included in the DFMEA and said that they should be identified when looking at other interfaces. Before John could elaborate on his answer to Wolfgang, Yumiko asked how team members would know which interfaces should be analysed in identifying causes of the failure mode being considered.

John thanked Yumiko for her question and said that the interface matrix can be used to help link interfaces with a particular failure mode through the Affected System Function. In referring team members to the propulsion system interface matrix (Figure A.6), John said that he would take the example of the cassette which had one internal interface, the chain/cassette that they were currently considering, as well as three external interfaces with the rear wheel, rider and external environment. John said that as they had identified, the function of the cassette was to *Convert force in chain to torque at Rear Wheel*. John explained that potentially any of the exchanges at an interface with the cassette might be associated with this function as an Affected System Function and hence a potential cause of a failure mode of this function. John added that it was a case of looking at the interface analysis in the internal and external interface tables for each of these interfaces and seeing which interfaces had the cassette function as an Affected System Function. Yumiko thanked John for his explanation.

John, in returning to the development of the DFMEA, said that since they had already identified the causes associated with the chain/cassette interface, he suggested they add the causes associated with the cassette/rear wheel as an example of an external interface with the same Affected System Function. In looking at the time, John realised that the meeting was going to overrun if they were going to cover sufficient content to allow Team B to take away the key assignment of starting to populate the propulsion system DFMEA. John shared his concern with the team, and Phil said that he was prepared to stay an extra half an hour. John said that he hated to do this but asked if others would be able, and prepared, to stay longer like Phil. Wolfgang, Andy and Sam said that they had other meetings to go to, and Yumiko said it was getting late for her, but she was used to catching up with Team B on work done outside of her normal day. After looking at their calendars, Sarah and Jennifer said that they could stay, and Sarah confirmed that the conference room was not booked for another meeting. John said that while he was hoping to answer Wolfgang's question by example, he would answer it in general terms and then conduct what he called a mini Warm Down for those who were leaving on time.

John asked Wolfgang to remind him of the two additional causes of temporary loss of function that he had mentioned, and Wolfgang said they were the chain slipping on the chainwheel and the cassette slipping on the rear wheel. John said that while these were important causes, they were associated with other interfaces than the ones they were currently considering and named these as the chainset/chain and cassette/rear wheel interfaces. John added that the reason that slippage of the chain on the chainset might appear at first sight to be a cause of the failure mode at the chain/cassette interface was because of the linear nature of the Drivetrain energy flow. Phil said that it was like the fact that his TV lost the sound while watching a programme the night before last, and he thought the fault was with the TV until he changed channel. John thanked Phil for his example and said that in the worst-case scenario, Phil would have send his TV back for repair in attempting to fix a failure mode at the wrong interface.

In facilitating the mini Warm Down, John asked Wolfgang, Andy and Sam to make a closing statement. They all made what John took to be generally positive statements, although Sam said that his head was beginning to spin with the amount of material they had covered. In view of Sam's feedback, John suggested, what he said might appear to be a contradictory recommendation, that those who were staying because there was more work to do took a 5-minute break before the team continued into what he called "extra time". Sarah said that she would benefit from a break, and Jennifer and Phil agreed with her.

After their second break, John asked the remaining team members to update the FMEA extract that they were currently considering by analysing the external cassette/rear wheel interface analysis. With Jennifer's assistance and some deft cutting and pasting from the external interface table, Jennifer displayed the required interface information alongside that for the chain/cassette interface that they had already considered. With team members' assistance, John then completed the additional potential causes column on the DFMEA extract for the intermittent function failure mode that they were analysing (Figure 7.19).

Propulsion System Internal Interface Table Extract

Interface Ref	Interface		Type	Impact	Interface Function / Requirement	Affected System Function
3-D	Chain	Cassette	E	2	Convert chain tension to torque at cassette	Convert force in chain to torque at rear wheel
			P	2	Engage contact between chain and cassette	
			M	1	Pass lubrication between chain and cassette	
			M	−1	Operate with specified level of dirt/debris	

Propulsion System External Interface Table Extract

Interface Ref	Interface		Type	Impact	Interface Function / Requirement	Affected System Function
4-E2	Cassette	Rear Wheel	E	2	Couple cassette to rear wheel	Convert force in chain to torque at rear wheel
			P	2	Locate cassette on rear wheel	
			E	2	Lock cassette to rear wheel	
			E	−1	Operate within specified level of vibration from wheel	

Propulsion System DFMEA Extract

Function	Potential Failure Mode	Potential Effect(s) of Failure	Potential Cause(s)/ Mechanism(s) of Failure
Convert force in chain to torque at rear wheel	Temporary loss of conversion of force in chain to torque at rear wheel	Rider experiences jolt in pedalling Increased effort required by rider	Poor contact between cassette and chain due to cassette teeth and chain wear - intermittent chain slippage
			Excess lubrication attracting dirt and dust causing premature chain and cassette wear and intermittent chain slippage
			Dirt and dust causing premature chain and cassette wear and intermittent chain slippage
			Cassette not located properly on wheel causing intermittent chain slippage over cassette teeth
			Cassette not locked to wheel causing intermittent cassette slippage on rear wheel
			Vibration from wheel causing intermittent loss of contact between cassette and chain

Figure 7.19 Propulsion system DFMEA extract with causes identified from interface analysis.

John pointed out that the DFMEA causes now included the cause that Wolfgang had spoken about of the cassette slipping on the rear wheel. John asked Phil, Sarah and Jennifer if they would have been tempted to include this cause when Wolfgang had mentioned it. The three remaining team members agreed that they probably would have done. John was impressed by the team members' honesty if not with their Failure Mode Avoidance discipline.

In expanding on the point behind his question, John said that interface analysis provided a coherent and disciplined way of developing a DFMEA. John added that his experience was that engineers using the process for the first time did not have full confidence in it and so added causes based on their engineering knowledge to a DFMEA alongside those gleaned from interface analysis. John said that when this happened, the development of the DFMEA suffered from what he called *butterfly engineering* with the team flitting from one interface to another and very quickly the coherent relationship between an interface table and corresponding DFMEA was lost. John explained that when the link between interface analysis and its corresponding DFMEA was disrupted, the systems structure inherent in interface analysis which is carried across to the DFMEA can also be lost with the DFMEA then including, what he termed, a mishmash of system levels. John continued by saying that disturbing the structural and logical relationship between an interface table and the DFMEA made it more difficult to ensure that all interfaces and their exchanges were considered. John added that it was also more difficult to maintain traceability between the documents when they need to be updated. In referring team members to the propulsion system interface matrix (Figure A.6), John said that as team members could see the cassette had two more external interfaces with the rider and external environment and added that he would leave it to Team B to include causes associated with the exchanges at these interfaces to the DFMEA extract.

John thanked the team for their help and said that he wanted to look next at potential causes due to mistakes. John said that in looking at causes due to mistakes he was thinking of past mistakes which should not be repeated and not every conceivable mistake since there were a myriad of such mistakes some of which might be realistic and other fanciful. John asked if team members could recall any past system-level design mistakes which had been made which resulted in a problem at the chain/cassette or cassette/rear wheel interfaces. Jennifer asked John what he meant by system-level design mistakes, and John said mistakes due to the manner in which the Drivetrain was integrated together into the eBike system rather than faults due to the Drivetrain components themselves. Phil recalled, what he termed a classic mistake, when the components of the Drivetrain for a particular model were chosen from different Groupset manufacturers and the cassette was found to be incompatible with the chain for particular higher speeds of chain and cassette once the components started to wear resulting in chain slippage, or worst still, chain breakage. Phil added that this problem was not seen during Design Verification. John was pleased that his new-found understanding of the bicycle industry was such that he knew that a Groupset comprised the Drivetrain and braking components and that the speed of a cassette represented the number of teeth on a sprocket. John thanked Phil and updated the FMEA extract accordingly, highlighting the cause due to a mistake (Figure 7.20).

Function	Potential Failure Mode	Potential Effect(s) of Failure	Potential Cause(s)/ Mechanism(s) of Failure
Convert force in chain to torque at rear wheel	Temporary loss of conversion of force in chain to torque at rear wheel	Rider experiences jolt in pedalling Increased effort required by rider	Poor contact between cassette and chain due to cassette teeth and chain wear - intermittent chain slippage
			Excess lubrication attracting dirt and dust causing premature chain and cassette wear and intermittent chain slippage
			Dirt and dust causing premature chain and cassette wear and intermittent chain slippage
			Cassette not located properly on wheel causing intermittent chain slippage over cassette teeth
			Cassette not locked to wheel causing intermittent cassette slippage on rear wheel
			Vibration from wheel causing intermittent loss of contact between cassette and chain
			Incompatibility of higher speed chain and cassette across groupset manufacturers causing chain slippage

Figure 7.20 Propulsion system DFMEA extract including cause of failure due to a mistake.

John asked if there were any further questions or comments, and Phil asked if they should include the potential causes associated with the other external interfaces that John had spoken about earlier, and John replied that this would have to be completed later as they had already overrun the meeting time badly. Since no one else spoke, John took advantage belatedly to start to move into the meeting Warm Down, although before doing so John asked the remaining Team B members if they would be able to complete the first three columns of the propulsion system DFMEA based on the information contained in the interface tables. John said that he would like to review a sample of the DFMEA at the next Oe375 team meeting. Phil said that, while he was speaking for himself, he felt sure they could and since there was no dissent from the other two Team B members, John was about to move into Warm Down when Jennifer reminded him about the Warm Up rota. John thanked Jennifer and asked if she would send an email out to all Oe375 team members and ask those members who have not yet conducted a Warm Up to sign up for the rota. John then thanked Sarah, Phil and Jennifer for staying on to complete the DFMEA extract for the two interfaces that they had considered.

The Warm Down process went smoothly with remaining team members being generally positive about what they had accomplished in the meeting, although

Jennifer said that, like Sam, she felt a bit overwhelmed by the amount of material that had been covered in the meeting. Phil expressed some concern about the team using a different approach to DFMEA to the accepted process in Oxton and said that some of his colleagues working on other programmes might wonder what was going on. Following the conclusion of the meeting, John caught Phil as he was leaving and thanked him for his comment in Warm Down and said that he would talk to Mo Lakhani on the subject he had raised.

On returning to his office and sitting down at his desk, John reflected on the Oe375 meeting that had just finished. John was pleased with the way in which the team were willing to participate in the meeting and thought that the use of the meeting software hand-up signal was becoming redundant. John was also impressed by the way team members had actively engaged in the application of the OFTEN framework both within and between meetings. John was disappointed in himself, however, in attempting to shoehorn too much into a single meeting. He reflected that he had succumbed to the pressure of trying to keep to the overall timing of the programme by having insufficient time in the short term which was not an unusual experience.

John then opened his email Inbox, and his attention was immediately taken away from the Oe375 programme on seeing that Vicky Harrison had sent him a blind copy of an email that she had sent on Jim Eccleston's behalf to Georgia Lee at Blade. The email was a cover note for a PDF copy of a letter to Georgia signed by Jim. The letter was noticeable for its brevity with Jim starting the letter by confirming his view that John Perry was the focus of the alleged accusation that Blade had made. Jim then summarised the actions that he had taken, including interviewing John and all the Oxton members of the programme team with which John had worked since joining the Company as well as talking to senior managers at Blade. Jim added that a search by the Oxton IT department of John's company laptop and flash drive had not revealed any Blade material. Jim concluded his letter by saying that, despite a diligent investigation, he had not found a scrap of evidence to support the allegation that Blade had raised and politely requested that, should a similar event occur in the future, Blade think twice about potentially wasting his time on what might also prove to be a fool's errand. John was pleased by the clarity of Jim's letter to Georgia and wondered what her reaction to it might be. He also pondered which Oxton senior managers Jim was referring to, knowing at least one was Joe Bowen and recalled that Jim had said that he had spoken to Joe in, what he termed, general terms. In closing the PDF file, John wondered if that represented an end of the matter or whether whoever had made the allegation had more mischief in store.

Vicky had followed up the email with another note saying that she had John's original laptop and flash drive and that he was welcome to pick them up from her as soon as he liked. Thinking there was no time like the present and that he might be able to glean some more information from Vicky, he got up and walked along to Jim's office. Vicky, looking up from her screen, greeted him with a smile and pointed to the laptop and flash drive on the corner of her desk. John was reluctant to ask Vicki straight out what she knew about the situation that had never been far from his thoughts over the past days and so, in thanking her for sending him a copy of Jim's letter to Georgia, merely commented that the letter had been brief and to the point. In response, all that Vicki said was that this was typical of Jim and returned to whatever task she was engaged on when John had interrupted her. John picked up his laptop and flash drive and asked Vicki if he should return his temporary laptop to the IT department, and she said if he returned it to her, she would see that it went back. John thanked Vicki, and in walking back to his office, he thought that he would have to wait until his next one-to-one with Jim to see if he would be any more forthcoming. Having texted Jane the good news, John spent the time up to lunch updating his laptop and flash drive with the new files that he had saved to his temporary devices since his laptop had been requisitioned, pleased that things seemed to be returning to normal.

Background Reading and Viewing

Speaking Skills	Syer, J. and Connolly, C. (1996) *How Teamwork Works: The Dynamics of Effective Teamwork*, McGraw-Hill Pages 228–243
Interface Matrix information flow to FMEA	Henshall, E., Campean, F. and Rutter, B. (2014) A Systems Approach to the Development and Use of FMEA in Complex Automotive Applications, *SAE International Journal of Materials and Manufacturing*, 7(2), https://doi.org/10.4271/2014-01-0740 Pages 280–284
FMEA	QualityQAI. *What Is FMEA*, YouTube, https://www.youtube.com/watch?v=IWPt6exPwjI&t=3s, Accessed 18 December, 2022 McDermott, R., Mikulak, R. and Beauregard, M. (2009) *The Basics of FMEA*, CRC Press Pages 1–14

Chapter 8

Branching Out

John arrived home that evening to find that Jane was not yet back from work. While officially Jane had a single shift day job, working life in a hospital was never entirely predictable, and she and John had this unspoken agreement that whoever arrived home first of an evening would start the preparations for the evening meal. To make this arrangement run smoothly, again without any formal spoken agreement, they had developed a schedule of daily meals that they tended to stick to and this evening was stir fry, aligned to this morning's weekly early morning supermarket delivery. John was busily chopping fresh vegetables and listening to the evening news on the radio when he heard Jane coming through the front door. He had made himself a cup of tea and, as was their practice, ensured that there was sufficient water for a second cup in the kettle which John had switched on to re-boil on hearing Jane arrive. Knowing that Jane would wish to sit down at the kitchen table and discuss his news of the day as a matter of urgency, he stopped what he was doing, made her tea in one of her favourite China mugs and placed it on the table at her seat just in time for her entry into the kitchen. He had set his laptop on the table earlier knowing that she would want to see Jim's letter to Georgia Lee, although he knew that she would be disappointed that he could not tell her a lot more over and above what she could read.

Jane duly sat down opposite John and gave him one of her looks which John knew meant that she was expecting him to initiate the conversation. In turning the laptop round and pushing it towards her so she could read it, John told Jane that the copy of the letter that Jim Eccleston had sent to Blade was displayed. Having read the letter, in what to John seemed like an instant, Jane asked him if that was it. John explained that Vicki Harrison had sent him a blind copy of the covering email to Georgia Lee at Blade which said no more than a PDF copy of the letter from Jim was attached with the original copy being sent through the post. Jane asked John if he had spoken to Jim or Vicki, and John responded by saying that Vicki had sent him another email saying that she had the laptop that had been taken away from him and he had used this opportunity to speak briefly to her. Jane asked him what Vicki had said, and John said that she had been unforthcoming and that he had judged from her body language that, either she had nothing to say or she did not want to say

DOI: 10.4324/9781003286066-8

anything. Jane asked John if he meant that Vicki had rudely ignored him to which he answered that she had not, but appeared either to be or made herself out to be, preoccupied with another task. John added that while he had spoken to Vicki, he had thought he would leave any discussion with Jim to their next one-to-one since that was the next day.

Jane stopped her mini-interrogation at this point, sat back and started to drink her tea. Having considered the matter, she told John that of course Jim's letter was good news because it had exonerated him and was very welcome. John agreed with her and said that while the original accusation was completely ridiculous, and Jim's investigation had come to the only logical conclusion available, it did not seem to him that Jim's letter represented the final act in the drama. Jane said that while she could understand John's position, they had to try to put it behind them and getting up she said to John, more as a command than a statement, that they should get on with preparing the stir fry. Jane added that they should open the special bottle of red that they we were saving for a rainy day and defrost two of the indulgent deserts they kept in the freezer for special occasions.

Following an enjoyable evening and good night's sleep, John felt in good form the next morning on arriving at work and was looking forward to his one-to-one with Jim, as much to bring him up to speed with all that had happened on the Oe375 programme since their last meeting, as to discuss the results of Jim's investigation. Vicki seemed her old self in chatting to John about last evening's Premiership football as he waited to be invited into Jim's office, which he was as soon as Jim had finished a phone call. John began the meeting by suggesting they develop an agenda for the meeting since it was a while since they had last spoken and the Oe375 team had done quite a bit of work in the intervening time, adding that he had a couple of other items that he would like to discuss. Jim said that was a good idea and moved the notepad on the table in front of him so that he could write on it and asked John what he would like to discuss. John said that the items he would like to discuss were the People Skills of Feedback and Verbal Communication in Team Meetings, the Oe375 propulsion system boundary diagram, function tree, interface analysis and design failure mode and effects analysis (DFMEA). In suggesting the agenda items, John grouped the People Skills together and Technical Skills together, even though this did not reflect the order in which he had introduced them in the two Oe375 team meetings that had taken place since Jim and he had last met. John thought that grouping the agenda items in this way would make it easier for him to stress the coherent information flow between the Oxton first-time engineering norm (OFTEN) Phase 1 and 2 tools. Jim wrote John's list down making sure his writing would be legible to both of them and then asked John what his additional items were, to which John responded by saying he wanted to talk about Mo Lakhani's involvement in the Oe375 programme and any updates on the Blade accusation. Jim tore the top sheet off his notepad and placed it, so John and he could read it and he could easily access it.

Jim suggested that they should tick off items as they covered them and said that they should start with the last item since that was probably one of the quickest. John said nothing waiting for Jim to speak. Jim quickly filled the silence by saying that he assumed that Vicki had sent John a copy of the letter he had sent to Blade summarising the actions that he had taken and the conclusions he had reached. John confirmed that Vicki had sent him a copy of the letter, and Jim asked John how he felt about it. John said that he was pleased by the clarity of the letter and the fact that it reflected the only rational conclusion that could have been drawn, adding that it had still left him wondering what was behind the accusation. Jim said that John would probably never know who had contacted Blade and added that his time as a professional cyclist had taught him never to underestimate the darker side of the human soul. Jim said that he had an excellent relationship both on and away from the road with some of his competitors from other teams and, while he never thought of any of them as being cheats, some had the occasional rush of blood in the heat of the moment. Jim continued by saying that there were times when he had been deliberately elbowed away from the line that he was intending to take in a race by people he knew well. He said that, although this type of cheating could have adversely affected his finishing position, he could not have afforded to dwell on such occurrences. He added that he had learnt to take such occasions in his stride and, because people had resorted to cheating in the hope of gaining an advantage over him, to look on them as a sign of his success. He advised John to do the same in the current circumstances, adding that he was intending to consign the incident to history as of that moment. John took what Jim had just said as an indication that they should move swiftly on to the next item on their agenda, and he showed Jim the team system slide on his laptop as a prelude to taking him through the use of feedback in team meetings.

Jim said he liked the emphasis John placed on using descriptive feedback and related back to what he had just said about cycling competitors seeking an unfair advantage. He said that using more descriptive feedback would probably have proved useful in helping him to take the emotion out of post-race confrontations which had sometimes led to physical aggression. In showing Jim the "Framing Information: The Spoken Word" slide, John quicky summarised the Speaking Guidelines that he had shared with the Oe375 team. John said that the straightforward set of recommendations contained on the slides also facilitated clear communication between individuals in team situations, including the types of interaction that Jim was talking about. Jim studied the slide and said he could see what John was saying, although adding that he thought that they seemed more valuable in a business context. Jim said that he liked people to take ownership of their own opinions and got frustrated when individuals projected their opinion onto other team members, with some having the arrogance to assume that their opinion was by definition the team opinion. Jim said he was thinking out loud but wondered about having a poster made up relating to the content of the slide to pin up in Oxton meeting rooms, saying that if he could add a bit of humour, it might make it more palatable. He said that he would discuss it with Vicki and asked John if the slide was on the Oe375 ShareDrive, and John confirmed that it was available.

Turning to the Technical Skills, John explained how the propulsion system boundary diagram and function tree had been derived from the propulsion system state flow diagram (SSFD). John was a bit surprised that Jim was aware that the boundary diagram and function tree were an inherent feature of the SSFD until he remembered that Jim had read the papers that he had recommended to him. John was reminded of Jim's ability to think insightfully when he commented that deriving the boundary diagram and function tree from the SSFD meant that these graphics were established through fundamental engineering principles. John summarised the thought processes Sarah Jones, Jennifer Coulter, Phil Eastman and Yumiko Izumi had gone through in establishing the structure of the propulsion system function tree. Jim said that he was impressed by what had been done and, after a brief pause, added that while he knew Sarah, Jennifer and Phil, he had forgotten who Yumiko Izumi was.

John realised that he had forgotten to tell Jim in their last one-to-one about the working arrangements that had been agreed between himself and the external members of the Oe375 PD team. Unsure of how Jim might react to what he was about to say, he explained that Yumiko worked for Agano and while the other two external representatives on the team from HJK and Teutoberg had said that they did not have time between team meetings to work directly on Oe375 design analysis, Yumiko's manager had given her permission to devote whatever time was necessary to Oe375. If Jim was surprised by what John had told him, he did not show it, other than pausing for a short while before responding. Jim said that he had mixed feelings about what John had told him, saying that on the one hand he applauded good relationships with suppliers but on the other hand he would not wish to compromise any proprietary information. John said that he assumed the non-disclosure agreement that Oxton had with Agano would protect them, and Jim concurred that it was pretty watertight. John said that because they were currently focusing on the propulsion system, Oxton were getting more from Agano than Agano were getting from Oxton, adding that Yumiko was proving to be a valuable member of the team. Jim said that he was happy to keep things as they were for the moment but that they would need to review the situation before the Oe375 team started to analyse other systems. John then added that collectively Sarah, Jennifer, Phil and Yumiko had decided they would like to be known as Team B and explained their reasoning for this name choice.

John moved the discussion on to the Oe375 propulsion system interface analysis by showing Jim the interface matrix and interface table templates. Jim said that he was familiar with the interface matrix but had not seen an interface table before but could see it was more comprehensive than the matrix and asked why the matrix would be required if all the information it contained was in the table. John said that he found the matrix to be useful in acting as a high-level map of the interface tables and, subsequently, the DFMEA structure providing the sequence of interfaces analysed in the tables followed the row- and column-wise structure of the matrix. John illustrated the point by showing Jim the completed propulsion system interface matrix (Figure A.6) and, in drawing Jim's attention to the internal interfaces with the battery, quickly found these interfaces in the propulsion system internal interface table using the interface references. In taking Jim through the detail of these interfaces, John reinforced the comprehensive nature of the table over the matrix mentioning the ability to distinguish between exchanges of different types, and document multiple exchanges of the same type, in the table.

John then said that he was going to move on to the subject of failure mode and effects analysis (FMEA) and noted that Jim had ticked off all their agenda items but this and Mo Lakhani's involvement in Oe375. John was aware that Jim would know, both from the discussion he had had with him and the papers he had read, that the OFTEN approach to DFMEA was based on the identification of function failure modes and felt sure that Jim would have raised this with him before now if he had a problem with this. Nevertheless, John felt that it would be useful to discuss with Jim the historical basis of Oxton's current approach to FMEA and raised the subject by saying that in reviewing one of Oxton's existing DFMEAs for his own enlightenment, he had come across the Oxton FMEA Manual and had noticed that it was copyrighted to an automotive OEM (original equipment manufacturer). Jim responded to John's comment by saying that in the early days after he founded the company, he had met a Quality Manager from the automotive company at a conference and that had resulted in Oxton being given the rights to use the automotive company's previous FMEA Manual. Jim added that, at the time, the automotive OEM had recently replaced their FMEA Manual and saw Oxton as a non-competitive company. Jim said that both hard and digital copies of the manual were not difficult to obtain anyway at that time since they had been freely available to automotive suppliers, and digital copies could be found on the internet for unofficial download. John thanked Jim for sharing this information with him and said he had been wondering what the provenance of the manual was. John then said that he would illustrate how a DFMEA could be developed from the information gained during interface analysis, adding that this was possible providing the functional failure approach to FMEA was used.

John quickly summarised the four types of function failure mode and emphasised the fact that function failure modes shared the same metric as their associated function. John then showed Jim the slide displaying the transfer and transformation of information from the interface table to the DFMEA. Jim said that he was impressed and that he felt that he was getting a deeper understanding of the way information flowed coherently in using the OFTEN framework. Jim added that while his direct experience of developing DFMEAs was restricted to the early days of the company, he recalled that the current Oxton form requires both the part and part functions to be documented. Jim said that, as he recalled, the listing of functions led to endless debates as to what functions of the bike parts should be listed and gave the example of handlebars having functions associated with steering, braking, gear change, balance, rider support and so on. Jim said that he was struck by the clarity and objectivity of the OFTEN approach. In looking at the time, John noticed that they had 5 minutes left and so he moved on to the last agenda item by saying that he had recently had a meeting with Mo Lakhani.

Jim asked John if he had asked to see Mo or had Mo asked to see him, and John explained that Mo had asked to see him having heard good reports from Phil and Sarah about what was happening on Oe375. John said that Mo's initial interest was to explore the OFTEN framework's capabilities in terms of problem solving and added that Mo was interested to learn that the main thrust of OFTEN was avoiding failure through design which had led them on to a discussion of the role of Design Verification. In smiling, Jim asked John if Mo had repeated his often-spoken catch phrase "there is no substitute for testing with the final product", which John said that perhaps surprisingly he had not. John said that Mo was supportive of what he was doing with Oe375, adding that Mo was happy for John to use the alternative approach to DFMEA. John said that he and Mo had agreed that John would keep Mo in the loop in terms of how Oe375 was progressing and added that Mo had said that he would directly support Oe375 Design Verification effort at the appropriate time. Jim said he was pleased that Mo and John had spoken to each other, and that Mo had taken a positive approach to what John was doing. Jim then said that he had another meeting, and, in wishing Jim a good day, John packed up his things and left.

On getting back to his desk and sitting down with a cup of coffee, John thought the meeting with Jim had gone well and even his neglecting to tell Jim about him agreeing on a working relationship with the external Oe375 team members did not seem to rankle him. John rationalised his omission by saying to himself that he had a lot on his mind at the time. John decided to set up another meeting with Mo to take him through the detail of the work that the Oe375 team had been doing and wondered whether to do it before or after the next Oe375 team meeting. He decided to do it after the Oe375 meeting as by then they would have looked at initial risk assessment in the DFMEA process. After sending Mo a meeting notice, John opened the Warm Up rota and was pleased to see that those who had not conducted a Warm Up yet had signed up for a future Oe375 meeting, with Andy volunteering for the next meeting. Recognising that the next Review/Preview meeting was that afternoon and as he did not have access to Andy's calendar, John gave him a call and was pleased that he answered rather than getting his voice mail. John asked Andy if he could attend the first 15 minutes of the Review/Preview meeting and, after asking when and how long the meeting was, Andy said he could attend for 15 minutes at the end of the meeting. John told Andy that this was fine and said he would send him a meeting notice to sign in with. John thought that Andy attending at the end of the meeting was useful since it would enable him to bring Sarah up to date with the Blade situation. John then directed his attention to the People Skills he wanted to introduce in the next Oe375 meeting so that he could suggest a related Warm Up exercise to Andy.

Sarah knocked on his open office door spot on time for the Review/Preview meeting and, without waiting for John to invite her in, went over and sat in her usual position at the table. John got up from behind his desk, exchanged greetings with Sarah and went over and closed his office door. Sarah sat silently and John assumed that she was giving him space to speak first which he did once he had also sat down at the table. John told Sarah that Jim had sent a letter to Blade after concluding his investigation into their allegation. John said that, although he did not think it appropriate that Sarah had sight of the letter, given its brevity and directness, he could relate its content to her which he did. After John had finished explaining what Jim's letter had said, Sarah said that she was very pleased that Jim had made it clear in no uncertain terms that John was so obviously innocent. Sarah asked John if she might tell Jennifer the good news and John said that she could, although he asked her to ensure that the matter was kept confidential at this stage. Sarah then asked John if he had any idea who had made the phone call to Blade, and John said he still had no idea, adding that Jim had advised him to put the matter behind him and that this was what he was intending to do.

In moving on to the substance of the Review/Preview meeting, John asked Sarah how she thought the last Oe375 team meeting had gone, and she immediately turned the tables by asking him what he thought of the meeting. John said that he was disappointed in himself in attempting to do what, with a bit of forethought, he could have seen was too much in the meeting. John added that he was appreciative of Sarah, Jennifer and Phil staying on. Sarah said that John should not worry as it was not the first time an Oxton meeting had badly overrun, and it certainly would not be the last. Turning to the meeting itself, John said that he was pleased with the meeting process and outcomes and added that he was impressed by the way in which team members were contributing to the meetings. John said that he was also pleased by the way that team members questioned the mechanics of the processes used within the OFTEN framework, and Sarah asked him why he would be pleased with this. John responded by saying that he assumed that people question the process because they were checking their understanding and added that he saw this as a good thing. John then asked Sarah for her view of the meeting, and she said that she had enjoyed it and was intrigued how, in her words, the DFMEA fell out of the interface table. Sarah added that she now understood the power of the function failure approach to DFMEA when used within the OFTEN framework.

John explained to Sarah that Andy would be joining them for the last 15 minutes of the meeting to talk about the Warm Up for the Oe375 meeting. John continued by saying that he wanted to introduce the People Skill of *Attitudes* and, in this context, he intended to suggest to Andy that he conduct a Warm Up based on a process that John called *See, Imagine, Feel,* saying that any discussion of this was probably best left until Andy joined them. Sarah asked what Technical Skills John wanted to include in the meeting, and John said one thing that he would like to do was to review the propulsion system DFMEA work that Team B was doing and asked Sarah how that was going. Sarah replied that it was going well and said that the DFMEA should be completed in time for the meeting. John said he would leave it up to Team B what interfaces to show. John continued by saying that he also wanted to look at the use of the *severity* and *occurrence* ratings in the DFMEA and introduce the tool of the function fault tree. In response to her question as to what a function fault tree was, John told Sarah that it was a fault tree based on function failure modes which, like the DFMEA, was developed using information generated in Phase 1 of the OFTEN framework. John asked Sarah if fault tree analysis was used within Oxton, and she replied that the only use that she knew of was a fault tree that Mo Lakhani's area had developed for fault tracing. John said he would ask Mo about it the next time he had a chance and said that he understood from Sarah's answer that fault tree analysis was not used as part of product design and Sarah confirmed that this was the case.

Looking at her watch, Sarah said that they had a short while before Andy was due to join them and that she would like to discuss something with John. John asked Sarah what was on her mind, and she said that it had occurred to her after the last Oe375 meeting that if the OFTEN framework was to be adopted for other programmes, the current Oxton FMEA Manual would be obsolete, and a new manual would be required. Sarah continued by saying that, having thought about the need for a new FMEA Manual, she had extended this idea to the development of a manual to cover the whole OFTEN framework. John told Sarah that he thought her idea was interesting and was something that he had not thought of. He said that his focus at the moment was on ensuring that the Oe375 programme was successful, adding that if it were not, then it was unlikely that OFTEN would be used on other programmes. In recognising that it might seem to Sarah that he was presenting reasons why her idea was not worthy of further discussion, John asked her to explain more about her suggestion. Just at this point, Andy signed in to the meeting, and before greeting him, John said that if Sarah had the time after their discussion with Andy, he would like to explore her idea further.

After John, Sarah and Andy had exchanged greetings, John welcomed Andy to his first Oe375 team Review/Preview meeting. John said that he intended to explore the subject of Attitudes as the next Oe375 team People Skill and proposed that the core of the Warm Up was based on an exercise that he called *See, Imagine, Feel*. John explained that in the first part of the exercise, an individual observes something about the behaviour of another, citing the fact that Sarah was currently clicking her pen as an example. John added that the first person then says what they imagine about the second person based on their observation and continued with his example of Sarah's behaviour by saying that he imagined in clicking her pen, Sarah was bored. John then said that in the last part of the exercise, the first person says how they feel about what they imagined and concluded his example by saying that Sarah being bored made him feel worried, explaining that this was because he thought she might be not interested in what he was saying. Sarah said that she was not aware of her clicking her pen, but it was probably because she was listening intently to what John was saying, adding that she was far from bored. John said that the aim of the See, Imagine, Feel exercise was for people to realise that our judgement of others is based on assumptions that we make based on our past experience which may or may not be correct. Having explained how the exercise worked, John said he was concerned that because the exercise was based on the observation of another, it would be difficult for those in the remote locations to participate, and Andy and Sarah agreed with him. John asked them if they had any ideas.

Andy asked if the exercise had to be based on an observation made in the meeting. John responded by saying that he had never thought about doing the exercise other than in real time, adding that he would like to explore Andy's idea and asked him to explain what he was thinking. Andy said that one of the things about his partner that wound him up was that he walked around eating his breakfast in the morning. Andy added that he was forever telling his partner to sit down because he thought that his partner would benefit from eating his breakfast in a relaxed manner. Andy continued by saying that when he and his partner had it out one day, his partner explained that he walked around because he had just spent 8 hours lying down, and he found the exercise helped him wake up. Andy revealed that his partner's explanation made him feel awkward because he felt that he had been nagging him. Andy said that he could frame his example in terms of See, Imagine, Feel and then ask team members to think about and share a similar experience. Noticing the time, Andy said that he would need to sign out from the meeting in a minute or so and asked John if he had any reading material that he could look at in preparing for the Warm Up. John said he would send him a copy of a paper, telling Andy that, although it was published 25 years ago, he thought it was very useful. Andy thanked John and said that he would read the paper at home that evening. As Andy was about to sign out from the meeting, John told him that if he needed any further help with the Warm Up to get in touch with him, to which Andy responded that he was sure that it would be fine.

After Andy had left the meeting, John asked Sarah if she had a few minutes to explore her idea for an OFTEN Manual further. Sarah said she had 10 minutes, and John asked her to say more about what she was thinking. Sarah repeated what she had said before that rather than just develop a new function failure based FMEA Manual, a manual describing the use of the OFTEN framework be developed. John asked who would develop the manual, and Sarah responded that it could only be John and the Oxton members of the Oe375 team. Sarah said that, while she realised that everyone was very busy working on the programme, she thought that the slides that John was developing would form a useful foundation for such a manual. John said he liked Sarah's idea and while, as he had said earlier, he thought that they need to confirm that the OFTEN approach was successful before developing a manual, he suggested they keep Sarah's proposal in mind. John added that, as a start, he would archive and catalogue the original slides that he was developing for the programme. Sarah said that she was happy with what John had proposed and said that she would keep the idea in mind and work on it if she had any spare time.

After Sarah had left, John thought about what he had just discussed with her and decided to review his Oe375 slides there and then, knowing that if he left it, it might slip through the net and so not get done at all. The fact that he needed to organise his slides was something that had occurred to him before Sarah had mentioned the idea of an OFTEN Manual since they were currently all contained within a single file. Given that this file included both the slides he had used with the team and those he had developed but not used, the slide count was approaching one hundred which meant the file was becoming unwieldy when he was looking for a certain slide. He decided to catalogue the slides by OFTEN phase in the first instance and then by the major Technical and People skills. John numbered the separate files, so they remained in the order in which he had used the slides with the Oe375 team, so retaining the linkage between particular People and Technical Skills. John then spent the rest of his day preparing for the next day's Oe375 team meeting.

John found that he was the first to arrive at the conference room for the Oe375 team meeting. He had just set up his laptop and signed into the meeting as the host in case any remote members were a few minutes early when Phil arrived. Phil and he rearranged the room furniture to what was now their standard layout. Before taking her usual seat, Jennifer came up to John and said that she had got to leave after an hour as she had an emergency dental appointment. John thanked her for letting him know and asked what the trouble was. Jennifer said she had a raging toothache in the night which she had managed to control with strong painkillers and had been lucky in getting an early appointment at the dentist. By this time, everyone had arrived or signed in and so John handed the meeting over to Andy for the Warm Up. Andy began by saying he noticed that the room at Oxton was all set up. He then said that he would like everyone to spend a few minutes in identifying an experience in which they had observed a behaviour in another person and had imagined that the behaviour was the result of one thing, but it turned out that it had been the result of something else. Andy continued by asking team members to say how they felt both as an immediate result of what they observed and then afterwards when they discovered that what they imagined was incorrect. Andy gave the example of seeing his partner walking around eating breakfast and explained his reaction to both his observation and subsequently learning that his assumption of why his partner acted in this way was wrong.

John was impressed by the way Andy used the See, Imagine, Feel framework without making this explicit, and that he had asked people to think about their feelings both immediately on observing a behaviour in another, and after they had found that their judgement had been incorrect. After giving the team a few minutes, Andy checked if they all had an example that they could tell the other team members which they said they had. John thought that the *Others* part of the Warm Up was going well and he particularly liked Yumiko's example of going to buy a drink in a bar for her friend and her friend saying she wanted a soft drink. Yumiko said that she assumed that her friend had had a heavy night the evening before and felt sympathy, adding that she felt surprise and joy when her friend told her that she was pregnant. Andy came to John last, and he related the example of when playing football and an opposition player, who had a reputation for play-acting injury, went down under a strong tackle and stayed down moaning. John said that he imagined the player was play-acting and felt so annoyed that he told him to get up and went to assist him up only for the physio to say that it looked like he had broken his leg, which John said made him feel foolish.

Andy then explained how the exercise that the team had just completed could be set in the framework of See, Imagine, Feel and explained that we all make judgements based on assumptions that are founded on our past experiences. Andy added that because people have different experiences, they may make different judgements in observing the same behaviour in others, and hence feel differently, with a particular feeling only being appropriate if the judgement was correct. Andy asked for comments both on the exercise and on what he had said. Sarah said that, in thinking about what Andy had said, she was now aware that she made, what she described as almost subconscious, judgements about people when she saw them for the first time. Wolfgang said that while it was all very interesting, unlike the visualisation and restatement exercises that they had done in previous Warm Ups, he was not sure of what it had to do with applying the OFTEN framework. John was prepared to intervene if required, but Andy addressed Wolfgang's comment head on by giving the example of an engineer who on being asked a question about a particular state flow might sit with a puzzled look on their face, at which point Andy held his chin, looked up and screwed his face up to demonstrate what he meant by a puzzled look. He continued by saying the puzzled look might be associated by some people with someone struggling for an answer and therefore these people might dismiss the credibility of any response the engineer gives. Andy continued by saying that other people might associate the puzzled look with someone who is thinking deeply which might give these people a high confidence in the credibility of the answer that the engineer gives. Andy concluded the example by saying that being aware of the associations we make with the behaviours we observe helps us to see why we make the judgements we do and give us the chance to adjust our reaction accordingly. Wolfgang thanked Andy and said he looked forward to seeing if what he had said would help him. Seeing John place one hand perpendicularly below the other in a time-out signal, Andy quickly went through the meeting purpose and agenda and handed the meeting over to John.

John thanked Andy for what he said was an interesting Warm Up and said that he wanted to look further at what Andy had just said about the fact that we all make judgements based on assumptions formed from our past experiences. John explained that when our behaviour is directed by our past experiences, then we are responding to an *Attitude*. John then shared a slide (Figure 8.1).

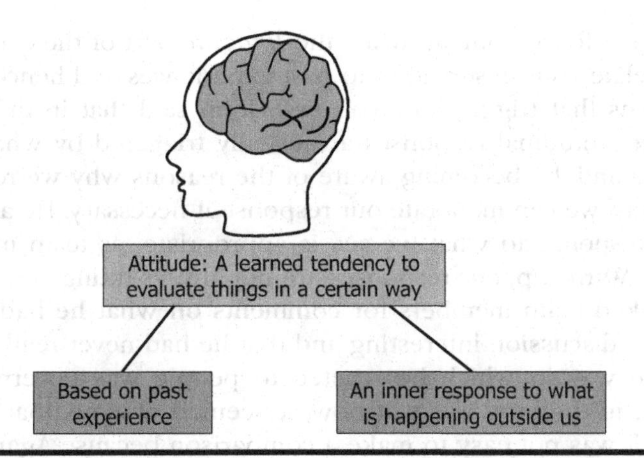

Figure 8.1 Attitudes.

In explaining the slide, John said we all have attitudes which we learn over time with some of those formed in childhood staying with us for the rest of our lives. John said that an attitude means that we tend to respond in a certain way when presented with a situation with which we are familiar and he linked back to the Warm Up by saying that the Imagine step of the See, Imagine, Feel process is where we associate what we observed with our past experience. John asked the team members if they behaved in a particular way in a given situation because they had been taught to do so by their parents. Sarah said that when she tells her children to tidy their bedrooms, she can hear her mother telling her to do the same. In picking up Sarah's example, Phil said that he pictured his Dad speaking when he used to tell his children off at the meal table when they were young for not holding their knife and fork, in a way that his Dad would call, properly. John thanked Sarah and Phil for their examples and continued by saying that attitudes also develop when we observe what others do, and said that in his youth, he felt strong peer pressure to dress and act in the same ways as his friends. Wolfgang said that he had experienced the same thing as John, adding that as he saw it, each generation wanted both to be seen to be different to their parents and to conform to their own norms. In agreeing with Wolfgang, John said as well as being formed through peer group pressure, attitudes are also formed by association, adding that this was well known to the marketing industry. John explained that manufacturers of clothing and sports equipment pay top sportsmen and women to endorse their products which people then buy to associate themselves with a particular athlete.

John said that we are all influenced by the society in which we live, with the people with whom we interact shaping what we believe and think, and how we express ourselves. John continued by saying that, while we all need attitudes to be able to cope with the myriad of information that we receive all of the time, problems can occur when we let our attitudes take control rather than using them as guidance. John explained that if our behaviour in particular circumstances is not to be dictated solely by our attitudes, then we need to be aware of the influence that our attitudes have on our behaviour. Relating back to the See, Imagine, Feel exercise, John said that stating what you notice in the *See* part of the exercise allows you to suspend judgement for a moment, while the *Imagine* part of the exercise helps you to consciously relate your observation to past experiences and hence become aware of the associations that trigger how you *Feel*. John said that in the context of the Oe375 team, our attitudinal responses are usually triggered by what another team member is doing and, by becoming aware of the reasons why we react to a person in a particular way, we can moderate our response if necessary. He added that while sometimes our response to what we see is appropriate, as team members had all illustrated in the Warm Up, our responses are not always fitting.

John then asked team members for comments on what he had said. Sam said that he found the discussion interesting and that he had never really thought about the fact that the way in which he reacted to people was governed by his past experiences but, in thinking about it now, it seemed obvious that it was. Yumiko said that, while it was not easy to make a comparison because Agano was a bigger company than Oxton, her impression was that Agano was a more hierarchical company than Oxton and this probably affected the way in which employees in each company behave towards their superiors. John agreed with Yumiko that company culture can have a strong influence on an employee's attitudes. As there were no more comments, John said that he would like to move to the next agenda item and confirmed with Phil that Team B were prepared to show the whole team a section of the propulsion system DFMEA that they had developed. Phil said that he was going to look at two interfaces and asked Jennifer to share the worksheet showing both the extract from the DFMEA and the associated internal and external interface table extracts for the interfaces that Team B had decided to review with the team (Figure 8.2).

After apologising for having quite a lot of information in the graphic he was sharing, Phil explained that he wanted to discuss the way in which Team B had used the different OFTEN tools. Phil said that as team members could see the partial failure mode *Loss in converting controller AC power to torque at chainset* for the function of the Motor of *Convert controller AC power to torque at the chainset* had been analysed on the DFMEA extract. Phil explained that while some degree of loss was to be expected, the words in brackets after the failure mode description made it explicit that the loss was measurably greater than would normally be expected.

Phil continued by saying that as had been explained in the previous meeting, the two internal interfaces *Controller/Motor* and *Motor/Chainset* and the three external interfaces *Motor/Frame*, *Motor/Rider* and *Motor/Environment* shared the same Affected System Function as the function being analysed in the DFMEA extract. Phil said that by way of example, the graphic Jennifer was sharing included one internal and one external interface extract from the respective interface tables. Phil explained that Team B had identified the causes documented on the DFMEA extracts by considering the exchanges and interface functions/requirements listed on the interface table extracts. Phil asked if there were any questions.

Andy said that the causes in the FMEA extract did not specifically include the motor itself being faulty, although he admitted that what he called the "*Coverall Cause*" listed as the first cause could cover this. Phil asked John if he would address Andy's point. John said that in a system-level DFMEA, it is assumed that the subsystems like the motor all operate to the required specification. John added that Oxton would expect the motor to be reliable and for the motor supplier to use DFMEA in

Propulsion System Internal Interface Table Extract

Interface Ref	Interface		Type	Impact	Interface Function / Requirement	Affected System Function
2-H	Motor	Chainset	E	2	Transmit torque from motor to chainset	Convert controller AC power to torque at chainset
			E	2	Clamp chainset to motor	
			P	2	Connect motor physically to chainset	
			E	-1	Conduct heat away from motor via chainset	
			E	-1	Operate within specified level of vibration	

Propulsion System External Interface Table Extract

Interface Ref	Interface		Type	Impact	Interface Function / Requirement	Affected System Function
8-E7	Motor	Environ-ment	E	-1	Convect heat away from motor	Convert controller AC power to torque at chainset
			M	-2	Seal motor from liquid ingress	
			M	-1	Seal motor from dust ingress	

Controller, Motor, Chainset Triad

Propulsion System Function Tree Extract

Propulsion System DFMEA Extract

Function	Potential Failure Mode	Potential Effect(s) of Failure	Potential Cause(s)/ Mechanism(s) of Failure
Convert controller AC power to torque at chainset	Loss in converting controller AC power to torque at chainset (greater than expected)	Increased effort required by rider	Insufficient torque transmitted from motor to chainset
			Motor not fully attached to chainset
			Motor not in full physical contact with chainset
			Motor overheats causing loss of efficiency
			Vibration transfer chainset-motor; loss of motor efficiency
			Water ingress to motor due to exposure
			Water ingress to motor due to inappropriate IP rating
			Dust/dirt ingress to motor due to inappropriate IP rating

Figure 8.2 Associated extracts of propulsion system DFMEA, interface analysis, SSFD triad and function tree.

designing the motor. Yumiko confirmed that Agano used DFMEA and added from their perspective the motor DFMEA would be at system level since the motor was one of their systems. Yumiko added that Agano also developed subsystem/component-level DFMEAs for the parts they designed where it was felt appropriate and cited the example of a recent innovation in the design of a brushless direct current motor (BLDC) rotor to reduce its magnetic flux leakage being supported by the development of DFMEA. John thanked Yumiko for her input and then said that he was interested in Andy's description of a "Coverall Cause" of *Insufficient torque transmitted from motor to chainset*. Andy responded by explaining that he saw the statement of this cause as nothing more than a restatement of the failure mode. John said that he agreed with Andy's observation and asked Andy if he was implying that stating this cause was not helpful, and Andy replied that he was.

Before addressing Andy's point, John apologised to Phil for interrupting his presentation. Phil said that this was fine, adding that he was finding the discussion interesting. John told team members to assume that the motor performed to specification and asked why the motor might fail to deliver sufficient torque for a reason other than those due to the causes already listed on the interface table extract. Sam said that this might be because the specification was inadequate. Sarah said that this was Oxton's experience with the last eBike which was generally recognised by both press reviews and feedback from customers to be underpowered. John asked what type of cause this amounted to, and Sam said it was an error of engineering judgement in that the design team for the previous eBike should have determined the power requirements analytically. John agreed with Sam and said that in the language of Failure Mode Avoidance, he would call this a mistake. The team then agreed on revised wording for this cause to "*Insufficient torque transmitted from motor to chainset – motor underspecified*".

Phil thanked John and Andy for their input and asked if there were any further questions. Sam asked what an IP rating was. Phil said it stood for ingress protection rating. John explained that the letters IP were followed by two numbers with the first representing the protection against solids and the second the protection against water. Sarah said that IP was also taken to mean international protection rating. Phil pointed out to team members that there were two causes listed relating to water ingress with the first relating both to the design of the motor and its position on the eBike. Phil added that the second cause related to water ingress of choosing a motor with an inappropriate IP rating corresponded to a mistake since the IP rating was known to be an important design factor. Wolfgang then observed that the connecting and clamping of the motor to the chainset was a manufacturing task, adding that his understanding was that manufacturing causes should not be included on a DFMEA. Sarah intervened saying that she agreed with Wolfgang that manufacturing causes should not be included on a Design FMEA and added that the connecting and clamping here referred to the design of the interface between the motor and chainset.

In thanking Team B for their efforts, John noticed Jennifer getting up and remembered that she had her dental appointment. As he hoped she would, Jennifer explained where she was going and why and left to several team members wishing her good luck. John said that he wanted to talk about the assessment of risk in the context of DFMEA next and added that he assumed team members would be familiar with the use of the *Severity and Occurrence ratings* in developing a DFMEA since these were used irrespective of whether the DFMEA took the functional or hardware approach. Sam said he was not sure what these ratings were, and John explained that severity was a measure of the seriousness of the effect of a failure mode while occurrence was the chance that a cause would occur while the product was in use. John said that both severity and occurrence were rated on a scale of 1 to 10 and shared a worksheet showing two tables (Tables 8.1 and 8.2).

Table 8.1 Severity Ratings

Severity Ratings		
Rank		**Severity of Effect**
1	Minor	Customer does not discern the effect
2, 3	Low	Some customers slightly annoyed; minor deterioration in system performance
4, 5, 6	Moderate	Customer annoyed or uncomfortable and notices some deterioration in system performance
7, 8	High	Significant customer dissatisfaction. System may fail to operate but does not affect safety
9, 10	Very High	Effects on safe operation of bicycle, non-compliance with government regulations

Table 8.2 Occurrence Ratings

Occurrence Ratings		
Rank		**Chance of Failure**
1	Remote	Failure not likely
2, 3	Low	Failures rarely happen
4, 5, 6	Moderate	Infrequent failures
7, 8	High	Numerous failures
9,10	Very High	Failure very likely

John said that he had based the tables on those in the Oxton FMEA Manual with modifications to take account of their usage in bicycle and eBike design. John reminded team members that, as the descriptions in the severity table indicated, within a system-level DFMEA the severity rating was a measure of the effect of a failure mode as experienced by the customer and/or the effect on bicycle or eBike operation. John said that he thought that the easiest way of understanding how the ratings were used was to apply them and suggested that the team add them to the DFMEA extract that Phil had just gone through (Figure 8.3).

Function	Potential Failure Mode	Potential Effect(s) of Failure	Severity	Potential Cause(s)/ Mechanism(s) of Failure	Occurrence
Convert controller AC power to torque at chainset	Loss in converting controller AC power to torque at chainset (greater than expected)	Increased effort required by rider	6	Insufficient torque transmitted to chainset - motor underspecified	7
				Motor not fully attached to chainset	1
				Motor not in full physical contact with chainset	1
				Motor overheats causing loss of efficiency	5
				Vibration transfer chainset-motor; loss of motor efficiency	3
				Water ingress to motor due to exposure	3
				Water ingress to motor due to inappropriate IP rating	4
				Dust/dirt ingress to motor due to inappropriate IP rating	3

Figure 8.3 Extract of propulsion system DFMEA including severity and occurrence ratings.

The team agreed on a moderate severity rating for the effect of the failure mode that they were considering because, while the partial function failure mode associated with less power converted to torque at the chainset than a rider might expect would require more pedal effort from the rider, it was not as severe an effect as a total failure of the Electric Drive. Phil made the point that, given the nature of some of the causes, the torque loss might build up slowly over time and hence not be as noticeable to the rider. Sam observed that this would make the type of the failure mode a degraded function, although conceded that this was still a partial function failure mode.

The team based the occurrence ratings on the Oxton experience with the last eBike, and in that context, they rated the failure mode cause associated with an under-specification of the motor as high. There was initially some disagreement within the team about the failure rates associated with fixing the chainset to the motor. This discussion was concluded when Sarah pointed out that the vast majority, if not all, of the failures at the chainset/motor interface that Oxton knew about were due either to manufacturing issues or to post-sale maintenance issues. Given this information, the team agreed to rate the two failures due to the interfacing components listed on the DFMEA extract as not likely to happen. Yumiko concurred that this was also Agano's experience. The cause of motor overheating and water ingress to the motor were both rated as moderate after several associated failures with the current eBike. The team agreed that the remaining causes had a low chance of occurring. Having agreed on the severity and occurrence ratings for the DFMEA extract, John asked Team B members if they would take away the task to add the severity and occurrence ratings to the rest of propulsion system DFMEA, and Sarah and Phil agreed in unison that they would do this.

John said that having established the severity and occurrence ratings, the team could use these to help identify the higher-risk items. John said that he had developed what he called a priority matrix to help with this assessment (Table 8.3).

Table 8.3 FMEA Priority Matrix

Priority Matrix

Occurrence	Severity				
	9 - 10	7 - 8	4 - 6	2 - 3	1
9 - 10	H	H	H	H	L
7 - 8	H	H	H	H	L
4 - 6	H	H	H	M	L
2 - 3	H	M	M	M	L
1	H	L	L	L	L

Key	
H	High Risk
M	Moderate Risk
L	Low Risk

John explained that the matrix was intended to be used in selecting the high- and moderate-risk items on an FMEA for further consideration. John stressed that the matrix was only intended for guidance and explained that since it was based on numbers, there was a danger that some people might treat it like a science and base the priority for further action solely on the matrix without using either their engineering knowledge or common sense. Sarah said that she had seen items on a Design FMEA prioritised by classifying them as Special. John said that he was aware of such a categorisation and added that whatever the exact detail of the different classification methods, the intent was the same in that it was to identify the higher-risk items on a Design FMEA so that they can be paid more attention.

In applying the priority matrix to the FMEA extract, they were currently considering the team decided that they would need to follow up on all the potential causes on the FMEA extract apart from the two causes with an occurrence rating of 1 (Figure 8.3). Sarah, who had taken over Jennifer's role in her absence of managing documents, asked John if she should delete these two causes. John said that the rating of the causes was valuable information both as a reminder of the discussion that Team B members had in allocating the ratings and to anyone else reading the DFMEA subsequently. John then said that he would explain how higher-risk items should be treated in the next Oe375 team meeting.

John suggested the team take their break, and as the Oxton members were getting up from their tables, Sarah called over to John and told him that she had promised to let Jennifer know what happened in the rest of the meeting. John took the opportunity during the break to send Andy a Chat message telling him that he liked his Warm Up and said that he was particularly impressed by the way he framed See, Imagine, Feel, both before and after the exercise. John added that he also liked the way in which Andy handled the comments and said that it was apparent to him that Andy had put a lot of thought into his facilitation of the Warm Up. Andy responded immediately by thanking John for his note.

Following the break, John said that for the next item on the agenda, he would like to talk about a graphical tool which was based on fault tree analysis (FTA) and complemented the DFMEA in analysing function failure modes. John asked how many team members were familiar with FTA, and Sam replied that he had learnt about it at University but not used it since. Wolfgang said that he had used FTA with his previous company. John asked Wolfgang how it was used, and Wolfgang replied that repair engineers used it in fault finding. John asked Wolfgang if he had been involved in developing fault trees and, when he said that he had, John asked him how the fault trees that he had worked on had been developed. In reply, Wolfgang said that having identified the top event, the causal events were identified through brainstorming based on engineering experience. John thanked Wolfgang and said he would explain the terms top event and causal event that Wolfgang had used shortly.

John then asked Sam if he would explain more about his experience with FTA, and Sam replied that they used it to verify software logic, and in doing so, like Wolfgang, they started with a particular failure event and looked at the possible failure paths that might cause that event. John thanked Sam and said that while conventional FTA was based on hardware or software failure events, he wanted to look at how the technique could be used to analyse function failure modes, adding that this paralleled the function failure mode approach to DFMEA. John continued by saying that, as Wolfgang and Sam would know, the development of a fault tree was based on the use of logic gates to connect failure events with the two most common gates being the *AND* and *OR* gates. In explaining the AND gate, John said he would look back to the Oe375 propulsion system function tree and shared a slide (Figure 8.4).

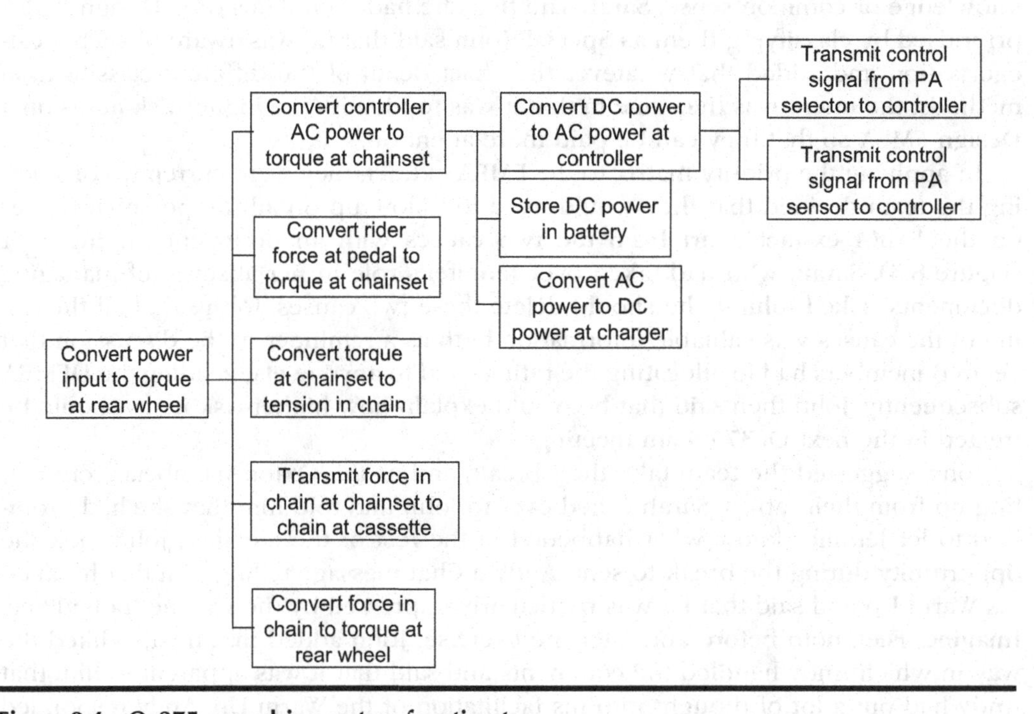

Figure 8.4 Oe375 propulsion system function tree.

John explained that if team members compared the function tree shown on the current slide to the original that they had developed (Figure A.5), they would see that the structure and content of the tree remained the same with the only differences being that the functions were written in smaller black font and placed in boxes and the lines joining the functions were thinner. John added that the reason for these changes should become clear shortly. John said that the logical link between functions in a function tree was an AND relationship, citing the example that the function *Convert power input to torque at rear wheel* when the eBike was in pedal assist mode was dependent on all the functions immediately beneath this function being achieved. John expanded his explanation by saying that the functions *Convert controller AC power to torque at chainset* AND *Convert rider force at pedal to torque at chainset* AND *Convert torque at chainset to tension in chain* AND *Transmit force in chain at chainset to chain at cassette* AND *Convert force in chain to torque at rear wheel* must all be achieved if the function above them was to be realised. John continued by saying that, although this was not normally done, it was perfectly legitimate to show the links between functions on the function tree using AND gate symbols and shared a further slide (Figure 8.5).

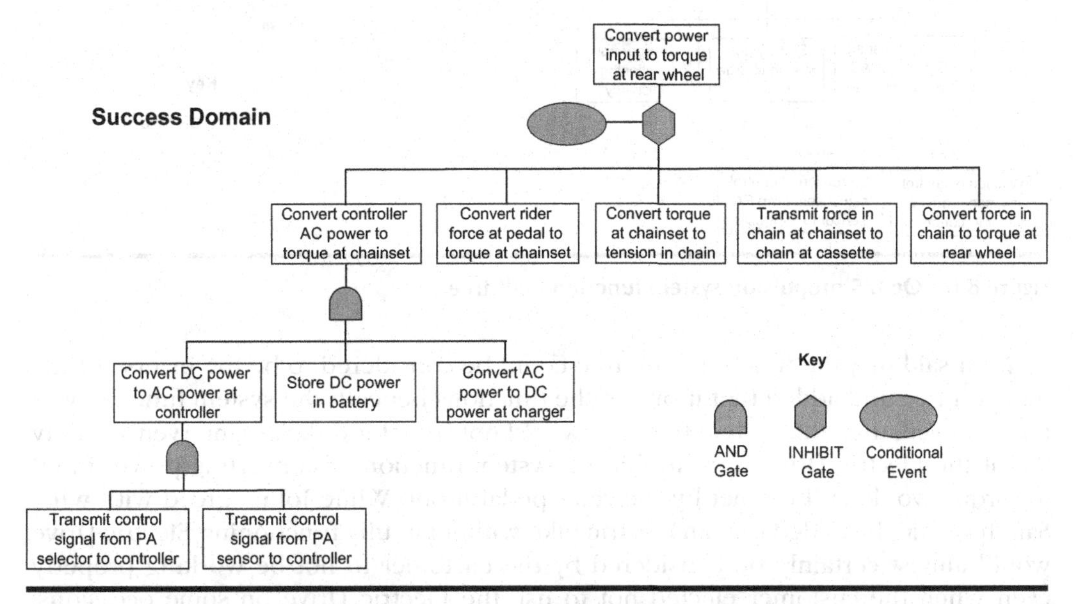

Figure 8.5 Oe375 propulsion system function tree with AND gates.

John said that as team members would have recognised, the orientation of the function tree was now top-down for reasons that would also become clear shortly. John referred team members to the symbol for the AND gate and explained that a function tree might be said to exist within what he called the Success Domain since the system function at the top of the tree would be successfully achieved if all the functions beneath the system function are successfully realised. John said that the hexagonal symbol represented an INHIBIT gate which he explained was a special type of AND gate which included a Conditional Event shown by the oval symbol. John explained that the condition in this case was determined by whether the eBike was in pedal assist mode or not. John added that in pedal assist mode, all of the events immediately below the INHIBIT gate had to be met for the event above the gate to be fully met, while when the eBike was not in pedal assist mode, the top event would be fully met without the Electric Drive events.

John continued by asking team members to assume that the Electric Drive and Drivetrain still both operated, but at a level of performance below that expected by the rider. John said that this corresponded to a partial function failure of the system which could be depicted as a failure event at the top of a fault tree and shared a slide showing such a situation (Figure 8.6).

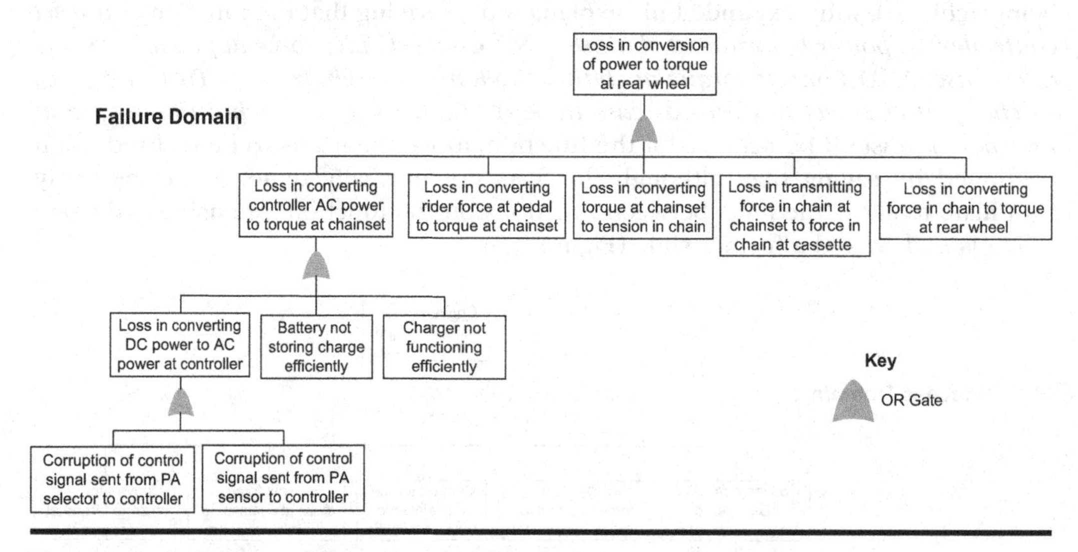

Figure 8.6 Oe375 propulsion system function fault tree.

John said that the function fault tree could be considered to be the inverse of the function tree and added that if one of the functions beneath the system function was not achieved, then the system function would not be achieved. Sam intervened to say that if the Electric Drive were to fail the system function of converting power input to torque would still be met by the rider pedal input. While John agreed with what Sam had said, he added that an electric bike without a fully functioning Electric Drive would almost certainly be considered by the customer to not be working properly even when the customer elected not to use the Electric Drive on some occasions. John explained that he was using the term "*Loss*" in the events in the tree in the same way that Phil had explained earlier in the context of the DFMEA extract he had taken the team through to mean a loss recognisably greater than that normally expected.

John explained that since fault trees were invariably drawn top-down, the failure event at the top of the Fault Tree was called the *Top Event* with the events beneath the top event being *causal events* since they could potentially cause the top event. John added that because all the events were expressed as functional failure modes, this fault tree was called a function fault tree. John then asked team members to compare the content and structure of the propulsion system function fault tree (Figure 8.6) with the content and structure of the propulsion system function tree (Figure 8.5). After allowing a few minutes for team members to compare the two trees, John asked for comments. Yumiko said that the function tree and function fault tree had the same structure with all the functions on the function tree being replaced with a partial function failure mode of that function on the function fault tree. John thanked Yumiko and asked if there were other differences or similarities, and Sam said that the AND gates on the function tree had become OR gates on the function fault tree and, judging by what was stated on each slide, while the function tree was set in the context of the Success Domain, the function fault tree was based in the context of the failure domain. John thanked Sam for his observations.

John summarised the differences between the two types of tree by saying that the function fault tree represented a mirror image in the failure domain of the function tree in the success domain. John then asked team members if the transition of the function tree into the function fault tree made engineering sense or if it was just a graphical manipulation. After a pause, Andy said that the function fault tree seemed valid from an engineering standpoint since essentially what it depicted was that any single partial function failure mode, or combination of single partial function failure modes, identified on the system function fault tree could cause the top event. Andy gave two examples, firstly saying that if the battery was not storing charge efficiently that might cause the motor output to be less than expected and hence cause loss in the Electric Drive function and subsequently cause the top event, and secondly, saying that a loss of the tension generated in the chain by the chainset along with a loss in converting this tension to torque at the cassette could affect the Drivetrain performance detrimentally and cause the top event. John thanked Andy and said that he thought that he had captured the logic of the function fault tree nicely and added that what Andy had described in his two examples was called the *path* by which a lower-level causal event might cause the top event failure.

John asked if anyone had any more comments, and Wolfgang said that he had always developed fault trees down to *Basic* events. John said that the failure events shown in the boxes in the propulsion system function fault tree beneath the top event were what was called *Intermediate* events because there were other causal events that would appear lower down the fault tree that resulted in the intermediate events. John added that, as Wolfgang had said, these lower-level root causal events were termed *Basic* events and were denoted by writing text inside a circular box. John said that the propulsion system function fault tree was currently at the system level of analysis because this was the level at which the propulsion system SSFD and hence function tree, from which the function fault tree was derived, had been developed. John continued by saying that basic events would be associated with components and while they might be added to the tree as it currently stood his recommendation was the function fault tree be extended to the component levels based on extending the function tree through component-level function analysis. John added that the function fault tree would then reach a level of resolution where basic events could be readily identified. John said that adding basic events or further intermediate events to a function fault tree in an ad hoc manner was the same as adding causes to an FMEA irrespective of which interface was being considered. As with the DFMEA, such action tended to disrupt the structural relationship between the methodology and the data analysis on which it was based.

Sarah said that there was a problem in extending the function fault tree to component level as John had recommended, since the design responsibility for the components that Oxton used lay with suppliers. John agreed with Sarah and said that nevertheless the system-level function fault tree in its present form was useful when viewed alongside its corresponding DFMEA. John shared a slide (Figure 8.7).

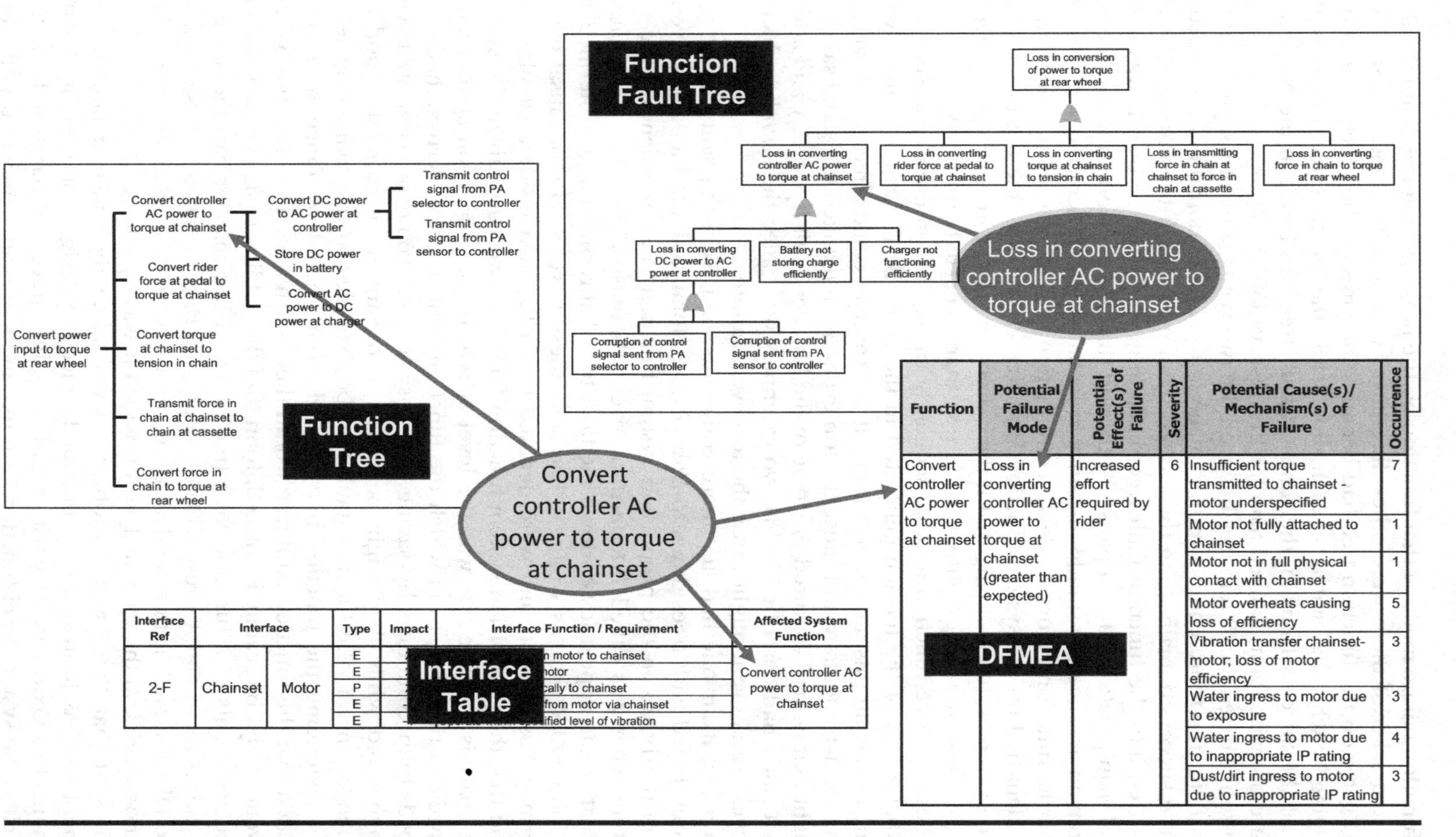

Figure 8.7 Relationship between OFTEN tools and function fault tree.

John said that, while much of the detail on the slide was illegible, team members would recognise the different types of documents with the key information being documented in the bubbles. John added that the slide showed extracts from the Oe375 propulsion system OFTEN Failure Mode Avoidance tools that the team had developed. John added that as the slide depicted the function fault tree and DFMEA, extracts were based on the same failure modes of the functions identified on the system function tree and more fundamentally on the system SSFD. John said that this meant that the higher-level content of the DFMEA matched the content of the function fault tree. John explained that in using the term the higher-level content for the DFMEA he meant the information documented in the two left columns of the DFMEA. John said that the commonality of content of both documents meant that they could easily be cross-referenced with the DFMEA providing detailed information on the causes and effects of the failure modes identified in both documents.

John said that the summing of partial failure modes was an important consideration and explained that while one partial failure mode acting alone might be insufficient to cause system failure, a number of partial failure modes acting together might cause a system to fail. John asked team members how using the propulsion system function fault tree and DFMEA alongside each other might help in identifying such situations. Sarah said that the graphical format of the function fault tree meant that it was easy to identify potential multiple causal events acting simultaneously that affected system performance such as losses at the chainset/chain and chain/cassette interfaces as well as a loss of the transmission of force within the chain itself. Sarah added that although the three partial function failure modes would be included in the DFMEA, they would appear as three separate line items which may well be spaced apart and hence it would not be as easy to identify them acting together without the function fault tree. John agreed with Sarah and said that what she had described about the function fault tree was its relative strength and asked team members what the relative strength of the DFMEA was compared to the function fault tree. Sam responded that the DFMEA contains a lot more information than the function fault tree, adding that he thought that the two tools complemented each other well.

John asked if anyone had any further questions or comments, and Wolfgang responded by saying that he was impressed by the manner in which the structure and content of the function fault tree had been derived from its associated function tree. Wolfgang added that his recollection of developing fault trees in the past was that it was a time-consuming process based on brainstorming which often led to a prolonged discussion between those developing the tree over what caused what. Wolfgang said that the fault trees that he had helped to develop tended not to be very tree-like in their structure, either being flat or uneven in shape. Wolfgang then observed that the propulsion system function fault tree had a partial function failure as a top event with all the intermediate events also being partial function failures and asked John if this was always the case. John asked Wolfgang if he could answer his own question. After some thought, Wolfgang said that he could envisage the propulsion system having a total function failure as a top event which was caused by a partial function failure at an intermediate event, quoting the example of a partially charged battery supplying insufficient voltage to operate the motor through the controller.

Phil said that he assumed that there were many different feasible combinations of partial function and total function failure events as well as intermittent and unintended function failure events that could be depicted on the function fault tree. In agreeing with what Phil had said, John explained that in his experience, it was not good use of time to try to develop a separate function fault tree for each feasible combination of the types of function failure mode. John added that he would construct a single function fault tree for each type of function failure mode as the top event that was both feasible and was likely to occur with the intermediate events on each tree generally being of the same type of function failure mode as the top event. John continued by saying that in analysing the function fault tree alongside the corresponding DFMEA, team members should be able to identify likely combinations of the same or different types of function failure mode that might occur even if not all the feasible and likely failure modes types were included on the function fault tree.

Realising that it was time for Warm Down, John confirmed that Team B members would complete the severity and occurrence ratings on the propulsion system DFMEA. He suggested that, if Team B members had the time after doing this, they might like to extend the propulsion system function fault tree to include events related to derailleur gearing after they had extended the propulsion system SSFD to include derailleur gearing. Phil said that he was sure that Team B could extend the SSFD at the second time of asking and so include derailleur gearing on the function fault tree. John found the Warm Down feedback, encouraging and particularly pleased when Sarah and Yumiko both said that they now appreciated the strength of a function-based approach to product design analysis. John was slightly put out to hear Andy say that he was not sure if what they were doing on Oe375 had any direct relevance to HJK but felt that he might change his perspective once they had completed all four phases of the OFTEN framework for the propulsion system.

On returning to his office, John made a mental note not to forget the two items that he promised Sarah and Phil that he would discuss with Mo Lakhani in his next meeting with him which he had scheduled for that afternoon about the use of FTA and the two different approaches DFMEA being used in Oxton. John then busied himself by considering what he should do in introducing the next phase of the OFTEN framework to the Oe375 team and bought sandwiches in the canteen to eat at his desk so he could continue working over lunch. After lunch, John prepared for his meeting with Mo by cutting and pasting the relevant interface table and DFMEA extracts from the work the Oe375 team had updated that morning to use by way of example in the meeting. Having done this, John was surprised to see it was nearly time for the meeting with Mo. As John expected, Mo greeted him with a broad smile as he knocked on his open door and ushered him to sit alongside him at the table in his room. After exchanging small talk, Mo walked over to his clean whiteboard and asked John what he wanted to discuss. John said that he would like to take Mo through the OFTEN work that had been done on the Oe375 programme and added that he thought that this would lead into the other points that he wanted to raise which were the use of DFMEA and fault tree analysis in Oxton. Despite the agenda being short, Mo still wrote it on the whiteboard and said that while he had some things that he wanted to talk to John about, they fell under the same headings (Figure 8.8).

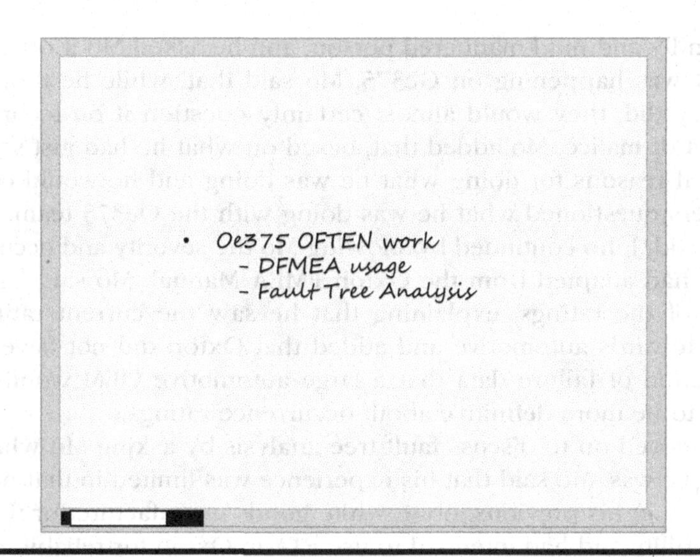

• Oe375 OFTEN work
- DFMEA usage
- Fault Tree Analysis

Figure 8.8 Agenda items.

Without lingering on fine detail, John proceeded to take Mo through the Oe375 propulsion system function analysis as a prelude to discussing the development of the DFMEA. Mo seemed intrigued with the SSFD and its use in deriving both the system boundary diagram and function tree and told John that while he had come across state diagrams before, he had never seen the SSFD. In moving on to the Oe375 propulsion system interface analysis, Mo said that he was familiar with the use of both the interface matrix and interface tables, although it seemed to John from the questions he was asking, and the comments he was making, Mo was impressed with the detail and rigour of the OFTEN approach. John then showed Mo the spreadsheet extract showing the development of the propulsion system DFMEA based on the interface table extracts for the three Electric Drive interfaces that he had cut and pasted together earlier. Mo said that he was impressed by the way in which the structure and information from the interface analysis was used in developing a DFMEA. John was pleased that Mo had appreciated both the structural and content relationships between the two types of document and took the opportunity to raise the issue that Phil had mentioned to him about how the use of a different approach to FMEA on the Oe375 programme might be viewed by those using the conventional Oxton approach on other programmes.

Mo said that someone else had raised that very point with him the other day and went on to say that Paul Clampin had been bleating in his ear on the subject. John asked Mo what his response to Paul had been, and before answering John's question, Mo said that the incident was typical of Paul and explained that Paul had spent a significant amount of time lately trying to ensure that the problems with the latest off-road bike were laid at Mo's door rather than his own. Without further prompting, Mo said that he did not have a lot of time for Paul and had told him if he had concerns about what was happening, he should speak to Jim about it. Mo added that he knew Paul was just trying to stir up trouble and would not speak to Jim. John was somewhat surprised by this minor outburst from Mo since he seemed to John to be such a friendly and mild-mannered person, and he asked Mo if others might also question what was happening on Oe375. Mo said that, while he thought this was unlikely, if they did, they would almost certainly question it on technical grounds rather than out of malice. Mo added that, based on what he had just seen, John had sound technical reasons for doing what he was doing and he would certainly back him up if others questioned what he was doing with the Oe375 team. Pleased with what Mo had said, John continued by showing Mo the severity and occurrence rating tables that he had adapted from the Oxton FMEA Manual. Mo said he liked John's simplification of the ratings, explaining that he saw the current ratings as being overly biased towards automotive and added that Oxton did not have the quantity and sophistication of failure data that a large automotive OEM would have which allowed them to be more definitive about occurrence ratings.

John then moved on to discuss fault tree analysis by asking Mo what his experience of the topic was. Mo said that his experience was limited in that he had helped to develop a few in his previous job in white goods manufacture for the assessment of system reliability and had intended to use FTA in Oxton for reliability assessment. Mo explained that this attempt came unstuck when he had difficulty in obtaining Oxton bicycle parts failure rates. John agreed that using FTA for system reliability assessment was a well-known practice, but it did rely on having good part failure data. John said that they were not using FTA on Oe375 for system reliability prediction but as an integral part of the design process to help with the identification of functional failure.

John then took Mo through the development of the Oe375 propulsion system function fault tree pointing out the similarities and differences with conventional fault tree development. Again Mo seemed to John to be genuinely interested in the development of the function fault tree and made a couple of what John thought were insightful comments. Firstly, Mo said that he knew that fault tree development had a reputation of being time-consuming and was aware that quite a lot of people had developed ways of automating the production of fault trees. Mo added that it seemed to him that the function fault tree lent itself to such automation, given that information on its structure and content was readily available from the function analysis work. Secondly, Mo said that developing the Oe375 function fault tree fully to basic event level would require the input of component suppliers since component design was their responsibility. John agreed with both of Mo's comments and said he was not sure himself at this moment how far they would take the propulsion system function fault tree because of the reason Mo had cited. John added that he thought that even in restricting it to system level, the function fault tree would be useful in defining the requirements that Oxton would cascade to component suppliers. When John then said that he had brought Mo up to date with what he was doing with Oe375, Mo concluded the meeting by thanking him for his time saying that he was supportive of what John was doing.

John came away from his meeting with Mo feeling that it had been a good use of his time. He realised that one key thing that he missed in joining Oxton from Blade was the ability to have in-depth discussions with his peers. John had valued this peer-to-peer discussion which had enabled him to get a fresh eyes objective perspective on what he was doing. While he recognised that he had developed a good working relationship with Sarah, John looked forward to continuing his discussions with Mo.

Background Reading and Viewing

See Imagine Feel	Brown, J., D'Emidio-Caston, M. and Benard, B. (2001) *Resilience Education*, Crown Press Inc Pages 75–76
Attitudes	Desai, S. (2014) *Components of Attitudes*, Khan Academy, YouTube, www.youtube.com/watch?v=cDq1_R-J51w, Accessed 18 December, 2022
FMEA: Severity and Occurrence Rating Scales	McDermott, R., Mikulak, R. and Beauregard, M. (2009) *The Basics of FMEA*, Productivity Press Pages 26–36 Automotive Industry Action Group. (2019) AIAG-*VDA Failure Mode and Effects Analysis (FMEA) Handbook*, Automotive Industry Action Group Pages 56–61
FMEA: Priority Matrix	Automotive Industry Action Group. (2019) AIAG-*VDA Failure Mode and Effects Analysis (FMEA) Handbook*, Automotive Industry Action Group Pages 62–64
Function Fault Tree	Campean, F. and Henshall, E. (2008) *A Function Failure Approach to Fault Tree Analysis for Automotive Systems*, SAE Technical Paper 2008-01-0846, https://doi.org/10.4271/2008-01-0846

Chapter 9

Investigating Noise

Driving home from work after a busy week, John was looking forward to going to Mike Holmes' retirement celebration at the Vertemont Hotel that Friday evening. John knew the hotel well despite never having stayed there as it was local to the Blade UK Headquarters and was frequently used for retirement and other company events, including department Christmas lunches. John was pleased that partners had also been invited since Jane deserved a night out, although John wondered how many people she would know. On arriving home, the first thing Jane said to him was to ask if they would need to eat before leaving. John replied that as far as he remembered, the invitation mentioned that there would be a buffet and added that he would check. After a brief search, John found the email containing the invitation on his phone and confirmed that there was a buffet. Following a short discussion, Jane and he decided to have a snack with their evening cup of tea, not knowing when the buffet would be served nor how substantial it would be, after which they watched the local TV news and then went upstairs to change. Jane was coming down the stairs while John was in the hall putting on his shoes and, looking up, John thought how lovely she looked and told her so, also asking if she was wearing a new dress. Jane thanked him and told him that he did not look so bad himself and added that she had bought the dress last year but had not had a chance to wear it before.

After eventually finding a space in the Vertemont's car park, in walking towards the hotel's main doors, John recognised that Bob Jenkins and a lady whom he took to be his partner were ahead of them. Bob was in the same apprentice intake as John, although John had lost touch with Bob relatively early in his Blade career when he had moved out of manufacturing. On entering the hotel, John noticed a free-standing board in the foyer directing those who were attending Mike Holmes' retirement event to the Eagle Conference Room. Looking round, John saw that Bob was standing in a queue at reception and assumed that he had not seen the notice and so went over to him to renew their acquaintance. After introducing their partners to each other, John asked Bob if he was going to Mike Holmes' retirement gathering and, when Bob said he was, John suggested they walk along together as he knew where it was. On the way, John asked Bob what he was doing now, and Bob replied that he was a senior toolmaker working in the Engineering Centre. In response to a reciprocal question

DOI: 10.4324/9781003286066-9

from Bob, John sad he had left Blade and was now working for Oxton, the bicycle and eBike manufacturer, as the Engineering Teams Manager. By this time, they had reached the entrance to the conference room and, after hanging up their coats on the coat rails standing near to the room entrance, joined a small queue of people being greeted by Mike Holmes. When their turn came, Mike said he was very pleased to see both John and Jane and shook their hands warmly, adding that they must catch up with each other, and telling them to get themselves a drink from the free bar.

Having got Jane and himself a drink, John looked round the room to see who he knew and spotted a group of fellow engineers from the area in which he last worked at Blade and went over with Jane to speak to them, tapping one of them that he knew well on the arm. John's former colleagues invited John and Jane to join them, and John introduced those that he knew to Jane leaving the others to introduce their partners and themselves. John's previous colleagues seemed keen to catch up with him and to learn how things were going in his new job, with John being equally keen to learn of the happenings in Blade since his departure. Some way into their discussions, Mike Holmes joined the group and the conversation changed to asking him about his retirement plans. Mike said that he intended to continue with several outside interests citing his school governorship, membership of the local horticultural society and playing a lot more golf by way of examples. He added that he had no other firm plans other than to go on several cruise holidays a year with his wife. Mike explained that his decision to retire was something that had happened recently when the company had offered, what he termed, an attractive package to those who like him had completed in excess of 30 years of service. After a while, Mike said he had to mingle, and everyone wished him well for the future. On leaving the group, Mike said that the buffet was available and that everyone should help themselves reminding them not to forget about the open bar.

The members of the group that John was with took it in turns to get food from the buffet and replenish their glasses, leaving sufficient numbers at any one time where they were so as to not surrender the territory they had claimed as their own in the room. The conversation continued between mouthfuls of food and drink with several of the partners unsuccessfully asking at intervals that the Blade engineers stop talking shop. Eventually, the partners of the Blade engineers and Jane formed their own satellite group to talk about things they found more interesting. After nearly an hour, John felt he should renew other acquaintances and in looking around for Jane, and not seeing her anywhere, he told the group members that he had been talking to that it had been great to catch up with them and he was now going to circulate. John quickly came across other people that he knew well and spent an enjoyable time speaking to such people before moving on to others. John found that he had nearly moved around the whole room when he finally spotted Jane with a group and joined her. In response to Jane' question, John told her what he had been doing and to whom he had been speaking before asking her if she had been OK. Jane responded that she had spoken to several interesting people and that she was fine. Just then a voice called out asking for silence in the room.

Mike's boss, Alicja Haegar, the Director of HR, was the owner of the voice, and standing on a plinth, she took charge of the proceedings by saying that she wanted to say a few words. She started by giving a brief resume of Mike's career in Blade and included the obligatory couple of tales of embarrassing moments that Mike had suffered during his time in the company which raised laughter in the room. Alicja then explained how Mike had taken the Engineering Apprentice School forward, how well-respected he was in the company and how supportive of his apprentices he had been, several of whom had risen to what she called dizzy heights in the company. Several "hear, hears" from the room accompanied what Alicja was saying at this point. Alicja then presented Mike with a number of presents. John was impressed most by what seemed to him to be an expensive-looking set of golf clubs and the model made by current apprentices of Blade's first global mass-produced vehicle, the Gox C. Alicja concluded her presentation by wishing Mike the very best for his retirement, and once the applause had died down, Mike replaced her on the plinth and thanked her for her kind words. Mike then said how much he and his wife appreciated everyone's kind wishes, adding that he very much appreciated the great presents that everyone had contributed to. Mike continued by saying that while he would miss the day-to-day interaction with colleagues, he was looking forward being able to spend more time with the family and using the golf clubs in earnest. Having concluded his short speech, he stepped back from the plinth and was greeted by further applause and some cheering.

John and Jane managed to speak to a few other people that John knew relatively well after the formal part of the evening before saying their goodbyes. Prior to leaving, John made a point of going up to Mike, who was deep in conversation, and waited patiently for the opportunity to speak with him. When Mike stopped speaking, John interjected and told him that he always looked back fondly on his time as an apprentice and said he would be forever grateful for being able to use the foundation of what he had learnt both technically and as a person at the Apprentice School as a springboard to his career. John said that he was grateful for Mike's support and personal example which had a strong positive influence on him and wished him well in his well-deserved retirement. John could feel his eyes beginning to well up, and in shaking his hand, he bid him a good night.

John was glad he had decided to book a taxi home, leaving his car to be collected over the weekend, and in looking at his watch, he saw that Jane and he would not have long to wait after collecting their coats. Once in the car, John asked Jane if she had enjoyed herself and was pleased when she said that she had. Jane then said that she had spoken to Bob Jenkins' wife Maria again and said that she seemed very nice. Jane said that in conversation Maria had told her she knew both of your move to Oxton and when it had happened. Jane said that she had expressed surprise to Maria at her knowing this, since you and Bob had lost touch with each other. Jane continued by saying that Maria had told her that Bob keeps an exercise book in which he notes down all the career moves of those in the apprentice intake that he and John had been a part of. Maria said that she had looked through the book with Bob before coming to Mike's retirement event assuming that some of Bob's contemporaries would be there. John and Jane agreed this seemed strange behaviour to them, but there was no accounting for what other people might do. On arriving home, Jane asked John if he wanted a hot drink before retiring, and John declined saying that he had enjoyed the evening and was looking forward to his bed.

John reflected on his way into work on Monday morning that their Friday night out had set Jane and him up nicely for a relaxing weekend together during which time, unusually for him, he had hardly thought about work and concluded that they ought to break their routine more often. John had suggested to Jane a couple of times that Bob Jenkins might be the person who had made the anonymous phone call to Blade. On the first occasion that he raised the subject, Jane said that it was not worth discussing it as there was not, and probably never would be, any evidence to support the supposition. Jane was firmer with him when he mentioned Bob as a suspect for the second time telling him to take Jim Eccleston's advice and forget about it.

John's first meeting of the morning was the Review/Preview meeting with Sarah. John was pleased that Phil was doing the Warm Up in the next Oe375 meeting and therefore also attending the meeting as he wanted to speak to Sarah and him about the alignment between aspects of Phase 3 of the Oxton first-time engineering norm (OFTEN) framework and current practice in Oxton. Phil and Sarah arrived together for the meeting, cups of coffee in hand, with Phil bringing a cup for John. In putting John's cup down on the table, Phil said that he thought John took milk but was not sure if he took sugar so had brought a couple of packets. John thanked Phil and said that his coffee preference was white without. Once everyone was settled, John said he was glad that Phil was there because he wanted to talk to him and Sarah about introducing Phase 3 of the OFTEN framework. John added that he was getting ahead of himself, and turning to Phil, asked him how he thought the last Oe375 meeting had gone. Phil said that he had made a few notes thinking that John would ask him that. Phil said that he found the Warm Up interesting and added that he knew that he tended to speak before he thought. Phil continued by saying that the See, Imagine, Feel exercise had made him realise that he responded to some people in a certain way because of his preconceptions about them. Phil said he was not sure if this would make any difference to how he responded to them in the future, but at least, he was more aware of how his presumptions about people influenced how he interacted with them.

Looking down his notes, Phil said that he liked John's revision of the severity and occurrence ratings, adding that Team B found them quite easy to use. Phil paused and ran his finger down his notes and then looking up said that he found the development of the function fault tree from the function tree interesting. In concluding his feedback on the previous Oe375 meeting, Phil said that he liked the way that information contained in the system state flow diagram (SSFD) was used in different but interrelated ways. John thanked Phil for what he called his detailed feedback, and turning to Sarah, asked if she had anything to add to what Phil had said. Sarah said that Phil had covered things well and that she had nothing further to add.

In looking at the agenda for the next Oe375 meeting, John said that he was intending to cover the People Skill of Attribution and explained that people tend to attribute behaviours to different causal factors depending on whether they are looking at that behaviour in themselves or in others. By way of example, John said that he liked to think that he got his job at Oxton because of his ability, but others might see it as him being lucky by being at the right place at the right time in meeting Jim Eccleston when Jim was looking to recruit someone. John asked Phil and Sarah if they had any ideas about a Warm Up exercise that might explore attribution in a similar way to his example. Phil said what they might do is ask team members to think of something that they were proud of accomplishing and explain why they thought they had accomplished it well. John said he liked the idea and said that attribution works the other way round too in that when people are not successful in accomplishing something, they tend to blame other factors such as bad luck rather than themselves. John added that team members might also be asked to think of an occasion when they had been unsuccessful at something and to explain why they thought this was the case. Sarah, Phil and John agreed that they had the core of an interesting Warm Up, and Phil said that he was happy to put the Warm Up exercise together based on what they had discussed. Phil then asked John how he thought he should handle any discussion following the exercise. John replied that if he were facilitating the exercise, he would firstly summarise any common themes in what people had said, adding that hopefully these would be along the lines that he had just indicated, and then he would ask people for their comments.

Turning to the Technical Skills content of the Oe375 meeting, John asked Phil how Team B were doing with including the severity and occurrence ratings in the propulsion system design failure modes and effects analysis (DFMEA), and Phil said that they had finished it and he would be happy to go through some examples in the meeting. John acknowledged Phil's offer and said that, since he wanted to consider high severity and/or high occurrence failure modes in the Oe375 meeting, they would be looking at the latest work that Team B had done as a matter of course. Phil said that this would be fine. John continued by saying, that, as Sarah and Phil would know, the next step described in the Oxton FMEA Manual following the identification of high-risk failure modes was Design Verification. John asked Sarah and Phil what influence the DFMEA had on the Design Verification that was conducted in Oxton. Phil said that, as he and Sarah had explained previously, the development of a DFMEA in Oxton had little to do with the design process and added that it was purely coincidental if failures that occurred during Design Verification were identified on the DFMEA.

In looking at the next phase of the OFTEN framework, John reminded Sarah and Phil that Phase 3 was entitled Countermeasure Development. John said that having identified high-risk failure modes, he wanted to cover the next step in the DFMEA process of ascertaining what measures were currently in place to mitigate each high-risk failure mode and to determine if these measures were adequate or needed to be improved. John said that he assumed that this analysis was not currently conducted in Oxton, and Sarah confirmed that this was the case. John then asked if the design changes that were put into place to deal with failures identified through Design Verification were documented, and Sarah said that they usually would be but in various places, citing the examples of revised component specifications and problem-solving reports. John asked how easy it might be to find this information, and Phil said that some information could be found easily while other information might require a search. John then asked Sarah and Phil if they had both been involved in looking at Design Verification failures for the last eBike, and they both said that they had. John thanked Phil and Sarah saying that what they had discussed was very useful information for him, adding that he would be needing their experience and expertise in the next Oe375 meeting. John then asked if there was anything else that they needed to include, and Phil said that at last Team B had extended the propulsion system SSFD and function tree to include the derailleur gearing and subsequently used the function tree to extend the function fault tree. John added that to the agenda before asking if they had had any difficulty in doing this task. Phil replied that extending the function tree had proved a bit tricky but once they had done that revising the function fault tree had been very straightforward. John said he looked forward to seeing Team B's analysis. John then noticed that it was getting close to the meeting finish time, and in concluding the meeting, he thanked Phil and Sarah for their help.

Walking from his car into the office the next day, John realised he was really looking forward to the Oe375 meeting and testing out his ideas on the approach to DFMEA. One thing that frustrated John about working in Blade was that in being an individual in a large global company, he felt powerless to accomplish any changes to the design methodologies used in the company. Admittedly, he had piloted the first-time engineering design (FTED) approach in Blade, but he felt that this work was only tolerated because it was part of his MSc research, and he would have struggled to promote the approach more widely in the company. John found working in Oxton to be a breath of fresh air since he had the freedom, subject to Jim Eccleston's approval, of doing things his way. This had meant that he was able to adapt the current Oxton DFMEA methodology in what he felt were relatively small but significant ways.

On walking into the conference room, John found that Phil had set up the room and was displaying a slide on the large screen. On studying the slide, John saw that it went to the core of the Warm Up, and he told Phil that he liked it. Phil thanked John, walked over to his laptop, and blanked the screen display. John said to Phil that he seemed to have everything under control and told him that he would leave him to start the meeting which Phil duly did once everyone had arrived or signed in. After welcoming team members to the meeting and explaining that he was doing the Warm Up, Phil shared his slide (Figure 9.1).

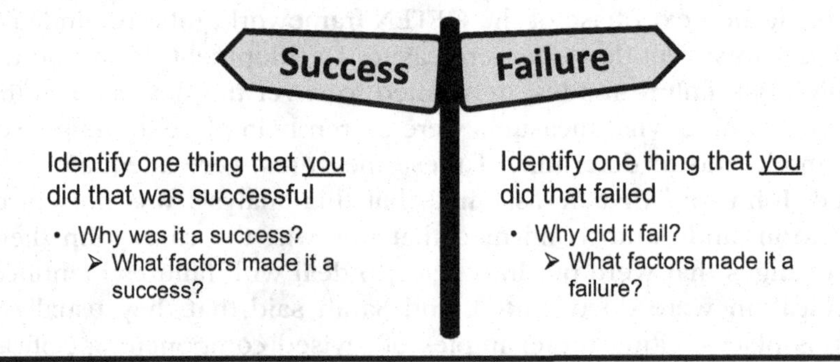

Figure 9.1 Warm Up slide – success and failure.

In explaining the slide, Phil said that he wanted each team member to think of something that they had done as an individual that was successful, and by way of example, Phil said that he recently built his own PC from a kit and to his amazement it worked first time. Phil said that in addition he would like team members to think of something they had done which was not successful, adding that he recently tried to change one of the presets on his kitchen radio and even though he pressed what seemed to him every possible combination of buttons the new station did not save. Phil then asked team members to identify the factors that made the first thing that they did successful and the factors that made the second thing they did unsuccessful. Phil added that he would give team members 5 minutes to identify both the success and the failure along with the factors that made the success and failure before they shared their experiences with each other.

After the 5 minutes was up, Phil asked team members to share both of their examples saying that they could do this in any order. John listened carefully to what was being said and particularly liked Jennifer and Andy's examples. Jennifer's success was that she had recently taken up crocheting and had crocheted what she described as a beautiful multi-coloured blanket which she said took a lot of patience, endurance and, if she had to say it herself, skill. Jennifer also said that she had done some tiling in her bathroom recently which turned out not to be as good as she would have liked, adding that she had managed to hide most of the worst area by repositioning the bathroom cabinet. Jennifer explained that the problem was that the wall was not true and the tile cutter she had borrowed from her father had seen better days. Andy said that he had recently won a 10k race organised by his company and while he had never run a competitive race before, he was always good at cross-country running at school, jogged regularly and had trained hard for this race. Andy then explained that his partner and he had his partner's parents over for Christmas Eve last year, and he cooked dinner which he described as a bit of a disaster. Andy said that the problem began when he found that the turkey was not fully defrosted even though he had followed the defrosting instructions on the wrapping to the letter. He continued by saying that even bringing the turkey into a warmer room meant that he was a couple of hours behind the clock. Andy added that things seemed to go from bad to worse with him over-cooking some of the vegetables because he had mixed up the controls on the cooker hotplates since these were finicky to operate and illogically laid out. Andy said that his partner's parents were polite about the meal, but he was sure that he had gone down a couple of notches in their estimation of him.

When all the team members, including John, had explained their success and failure examples, Phil asked team members if they could identify a common theme between the factors that individuals attributed to their success. Jennifer said that in the success examples, everyone had credited their success to personal factors. Phil agreed with Jennifer and asked the team what was common to the factors associated with examples of failure. Yumiko responded by saying that while Wolfgang and Sam had admitted that their failure was their own fault, most people, including herself, blamed other things. Phil thanked Jennifer and Yumiko for their input and said that we all tend to blame our failures on things outside of our control while attributing our successes to our own abilities, adding that John was going to look at this in more detail shortly. Phil asked if there were any comments on the exercise, and Wolfgang said that sometimes he knew that a failure was his fault but that did not stop him from trying to find other factors to blame it on. Sam said that he found the exercise interesting and added that, like in using other People Skills, he had identified a behaviour in himself that was obvious now he was aware of it but had not occurred to him before. Phil thanked Sam and Wolfgang for their comments, and since no one else spoke, he summarised the meeting agenda and handed the meeting facilitation over to John for the next item after Warm Up which he said was the People Skill of Attribution.

John thanked Phil and said that he was going to build on the experience of what he described as an interesting Warm Up, by looking at Attribution. John defined attribution as "*The act of regarding something as being caused by a person or thing*" and added that attribution in teams was often focused on attributing reasons for the behaviour of ourselves and other team members. John said that in attributing causal factors to a behaviour, the behaviour must be both recognised either in ourselves, or in others, and seen as intentional. John explained that behaviours are attributed either to what is called *Internal* or to *External* factors and shared a slide to help explain these two terms (Figure 9.2).

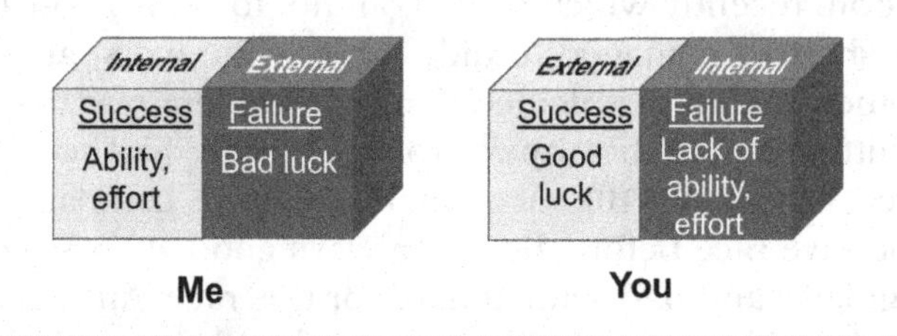

Figure 9.2 Dimensions of attribution.

John said that the success and failure examples given in the Warm Up fell into the graphic on the left-hand side of the slide since the success and failure events related to ourself or *Me*. John said that all of the success scenarios were attributed to factors internal to the person describing the behaviour, such as ability, effort, skill, patience, endurance, etc. John continued by saying that on the other hand, the failures were largely put down to external factors such as bad luck, incorrect or confusing instructions, worn tools, interference by others, etc. John said that the attributions often reverse when looking at behaviours in others or *You* as depicted on the right-hand side of the slide. John added that we might put the success of another down to good luck, being in the right place at the right time, while explaining any failures they might have as being due to a lack of ability or effort.

John said that while the slide might give the impression of a clear polarisation in attribution when considered in respect to ourselves and others, being based on human behaviours, the reality was more complex. John said that attribution in people with low self-esteem is often the opposite of those with normal or high self-esteem and gave the example of someone who blamed a lack of ability for their failures and good luck for their successes. John added that rather than dismissing our failures as due to circumstances beyond our control, they could often provide a valuable learning experience, and in understanding how we might have contributed to a failure, we can identify what we might do differently next time. John said such learning experiences extended to the technical aspects of Failure Mode Avoidance which is why they were identifying causes due to mistakes in developing the propulsion system DFMEA. John said that he hoped that mistakes were treated as learning opportunities in the Oe375 team with effort being directed to minimise the chances of the same mistakes being made again.

In returning to the model of Attribution that he was still sharing, John explained that the characteristics of internal and external were incorporated into what is termed the causal dimension of *Locus* or location. John added that the American Social Psychologist Bernard Weiner introduced two additional causal dimensions of *Stability* and *Control* and proposed that all three dimensions were linked to emotions. John explained that Andy might feel the emotion of pride because of his success in the 10k race, perceiving that it was due to the internal characteristic of ability, which was both stable, in that it did not fluctuate over time, and controllable because he had trained hard before the race. John added that being stable meant that Andy's performance was predictable. John asked Andy if his analysis was accurate, and Andy said that while he agreed that he was very proud to have won the race and he had trained hard, he was not sure he could repeat the success on another day when he might be feeling a bit under the weather so in that sense his skill might be unstable. In thanking Andy, John asked him if he, John, could tell the team his analysis of the Christmas Eve dinner, and Andy replied that of course he could. In response, John said that his perception was that Andy was frustrated by what he saw as his failure and attributed it to the unstable event of the turkey not defrosting fully which was out of his control. John added that the turkey defrost might be viewed as unstable because on another warmer day it would have defrosted, and the weather was beyond Andy's control. Andy said that on reflection, he thought the turkey defrost was controllable and he should have taken it out of the freezer sooner. John thanked Andy and made the point that the attribution of causes to an event was based on perception, and as Andy had nicely illustrated, this can vary from one person to the next.

John said that in the same manner that the way in which we attribute causes to something that we have done determines the emotion that is prompted in ourselves, the way in which we attribute causes to a behaviour we observe in a fellow team member will also prompt a particular emotion in ourselves. John asked team members to recall that he was late arriving at a meeting recently which he said was caused by heavy traffic due to an accident. John said that he assumed team members would see this as an external, unstable and uncontrollable event and have some sympathy for him. John then described an alternative scenario in which, on his late arrival, he had explained he had watched the Super Bowl on TV and had not got to bed until after 3:00am, resulting in him oversleeping. John explained that in this case team members would probably see this as an external, stable and controllable event and disapprove of his actions.

In concluding his presentation, John asked team members to recall the See, Imagine, Feel exercise that they did in the Warm Up for the last meeting. John said the important thing about Attribution as far as team work was concerned was that, just like in See, Imagine, Feel, we need to be aware that the way in which we attribute causes to a behaviour in others, which prompts an emotional response in ourselves, is based on our perceptions. John added that while our perceptions might sometimes be accurate, on other occasions, they may not. John asked team members if they had any comments on what he had just said, and Sam asked why John said that team members would see the Super Bowl event as stable since presumably it happened infrequently. John agreed that the Super Bowl only took place once a year, but if he watched it every year, it would be stable and added that, as Sam had reinforced, the nature of an attribution is dependent on the person doing the attributing.

In looking at the time, John decided that he should move on to the next agenda item, and after thanking the team members for their attention and comments, he said that he would like to review an extract of the propulsion system DFMEA to which Team B had allocated severity and occurrence ratings. John said that for continuity for Andy and Wolfgang, he suggested they look at the DFMEA extract that the team had developed in the meeting before last for the function associated with the cassette. Jennifer quickly found the extract John had mentioned and shared it, and Sarah said that she would go through it (Figure 9.3).

Function	*Potential Failure Mode*	*Potential Effect(s) of Failure*	*Severity*	*PotentialCause(s)/ Mechanism(s) of*	*Occurrence*
Convert force in chain to torque at rear wheel	Temporary loss of conversion of force in chain to torque at rear wheel	Rider experiences jolt in pedalling Increased effort required by rider	7	Poor contact between cassette and chain due to cassette teeth and chain wear - intermittent chain slippage	4
				Loss of lubrication on cassette teeth causing premature chain and cassette wear and intermittent chain slippage	6
				Excess lubrication attracting dirt and dust causing premature chain and cassette wear and intermittent chain slippage	4
				Dirt and dust causing premature chain and cassette wear and intermittent chain slippage	4
				Corrosion causing poor contact between chain and cassette causing intermittent chain slip	3
				Cassette not located properly on wheel causing intermittent chain slippage over cassette teeth	1
				Cassette not locked to wheel causing intermittent cassette slippage on wheel	1
				Vibration from wheel causing intermittent loss of contact between cassette and chain	3
				Incompatibility of higher speed chain and cassette across groupset manufacturers causing chain slippage	4

Rank		Severity of Effect	Chance of Failure
1	Minor/ Remote	Customer does not discern the effect	Failure not likely
2, 3	Low	Some customers slightly annoyed; minor deterioration in system performance	Failures rarely happen
4, 5, 6	Moderate	Customer annoyed or uncomfortable and notices some deterioration in system performance	Infrequent failures
7, 8	High	Significant customer dissatisfaction. System may fail to operate but does not affect safety	Numerous failures
9, 10	Very High	Effects safe operation of bicycle, non compliance with government regulations	Failure very likely

Figure 9.3 Extract of DFMEA including severity and occurrence ratings.

Sarah thanked Jennifer for sharing the combined severity and occurrence rating table alongside the DFMEA extract. Sarah said that since the last time the full Oe375 team had looked at the DFMEA extract, Team B had included causes associated with the external interfaces with the chain and cassette pointing out the causes due to the external environment in the extract that she was sharing. In looking at the severity rating, Sarah explained that Team B had assumed that the rider would experience a jolt as the pedals moved round with little reaction force as a result of the intermittent failure of the chain slipping on, or losing contact with, the cassette. Sarah said that such a jolt would feel uncomfortable to the rider, resulting in rider dissatisfaction. Sarah added that Team B members believed that the rider would still maintain control of the eBike in this situation and so the effect of the failure should not compromise safety and had rated the effect of the failure mode accordingly.

In directing team members' attention to the occurrence ratings, Sarah said that Team B members had assumed that the majority of eBike owners would conduct regular maintenance of Drivetrain components. Sarah added that in view of this, Team B allocated the causes on the DFMEA extract associated with maintenance as having a moderate or low occurrence rating. Sarah continued by saying that Yumiko had reminded Team B members that the cassette was normally located on the wheel using an asymmetrical spline so that it could not be located in the wrong position and was kept in position by a lock nut. Sarah said that in the light of these design features, Team B discussed whether to allocate an occurrence rating of 2 or 1 to the causes associated with these features. Sarah added that in the end, the team decided that such failures were not likely to happen because of the design of what were bought-in components to Oxton. In continuing, Sarah said that the team took account of the damping effect that the rear derailleur had on the sudden chain movement and gave the cause of the chain losing contact with the cassette because of a vibrational shock, a low occurrence rating. Sarah said that this left the rating for the cause due to the incompatibility of chain and cassette designs which she said was a mistake. Sarah said that this rating was also the subject of discussion in the team who agreed by what she called a majority vote to allocate a moderate occurrence rating. Sarah asked if there were any questions or comments, and Andy asked what the discussion on the last occurrence rating was about. Sarah said that to her and Phil's knowledge while the mistake had only happened once, they were both determined to ensure that it would not happen on Oe375 and so allocated a rating which was probably higher than what she described as "the statistics" warranted. Since there were no more questions, Sarah handed the meeting over to Phil to go through the extension of the Oe375 propulsion system function fault tree.

Phil thanked Sarah and explained that before being able to extend the propulsion system function fault tree to include derailleur gearing, Team B needed to extend the system SSFD to include the gearing which was a task hanging over from a few meetings ago. Phil said that Yumiko had reminded Team B members that mid-drive eBikes rarely had more than one chainring, and as a consequence, the team decided only to include the rear derailleur in the SSFD. Phil explained the group had little difficulty in identifying the derailleur gearing state flow but integrating this state flow into the propulsion system SSFD had proved a little more challenging. Phil shared a slide showing the three derailleur SSFD triads for the cassette derailleur (Figure 9.4) and reminded team members that *gear lever* was an alternative name for a *shifter*.

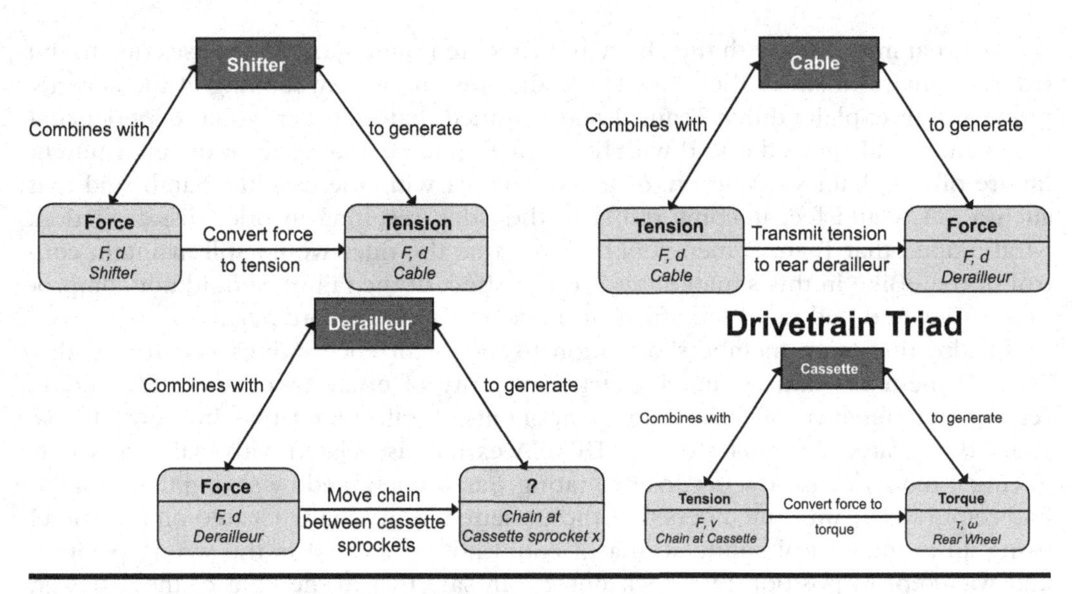

Figure 9.4 eBike cassette derailleur SSFD triads.

In explaining the slide, which initially did not have the Drivetrain triad, Phil said that, while the two triads at the top of the slide were straightforward to develop, the third triad proved somewhat more troublesome. Phil added that, while they agreed that the derailleur moved the chain from one cassette sprocket to another which they denoted as sprocket x, they were not sure how to describe the to-State. Phil said that, after they struggled for a while, he recommended to the group that they set the problem aside and proceed with adding the severity and occurrence ratings to the propulsion system DFMEA, which they were also working on at the time. Phil added that in his experience what appears to be a tricky problem often becomes less tricky if returned to after, what he described as, resting your brain. Phil said it was Yumiko who had a light bulb moment and asked her to tell the team what happened. Yumiko said that she was driving home from work visualising that she was the chain in the cassette derailleur, at which point Andy said "as you do" which prompted laughter. Yumiko interrupted the tale she was telling to point out that she normally visualised that she was a bicycle chain while driving. Yumiko then said that she realised that the state in question was that of the chain under tension at the cassette and what had happened as state flow was a change of location. Phil thanked Yumiko and said that the to-state that Yumiko had visualised was shown on a Drivetrain triad which then he added to the slide as an overlay, adding that this realisation enabled them to complete the propulsion system SSFD and shared a slide (Figure 9.5).

Phil asked if there were any questions or comments, and Andy said that there were now more than the recommended 10 states on the SSFD. Phil said that Team B felt that the addition of three more states to a model that was now familiar was acceptable. Wolfgang then said that there seemed to be a mix of what he called the quantification of attributes on the graphic. Phil asked Wolfgang what he meant, and Wolfgang said that while the main flow states of the Drivetrain and Electric Drive were quantified in terms of power, the derailleur Connecting Flow states were measured as energy. Phil thanked Wolfgang for his comment and said that Team B did

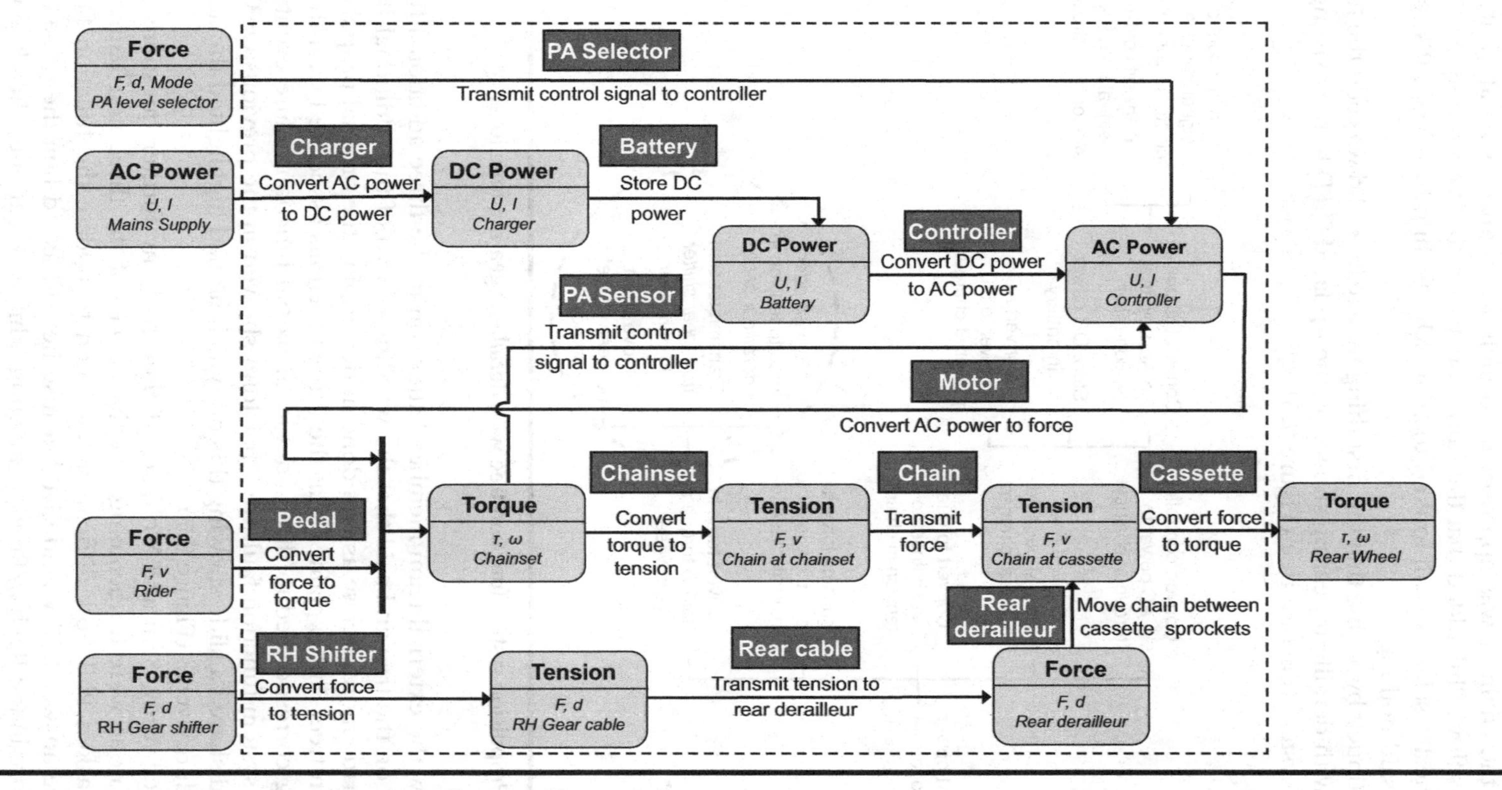

Figure 9.5 Propulsion system SSFD including derailleur gearing.

discuss this point and felt that, since the purpose of the derailleur was to change the location of the chain, it was appropriate to include distance as an attribute for the states in this flow. Phil added that the same was true of the Connecting Flow associated with the PA selector. Phil looked to John who feeling that Team B's reasoning was sound said nothing.

Phil continued by saying that the first thing that Team B did was add the functions associated with derailleur gearing shown on the updated SSFD to the existing Oe375 propulsion system function tree (Figure 9.6).

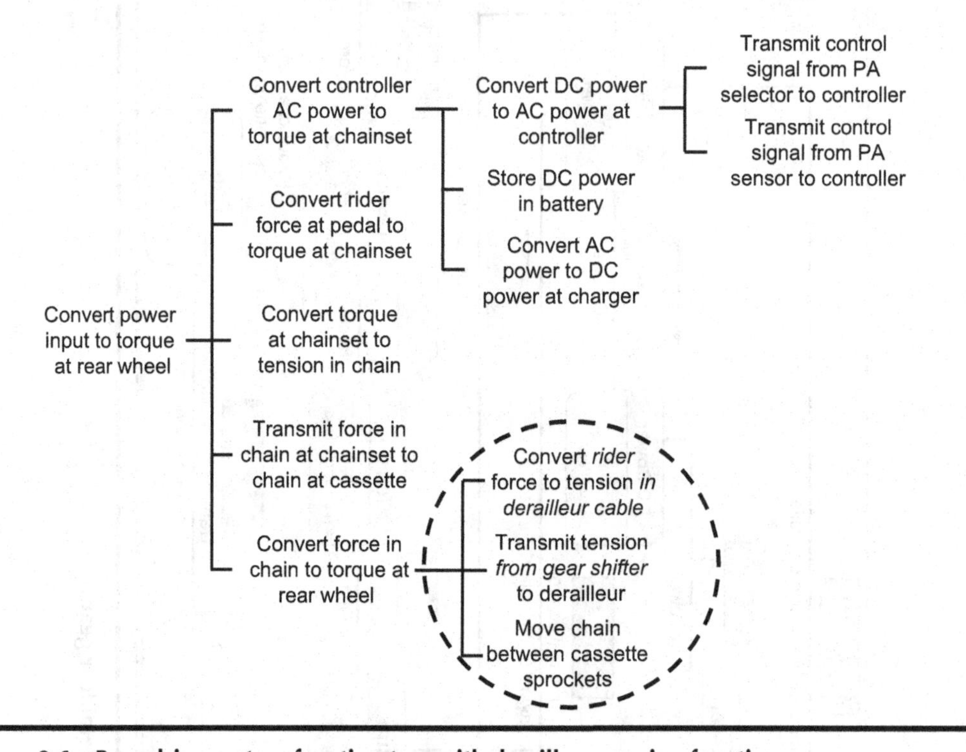

Figure 9.6 Propulsion system function tree with derailleur gearing functions.

Phil drew the external team members' attention to the three additional functions highlighted on the tree and said that the wording of two of the additional functions had been made more precise as was done in developing the original propulsion system function tree. Phil explained that the three functions associated with the derailleur gearing corresponded to three additional partial function failure events on the propulsion system function fault tree that John showed in the previous Oe375 team meeting and shared a slide showing this on which he had highlighted the additional function failure events (Figure 9.7).

Phil asked Andy, Sam and Wolfgang if they had any further questions or comments, and none were forthcoming. For the benefit of the three non-Team B members, John asked Phil how difficult it had been to update the function fault tree, knowing the answer Phil would give because he had asked him the same question the day before in the Review/Preview meeting. Phil said that once they had updated

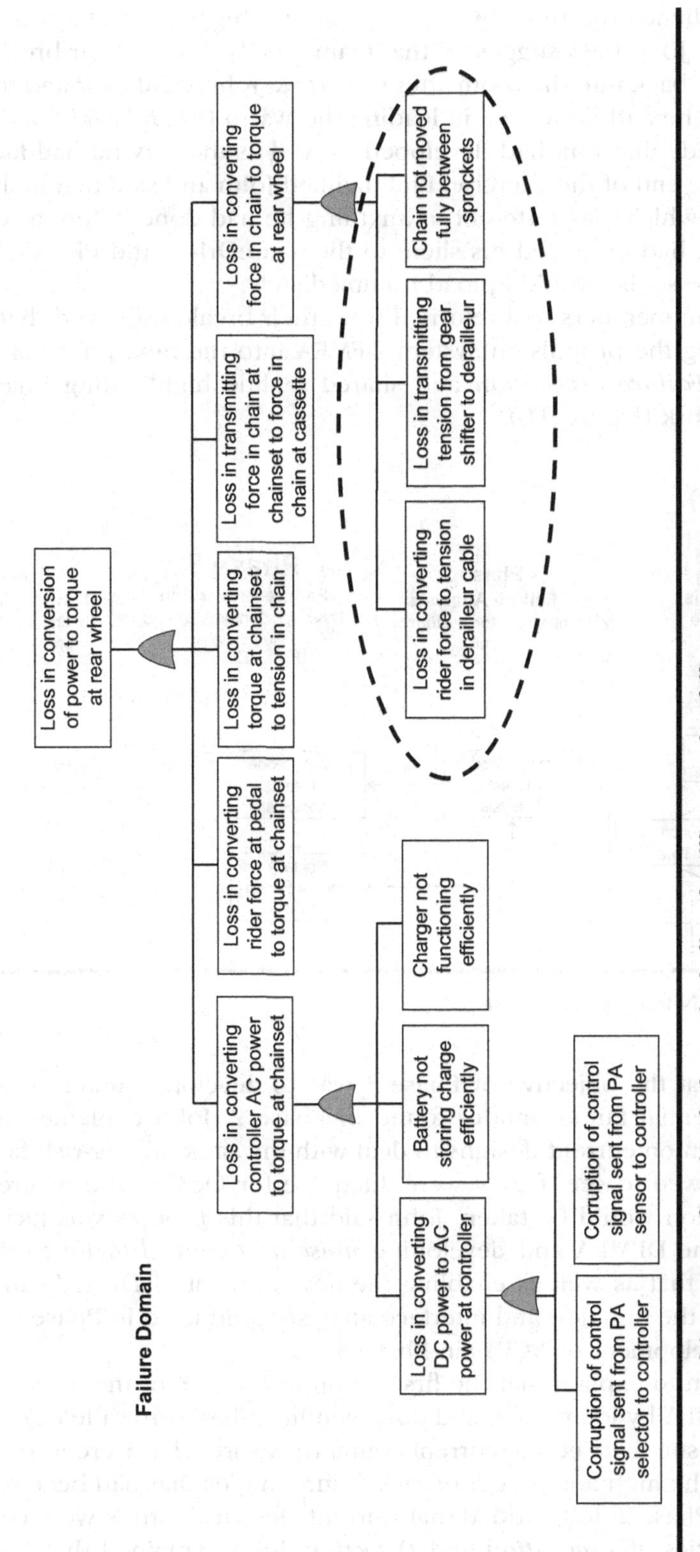

Figure 9.7 Propulsion system function fault tree including derailleur gearing failure events.

the SSFD, and hence the function tree, extending the function fault tree was very straightforward. John then suggested that team members take their break.

As Phil came back into the room after the break, John went over and told him that he thought that he had done well in leading the Warm Up and said that he liked the clarity of the slide that Phil had developed as well as the way he had facilitated the discussion at the end of the exercise. Phil thanked John and said that he had enjoyed the experience which was different to anything he had done before at Oxton. John asked Phil if he had uploaded his sliere to the ShareDrive, and Phil said that while he had not done so, he would upload it immediately.

Once all team members had returned from their break, John said that he wanted to look at taking the propulsion system DFMEA into the next phase of the OFTEN framework of *Failure Prevention* and shared a slide highlighting Phase 3 of the OFTEN framework (Figure 9.8).

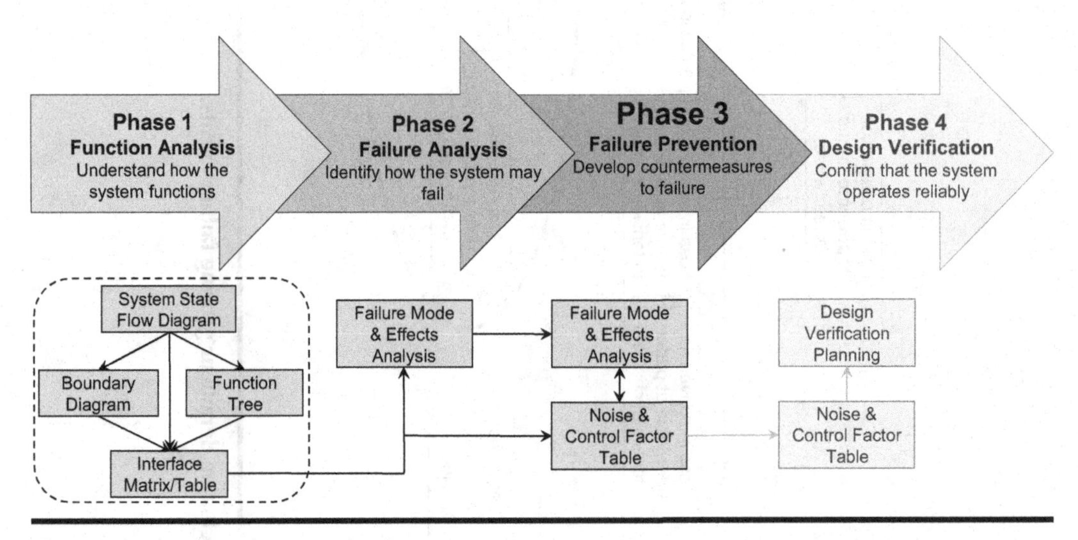

Figure 9.8 OFTEN framework: Phase 3.

John said that the objective of Phase 3 was to develop countermeasures to the higher-risk potential failure modes found in Phase 2. John explained that initially the actions taken on current designs to deal with the same higher-risk failure modes would be reviewed to see if they were adequate for Oe375, and where they were not, further action would be taken. John said that this process was facilitated both by extending the DFMEA and developing *noise and control factor tables* (NCFTs). John explained that as well as enabling the development of DFMEAs in Phase 2 of the framework, the function and interface analysis conducted in Phase 1 also underpinned the development of NCFTs in Phase 3.

John went on to explain that the first action in Phase 3 of the framework was to extend the DFMEA by identifying and documenting what were called *Current Design Controls* which summarised the current countermeasures that were in place for similar designs for the high and moderate-risk failure modes that had been identified on the DFMEA in Phase 2. John added that Current Design Controls were considered in the two categories of *Prevention* and *Detection*. John explained that, as the names implied, Prevention Controls were aimed at preventing a potential failure mode or its cause occurring, while Detection Controls looked to identify a failure mode before a design was released to production. John explained that the accepted wisdom was that the prevention of failure modes was the preferred option with the reliance on the detection of failure modes alone only being employed if a potential failure mode could not be prevented. John said that he would give some examples of both types of design control and shared a table (Table 9.1).

Table 9.1 Examples of Current Design Controls

Examples of Design Controls	
Prevention	***Detection***
Design standards	Durability test
Design robustness	Virtual tests
Analytical analysis	Design Verification tests
Poka-Yoke	Hardware-in-loop test
Proven technology	Designed experiment
Best practice	Electrical output/continuity test
Design parsimony	Prototype testing

John asked if there were any questions or comments on the example controls, and Jennifer asked what Poka-Yoke and design parsimony meant. John explained that the term Poka-Yoke was coined in Japan and meant mistake-proofing and asked team members if they could give him an example of the use of Poka-Yoke in bicycle design. Yumiko said that as she had mentioned earlier, the chainwheel and cassette are both designed to be assembled on an asymmetrical spline so that they can only be assembled in one orientation. Yumiko added that this design was used across major manufacturers. Andy said that what Yumiko had described was similar to the design of electrical connectors, including those on eBikes, which are designed to be asymmetric so that they can only be mated in one orientation and cannot be connected with the wrong polarity. Yumiko then explained that the term Poka-Yoke was originally Baka-Yoke which means fool-proofing and said that it was changed to mistake-proofing because of the negative connotations associated with thinking of a product user or assembler as a fool.

John thanked Andy and Yumiko for their input and said that, in answer to the second part of Jennifer's question, a parsimonious design meant it was one which used the least complexity required to achieve the desired function. Again, John asked team members if they could provide a bicycle example of a parsimonious versus a non-parsimonious design. Phil said that he was not sure if this is what John was after and said that the more modern integrated gear shifters used with dropped handlebars were more complex than historic shifters. John asked Phil to say more. Phil explained that historically gear change and brake application on bikes with dropped handlebars was achieved by moving separate levers, with the brake levers being located on the handlebars and the gear shifters on the frame down-tube. In an integrated shifter, the brake and gear levers are one and the same with the lever pulled towards the handle bar to brake and pushed sideways to shift gear. Phil added that the design varies between manufacturers, with some having a second lever beside the brake lever which again is pushed sideways to change gear in the opposite direction to that achieved by pushing the brake lever. Phil said that the integrated shifter design was significantly more complex with more moving parts, and hence, less parsimonious than the older separate shifter and brake lever designs. Sarah said that integrated shifters also had significant advantages over the older designs not the least of which was they allowed the rider to achieve more precise shifting without having to move their hands from the handlebars. Wolfgang observed that while integrated shifters had more moving parts than other shifter designs, they were reliable having been in use for several decades, but he added that they were more difficult to maintain. Sarah observed that the integrated shifter was a good example of an excitement feature that migrated to a basic feature over time.

As interesting as the discussion was, John decided that they needed to return to the subject in hand, and after thanking Phil, Wolfgang and Sarah for their contributions, he asked if there were any more questions on the list of examples of current design controls. Wolfgang asked what a hardware-in-loop test was, and John not wanting to open up the team discussion again at this point answered the question himself by saying that in a typical hardware-in-loop test, a software-driven controller is tested within a simulation of the system that it has to control rather than the actual system. John added that he was sure that the team would return to what he termed "HIL" testing later in applying the OFTEN framework. Wolfgang then asked why Design Verification was included as Detection Control in Phase 3 of the OFTEN framework when it was the core of Phase 4 of the framework. John explained that the same test might be used both as a Detection Control during the design process and later in conducting Design Verification with the main difference between its dual usage being the time in the design process, and hence, the maturity of the design, when the test was used.

John said that he thought it interesting that in the existing Oxton DFMEA process as defined in the Oxton FMEA Manual, Design Verification was documented instead of Current Controls. He added that the identification of Current Controls was now widespread practice. John explained that since the identification of Current Controls was something that had not been included in a DFMEA before in Oxton as far as Oe375 was concerned, it almost certainly meant looking towards those actions used on previous programmes to prevent or detect higher-risk failure modes. John reiterated that he believed strongly that the Oe375 design should focus on the prevention of failure modes in the first instant, adding that in his view, documenting past detection controls in a Oe375 DFMEA did not imply that they would be used on Oe375. Sam asked why they should bother identifying Detection Controls if they were unlikely to use them. John responded by saying that it was useful in assessing the effectiveness of Current Controls, and he was also sure that some of the tests that had previously been used for detection could be valuable as part of the design process.

John said that team members would remember that the team decided that they needed to follow up on the high- or moderate-risk items on the DFMEA extract that the team were looking at before the break. John explained that by follow-up action, he meant identifying the Current Controls Oxton had in place for each cause and assessing whether these actions were adequate or if further measures were required.

With John's guidance, the team started to identify the Current Design Controls Prevention for the causes on the DFMEA extract that the team had considered in their last meeting with Jennifer adding them to the document that she was sharing (Figure 9.9). When the team reached the two causes associated with interfacing the motor and chainset, Yumiko said that Agano either treated the chainset as an integral part of the Electric Drive or recommended a compatible chainset based on the design of the motor. Given this information and the low failure occurrence rating associated with these causes, the team decided that the mating of the chainset to motor was widely proven design and documented the Current Design Controls Prevention for the two causes associated with this design feature accordingly. The team decided that the remaining Prevention controls were associated with the engineering specification for the motor. As Jennifer was typing the last cell in the Current Design Controls Prevention column, John received a chat message from Sarah that she had sent to Phil and him saying that he should talk to her and Phil after the meeting about how specifications were developed in Oxton.

Function	Potential Failure Mode	Potential Effect(s) of Failure	Severity	Potential Cause(s)/ Mechanism(s) of Failure	Occurrence	Current Design Controls Prevention
Convert controller AC power to torque at chainset	Loss in converting controller AC power to torque at chainset (greater than expected)	Increased effort required by rider	6	Insufficient torque transmitted from motor to chainset - motor power under specified	7	Mid Drive motor specification
				Motor not fully attached to chainset	1	Proven design concept
				Motor not in full physical contact with chainset	1	Proven design concept
				Motor overheats causing loss of efficiency	5	Motor specification, Motor mounting location, Air flow
				Vibration transfer chainset-motor; loss of motor efficiency	3	Motor specification
				Water ingress to motor - exposure	3	Motor IP specification
				Water ingress to motor - inappropriate IP rating	4	Motor IP specification
				Dust/dirt ingress to motor - inappropriate IP rating	3	Motor IP specification

Figure 9.9 Extract of propulsion system DFMEA including current controls prevention.

Under John's guidance, the team then turned their attention to the Current Design Controls Detection. John said that including a detection rating alongside a Detection Control on the DFMEA to indicate the effectiveness of a test at detecting failure modes was valuable information. John then shared the Detection Rating Table that he had developed (Table 9.2).

Table 9.2 Detection Rating Table

Detection Ratings		
Rank		Likelihood of Detection
1	Certain	Design prevents failure; detection not applicable
2, 3	High	Virtual or physical test has very effective detection capability
4,5	Moderate	Virtual or physical test has good detection capability
6, 7	Low	Virtual or physical test has poor detection capability
8, 9	Remote	Available virtual or physical test not designed to detect specific failure mode(s)
10	Not feasible	No test or detection event available

The team agreed that the best detection event for assessing the conversion of controller AC power to torque at the chainset was the hardware-in-loop test designed to measure this outcome amongst other things. Although the causes associated with attaching the chainset to the motor had an occurrence rating of 1, the team decided to include this feature as a check during bicycle-level Design Verification testing. Given Oxton's reliance on Design Verification in previous programmes, the Oe375 team identified the tests currently used in this way as Detection Controls in the Oe375 programme for the causes of vibration and water ingress. Since no test was available to detect the cause of dust/dirt ingress and this cause had a low occurrence rating, the team decided that bicycle-level Design Verification testing was appropriate as the Current Design Controls Detection event. As the team identified each Detection Control, they assigned it a detection rating (Figure 9.10).

Function	Potential Failure Mode	Potential Effect(s) of Failure	Severity	Potential Cause(s)/ Mechanism(s) of Failure	Occurrence	Current Design Controls Prevention	Current Design Controls Detection	Detection
Convert controller AC power to torque at chainset	Loss in converting controller AC power to torque at chainset (greater than expected)	Increased effort required by rider	6	Insufficient torque transmitted from motor to chainset - motor power under specified	7	Mid Drive motor specification	Hardware in Loop test	2
				Motor not fully attached to chainset	1	Proven design concept	Bicycle DV test	4
				Motor not in full physical contact with chainset	1	Proven design concept	Bicycle DV test	4
				Motor overheats causing loss of efficiency	5	Motor specification, Motor mounting	Hardware in Loop Test	2
				Vibration transfer chainset-motor; loss of motor efficiency	3	Motor specification	Rig test	5
				Water ingress to motor - exposure	3	Motor IP specification	Water spray test	4
				Water ingress to motor - inappropriate IP rating	4	Motor IP specification	Water spray test	4
				Dust/dirt ingress to motor - inappropriate IP rating	3	Motor IP specification	Bicycle DV test	7

Figure 9.10 Propulsion system DFMEA extract including current controls detection and detection ratings.

In considering the need for improved prevention measures, the team agreed that Current Controls Prevention associated with causes with high and moderate occurrence ratings were clearly inadequate. They also decided to review the details of the prevention controls associated with the low occurrence rated causes to see if they required improvement. John reminded team members that the phase of the OFTEN framework that they were now in was entitled *Failure Prevention* and added that the two strategies for preventing the occurrence of failure modes were avoiding making mistakes during the design process and ensuring that the design was robust to noise. John said that mistakes were best avoided through vigilance. Jennifer asked what John meant by this statement, and John said that by vigilance he meant paying attention to detail during the design process and that this would include reviewing the adequacy of design specifications. John cited the example of choosing a motor with inappropriate IP ratings as a mistake and suggested that this failure mode could be avoided firstly by ensuring that the specification for the motor in terms of dust material and liquid ingress was appropriate and secondly adhering to the specification.

John said that he would like to consider the other prevention strategy of robustness improvement and reminded team members that a robust design was one that maintained its intended level of performance despite the effect of the noise factors which act to inhibit the functioning of the system. John added that while it was often possible to optimise a design through the selection of design parameter magnitudes, sometimes this was not enough and the design specification needed to be "beefed up", adding that this was a technical engineering term which produced laughter. John said that given the earlier comments about the current eBike being under powered, he suspected that Oe375 might need a more powerful or "beefed up" motor. John then added that he was getting ahead of himself because the required power output of the Oe375 motor would need to be confirmed later through design analysis and said that he would like to get back to looking at the development of countermeasures to failure by improving the robustness of a design and shared a slide (Figure 9.11).

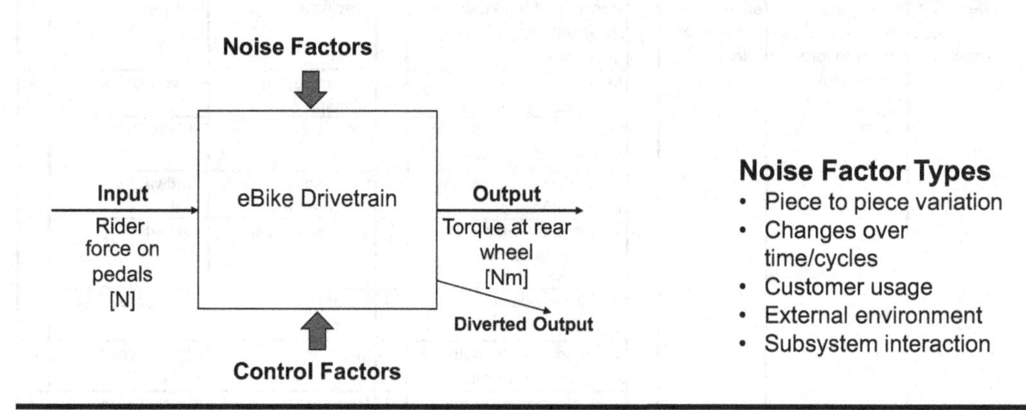

Figure 9.11 Oe375 drivetrain P-Diagram and types of noise factors.

John said that the left-hand side of the slide depicted what was called a *P-Diagram* for the Oe375 Drivetrain which was a block diagram representation of the Drivetrain system which included a graphical representation of the influence of *Noise Factors* on the system. John said that the graphic also illustrated the use of *Control Factors* which could be manipulated to minimise the disruptive action of noise factors on the system. Phil asked John if he would give examples of noise and control factors, and John said that if Phil would not mind waiting a few minutes, he would show the team some examples of noise and control factors. John added that as well as a system achieving its desired output, it may also generate non-useful or undesired output which he said was termed *Diverted Output* and pointed out that this was depicted on the P-Diagram. Jennifer asked John if he would give an example of a Diverted Output and followed up by asking him what the "P" in P-Diagram stood for. John responded by saying that it was inevitable with a physical system that not all of the input energy was converted into the desired output and gave the Drivetrain example of frictional heat produced as the chain moved over the chainwheel and cassette teeth. John explained that the term *Error States* was used as an alternative name for diverted output, although added that he was not keen on this term since to him it implied that something might be wrong with the system. In answering Jennifer's second question, John said that P stood for Parameter and explained that people often listed both noise and control factors as well as diverted output on a P-Diagram.

Turning to the right-hand side of the slide, John explained that it was useful to classify noise factors into five types to aid their identification and added that the team would look at these categories in more detail shortly. John said that when noise factors were listed on a P-Diagram, they were normally documented within the five categories and then explained that, because the amount and type of information that could be recorded on a P-Diagram were limited, his preference was to use a noise and control factor table (NCFT).

John said he would illustrate the way in which an NCFT was developed by continuing the team analysis of the Oe375 propulsion system. John added that he recommended that the team develop separate NCFTs for each of the Oe375 propulsion system subsystems and suggested that the team start with the NCFT for the motor by considering noise factors that act on this subsystem.

John said that it was important to appreciate that noise factors could only influence a system such as the motor by acting through the system inputs, outputs or system interfaces. Andy said that he was a bit confused since John had said that the motor was a system and a subsystem in almost the same breath. John agreed with Andy that his use of the term system might cause confusion and explained that in speaking of a motor as a system he was using the term in the general sense that a system was a collection of interacting items forming a combined whole. John explained that on the other hand, the motor was a constituent part of what was the combined whole of the propulsion system, and in the context of the propulsion system, the motor could be considered a subsystem. John asked Andy if this had helped him, and Andy replied that he thought that it had.

Anxious to move on, John shared a blank NCFT (Table 9.3) and said that he would describe the general layout of the table before starting to populate it.

Table 9.3 Blank Noise and Control Factor Table

Noise and Control Factor Table TITLE						Control Factors		
Interface Ref	Noise Factors by Type	Units	Range		Noise Factor Effect			
			Min	Max				
	Piece to Piece							
	Changes over time/cycles							
	Customer Usage							
	External Environment							
	Subsystem Interaction							
	Diverted Output							

John explained that noise factors were identified by reviewing the system interface analysis and documented on the table by type, along with the interface reference for the interface that each noise factor was associated with. John added that having identified a noise factor, the units in which it was measured were recorded together with an estimate of the maximum and minimum magnitudes of the noise factor that the system was likely to encounter in use. John said that the effect of noise factor on system performance was documented in the next column to the right and added that identifying the effect of a noise factor was equally, if not more, important than identifying the noise factor itself. In bringing team members' attention to the right of the table, John explained that the columns headed Control Factors formed a matrix between individual noise factors and individual control factors which the team would make use of shortly.

In beginning to look at how the table was populated, John said the fact that noise factors act through interfaces meant that the team could use the detailed information from the Oe375 propulsion system interface analysis to help them identify the noise factors influencing each subsystem of the propulsion system. Following Andy's comment, John was aware that his frequent use of the word system might cause confusion, but in setting this thought aside, said that he would look at the noise factors experienced by the motor by way of an example. John shared an extract from the propulsion system interface matrix to illustrate what he meant (Figure 9.12).

Oe375 Propulsion System		INT'L IFACE	EXTERNAL INTERFACES		
		H	E1	E6	E7
		Motor	Frame	Rider	Environment
2	Chainset	X			
7	Controller	X			
8	Motor		X	X	X

Figure 9.12 Extract from propulsion system interface matrix showing interfaces with motor.

John explained that the extract from the interface matrix only included interfaces with the motor, and as team members could see, the motor had two internal and three external interfaces which were detailed in the corresponding interface table extracts. In changing the content of his laptop screen, John said that he was now sharing an extract from the propulsion system internal interface table for the motor/chainset interface and the NCFT for the motor that he had started to populate (Figure 9.13).

Propulsion System Internal Interface Table Extract

Interface Ref	Interface		Type	Impact	Interface Function / Requirement
2-H	Motor	Chainset	E	2	Transmit torque from motor to chainset
			E	2	Clamp chainset to motor
			P	2	Connect motor physically to chainset
			E	-1	Conduct heat away from motor via chainset
			E	-1	Operate within specified level of vibration

Motor Noise and Control Factor Table Extract

Noise and Control Factor Table MOTOR				
Interface Ref	Noise Factors by Type	Units	Range	
			Min	Max
	Piece to Piece			
2-H	Variation in clamping torque chainset to motor	Nm		
	Changes over time/cycles			
2-H	Chainset to motor clamping force degrades	N		

Figure 9.13 Example of piece to piece and changes over time/cycles noise factors.

John said that he would select interface exchanges to illustrate how a noise factor of each type was identified. John explained that the first type of noise factor was *Piece-to-Piece* variation, adding that this was associated with variation of a particular feature of the product from one product to the next. John said that typically such noise was due to manufacturing variation and said that the example of this in the NCFT was variation in the torque applied to the fastener used in clamping the chainset to the motor (Figure 9.13). John said that this torque was related to the exchange with the interface function *Clamp chainset to motor* in the interface table extract for the motor/chainset interface. John added that on this occasion, the interface exchange relating to this interface function also indicated a second noise factor associated with the clamping force degrading over time within the category *Changes over time/cycles* (Figure 9.13). John asked team members if they recalled that he had spoken about treating time as an external interface when they were looking at causes of failure modes that occurred over time, and he had mentioned that he would introduce a practical way of looking at failure associated with time within Phase 3 of the OFTEN framework. John explained that at that time, he was thinking of the changes over time/cycles noise factor type.

John switched his laptop screen display to share the next two noise factor categories on the NCFT alongside two extracts from the interface tables for two other interfaces with the motor (Figure 9.14).

Interface Ref	Interface		Type	Impact	Interface Function / Requirement
7-H	Motor	Controller	E	2	Transmit AC power from controller to motor
			E	-2	Minimise electromagnetic interference
			I	2	Transmit motor speed and position signal
			E	-1	Minimise heat transfer from motor to controller
8-E7	Motor	Environ-ment	E	-1	Convect heat away from motor
			M	-2	Seal motor from liquid ingress
			M	-1	Seal motor from dust/dirt ingress

Interface Ref	Noise Factors by Type	Units	Range	
			Min	Max
	Customer Usage			
7-H	Duty cycle: Demanded motor operation	ω, Hz		
	External Environment			
8-E7	Variation in heat transfer coefficient of air	$Wm^{-2}K^{-1}$		

Figure 9.14 Example of customer usage and external environment noise factors.

John explained that the third noise factor category was termed *Customer Usage* which as the name implied related to the way in which a customer might use or interact with a product. John added that in the case of the propulsion system motor, the rider of the eBike would require different output speeds and power from the motor depending on the required speed of travel and external conditions like wind resistance and gradient which could be summed up within the phrase *Duty Cycle.* Sam said that he assumed from the graphic that John was sharing that this noise factor was associated with the motor/controller interface and added that he did not understand the association. John explained that the speed and power of the motor was controlled by varying the voltage applied to it and the variation in voltage was achieved through the controller. Sam thanked John for the explanation. John continued by saying that the fourth noise factor type was *External Environment*, and as the name implied, this was where noise factor exchanges with the external environment were listed. John added that the heat transfer coefficient, which was affected by factors such as air density and viscosity, would influence the cooling of the motor through convection and was an example of an external environment noise factor. Wolfgang said that he could not remember having come across the heat transfer coefficient before, and John replied that it was a standard characteristic used in the thermodynamics. Yumiko said that Agano used empirically derived values for the coefficient and added that she could provide typical values.

John then shared a graphic showing an example of the last type of noise factor of *Subsystem Interaction* (Figure 9.15) and said that vibration from the frame was an example of such a noise factor potentially affecting the motor performance.

Propulsion System External Interface Table Extract

Interface Ref	Interface		Type	Impact	Interface Function / Requirement
8-E1	Motor	Frame	E	2	Couple motor to frame so shaft can rotate freely
			P	2	Locate motor on frame so that shaft can rotate freely
			E	-1	Minimise vibration transfer between frame and motor
			E	-1	Conduct heat away from motor through frame

Motor Noise and Control Factor Table Extract

Interface Ref	Noise Factors by Type	Units	Range	
			Min	Max
	Subsystem Interaction			
8-E1	Vibration transferred to motor from frame	Hz		

Figure 9.15 Example of subsystem interaction noise factors.

John said that in identifying and documenting noise factors on an NCFT, the table should also include any *Diverted Output*. John explained that the electrical power input converted by the motor to the unwanted energy of heat was Diverted Output (Figure 9.16). John added that this Diverted Output was associated with the motor external environment interface 8-E7.

Propulsion System External Interface Table Extract

Interface Ref	Interface		Type	Impact	Interface Function / Requirement
8-E7	Motor	Environment	E	-2	Convect heat away from motor
			M	-2	Seal motor from liquid ingress
			M	-1	Seal motor from dust ingress

Motor Noise and Control Factor Table Extract

Interface Ref	Noise Factors by Type	Units	Range	
			Min	Max
	Diverted Ouput			
8-E7	Heat flow from motor	W		

Figure 9.16 Diverted outputs from motor.

John said that on this occasion, he had selected noise factor of each type from the NCFT and identified the interface exchange that each noise factor was associated with, adding that he would normally do things the other way round. He explained that he would usually identify noise factors by going through each of the relevant interfaces in both the internal and external interface tables in turn and document the noise factor or noise factors associated with each exchange in the appropriate noise factor category on the NCFT.

John asked if there were any questions or comments, and Wolfgang asked why manufacturing noise was included in the NCFT, but manufacturing causes were not included on a DFMEA. In telling Wolfgang that his question was a good one and pausing slightly, John bought himself enough time to sort out in his mind how he was going to answer. John said that he assumed that this was because historically the P-Diagram was developed independently of DFMEA and the NCFT was intended to be an extension of the P-Diagram. John added that as team members knew manufacturing also had its own version of DFMEA called the process FMEA. Jennifer then said that she had noticed that John had documented the noise factor units of measure but not their range. John responded that he did not know these and asked Sarah and Phil if they could help and added that what the team needed to document was their estimation of the maximum range of a noise factor the eBike could be expected to experience in use. Sarah said that she would identify those ranges that she could while John and the team carried on considering the NCFT.

John said that there were two more aspects of the NCFT he had mentioned in looking at the blank NCFT that he wanted to consider before going back to the interface exchanges with the motor that they had not yet considered and identify the noise factors associated with these exchanges. John explained that these two addition aspects related to identifying and documenting the *Effect* of the noise factors along with potential *Control Factors* that could be used to control the effect of the noise factors. John shared the NCFT which now included examples of the additional features he had just spoken about (Table 9.4).

Table 9.4 NCFT Including Noise Factors Effect and Control Factors

Noise and Control Factor Table MOTOR						Control Factors			
Interface Ref	Type	Units	Range Min	Range Max	Noise Factor Effect	Assembly control	Fasteners spec.	Motor specification	Air flow over motor
	Piece to Piece								
2-H	Variation in clamping torque chainset to motor	Nm			Reduced torque motor to chainset	X	X		
	Changes over time/cycles								
2-H	Chainset to motor clamping force degrades	N			Increased vibr'n reduces motor torque output	X	X		
	Customer Usage								
7-H	Duty cycle: Demanded motor operation	ω, Hz			Variation in torque required			X	
	External Environment								
8-E7	Variation in heat transfer coefficient of air	$Wm^{-2}K^{-1}$			Motor overheats - performance deteriorates			X	X
	Subsystem Interaction								
8-E1	Vibration transferred from frame to motor	Hz			Motor performance deterioration		X	X	
	Diverted Output								
8-E7	Heat flow from motor	W						X	X

John said that he assumed that the noise factor effects on the motor performance that he had documented on the NCFT were straightforward and there were murmurs of agreement. John explained that because they were currently doing system-level analysis, the control factors identified on the NCFT were system-level actions that would influence the system-level interface exchanges between subsystems and components. John added that since a noise factor might be addressed by more than one control factor, or one control factor might address more than one noise factor, as team members could see, each control factor was documented in a single column with an "X" denoting a relationship between a noise factor and control factor. John asked if team members had any questions, and Sarah asked why noise factor effects had been included in the NCFT. John said that this information can be useful both in identifying an appropriate design strategy to counter noise effects and in designing a Design Verification test and added that they would consider both aspects in more detail later.

Wolfgang then asked if the description of the control factors should be more specific. John asked Wolfgang if the statement behind his question was that he believed that the control factors should be more specific, and Wolfgang confirmed with a wry smile that this was what he intended by his question. John asked Wolfgang if he would give an example of what he meant, and Wolfgang replied that to merely reference a specification seemed vague. John told Wolfgang that he agreed with him that merely quoting a specification by name without knowing the detail of what was in the specification was inadequate. John added that what was important was that the specification gave reassurance that the motor would be robust to the relevant noise factors over the ranges quoted in the NCFT. John said that they would look at how this might be done shortly. Phil said that thinking about motor vibration had made him wonder about audible motor noise and added that he thought that this was an interface exchange between the motor and rider that had been missed during the interface analysis. John told Phil that he agreed with him and said that this was an example of why the OFTEN approach needed to be iterative. John thanked Phil for identifying the missing interface exchange, and Sarah said that Team B would update the interface matrix, table and DFMEA to include information related to the missing interface exchange.

John then suggested that the team populate the NCFT based on the interface exchanges with the motor that they had not yet considered to include the exchange of audible noise between the motor and rider. John was pleased how readily team members took to the task and how efficiently the table was completed (Table 9.5).

Table 9.5 NCFT for MOTOR

Noise and Control Factor Table MOTOR						Control Factors					
Interface Ref	Type	Units	Range Min	Range Max	Noise Factor Effect	Assembly control	Loom specification	Controller specificat'n	Fasteners spec	Motor specification	Air flow
	Piece to Piece										
2-H	Variation in clamping torque chainset to motor	Nm			Motor performance variation	X			X		
2-H	Variation in position assembling chainset to motor	deg, mm				X					
8-E1	Variation in position assembling motor to frame	deg, mm				X					
8-E1	Variation in assembly torque motor to frame	Nm				X					
7-H	Variation in contact resistance	$m\Omega$	1	3	Decreased voltage at motor		X				
	Changes over time/cycles										
2-H	Chainset to motor clamping force degrades	N			Increased vibr'n reduces motor torque output				X		
2-H	Motor to frame clamping force degrades	N							X		
7-H	Voltage loss due to corrosion	mV	0	800	Decreased voltage at motor		X			X	
	Customer Usage										
7-H	Duty cycle: Demanded motor operat'n	ω, Hz			Variation in torque demand			X		X	X
8-E6	Rider maintenance/cleaning	mm^3, ml			Variat'n in ingress into motor						
	External Environment										
8-E7	Variation in heat transfer coefficient of air	$Wm^{-2}K^{-1}$	20	40	Motor performance deterioration					X	X
8-E7	Ambient temperature	°C	–10	50						X	
8-E7	Liquid ingress to motor	ml	IP56	IP56						X	
8-E6,8-E7	Dust ingress to motor	mm^3	IP56	IP56						X	
	Subsystem Interaction										
8-E1	Vibration transferred from frame to motor	ω, Hz			Motor performance variation					X	
2-H	Chainset acts as heat sink for motor	W								X	
8-E1	Frame acts as heat sink for motor	W								X	
2-H	Vibration chainset to motor	Hz								X	
8-E1	Interference from frame to motor rotation	mm				X					
	Diverted Output										
8-E7,8-E1	Heat flow from motor	W			Motor performance deteriorat'n					X	X
8-E6	Audible noise emitted by motor	dB			Cust'r/Bystander annoyance			X		X	

There was some discussion in the team while developing the NCFT as to whether the heat transfer classified as Subsystem Interaction should be classified as Diverted Output. Since it appeared to him that the team was unlikely to reach consensus, John intervened to say that what really mattered was not whether a noise factor was categorised as one type or another type but that it had been identified and quantified. John added that for him an advantage of using interface analysis to identify noise factors was that it was a disciplined process which he contrasted with the conventional practice of brainstorming of noise factors and Diverted Output which he said in his experience was more hit and miss.

In quantifying the ranges of the external environment noise factors, Sarah recommended that the magnitude of the noise factors of dust and liquid ingress to the motor be specified by IP ratings of 56 and explained that 5 meant limited, non-harmful ingress of dust and 6 was protection against strong jets of water. Sam intervened and asked Jennifer if she would share the propulsion system DFMEA extract that the team were working on earlier in the meeting (Figure 9.10) which she did. After a pause, Sam said that the rating of 6 for liquid ingress meant that there was the potential for a mistake on the DFMEA. John asked Sam to explain his comment, and Sam said that the water spray test specified on the DFMEA was inadequate at testing motor protection from strong jets of water. John agreed with Sam's observation and asked him what he recommended, and Sam said that the spray test should be changed to a water jet test. Phil said that given the current Detection test that was used in Oxton was a water spray test, he thought it better to change the detection rating to 9, since the test was not designed to detect the failure mode. Sam said that he agreed with Phil, and Jennifer edited the DFMEA accordingly. John thanked Sam and Phil for their input and said that Sam's attention to detail was a good example of how mistakes can be avoided. Jennifer said that she would speak to Pre-Assembly and get information on the typical ranges of the piece-to-piece noises.

In picking up the point that Wolfgang had just made about the inadequacy of just quoting a design specification, John said that the final technical task he would like the team to do in the meeting was to carry forward the information on the NCFT extract relating to the motor as *Recommended Actions* on the propulsion system DFMEA that they had been looking at earlier. John asked Jennifer if she would share the document (Figure 9.17).

In looking at the cause at the top of the DFMEA extract of *Insufficient torque transmitted from motor to chainset*, the team decided that this was more than a robustness issue and was fundamental to the choice of motor in terms of properties like its torque-speed efficiency map. John recommended that once they had conducted interface analysis and noise factor analysis for the key systems associated with the eBike propulsion, they would be in a good position to model the propulsion system and assess the impact of the motor characteristics. The team decided that the other causes on the DFMEA extract were associated with noise factors and determined the recommended actions accordingly.

Once Jennifer had completed the DFMEA extract, John asked team members for comments. Phil said that, although he understood the rigour and discipline of the OFTEN approach, he wondered whether the team could have written the DFMEA extract without going through all the detailed work that they had. John said that Phil's observation from the perspective of hindsight was not unusual and certainly the team could have developed a DFMEA from scratch. John added that the questions the team should ask themselves when they had completed a DFMEA in that way was, would all the system functions and hence potential failure modes be identified and secondly would the DFMEA be arranged in a logical manner and at a consistent systems level. Phil observed that if the existing Oxton DFMEAs were anything to go by, then the answer to both of John's questions was a resounding "No" which drew smiles.

Jennifer asked why Oxton was doing all this analysis when they would not design or manufacture the motor that was used for Oe375. John said that the analysis that the Oe375 team had been doing was conducted at system level based on the interface exchanges that the motor would experience in use mounted to the Oe375 eBike. John added that this information could be used from two perspectives with the same purpose in mind. John said that the analysis would enable Oxton to select a motor which best matched the requirements of the Oe375 eBike and at the same time the requirements information could be cascaded to the motor supplier who could ensure that the motor as they designed it matches the Oe375 needs.

Wolfgang said he saw duplication of content between the DFMEA and the NCFT and wondered why both were developed. On being asked by John to say more, Wolfgang said the causes on the DFMEA were identified by considering interface

Function	Potential Failure Mode	Potential Effect(s) of Failure	Severity	Potential Cause(s)/ Mechanism(s) of Failure	Occurrence	Current Design Controls Prevention	Current Design Controls Detection	Detection	Recommended Action
Convert controller AC power to torque at chainset	Loss in converting controller AC power to torque at chainset (greater than expected)	Increased effort required by rider	6	Insufficient torque transmitted from motor to chainset - motor power under specified	7	Mid Drive motor specification	Hardware in Loop Test	2	Review motor specification based on eBike modelling
				Motor not fully attached to chainset	1	Proven design concept	Bicycle DV test	4	Review motor spec after modelling airflow
				Motor not in full physical contact with chainset	1	Proven design concept	Bicycle DV test	4	Review fixture specification
				Motor overheats causing loss of efficiency	5	Motor specification, Motor mounting	Hardware in Loop Test	2	Review motor spec after modelling airflow
				Vibration transfer chainset-motor; loss of motor efficiency	3	Motor specification	Rig test	5	Review Motor specification
				Water ingress to motor - poor shielding	3	Motor IP specification	Water spray test	4	Review motor spec and positioning
				Water ingress to motor - inappropriate IP rating	4	Motor IP specification	Water spray test	9	Motor IP56
				Dust/dirt ingress to motor - inappropriate IP rating	3	Motor IP specification	Bicycle DV test	7	Motor IP56

Figure 9.17 Extract of propulsion system DFMEA including recommended action.

exchanges identified on an interface table and the same exchanges were then reviewed in identifying noises on the NCFT and so it seemed to him that the causes documented in the DFMEA were the same as the noises documented in the NCFT. In responding to Wolfgang's observation, John said he agreed with Wolfgang that in general, noise is a cause of functional failure with specific noises causing particular function failure modes. John added that he saw the NCFT as a way of considering the data generated during interface analysis from a different perspective. In discussing the differences of perspective between the DFMEA and NCFT with team members, John summarised the key features on a worksheet as team members identified them (Table 9.6), with John adding the final three features.

Table 9.6 Key Differences of Perception Between DFMEA and NCFT

DFMEA	*NCFT*
Effect as perceived/experienced by customer	Effect as influencing system performance
Attention directed to failure mode/cause	Attention directed to noise factor
Causes grouped by failure mode	Noises grouped by type (same/similar effect)
Includes severity, occurrence and detection ratings	Units and range of noise factors identified
Includes current controls prevention	Includes control factors by multiple noise factors
Includes current controls detection	Helps identify robustness strategy (tbc)
Includes recommended action	Helps design efficient Design Verification tests (tbc)
Developed within subsystems	Can be developed across subsystems (tbc)

John said that two key strengths of the NCFT were facilitating the identification of the most appropriate robustness strategy and designing efficient and effective Design Verification testing, adding that since the team had yet to consider these two aspects, he designated these as "to be confirmed" (tbc) in the worksheet table. In explaining the third tbc item, John said that while his recommendation was that separate DFMEAs were developed for each of the eBike subsystems such as the propulsion system, an NCFT might be developed across subsystems which would be the case if an NCFT was developed to facilitate the development of model of eBike propulsion. John asked Wolfgang if he felt more comfortable that the DFMEA and NCFT were complementary rather duplicative methodologies, and Wolfgang responded that he would like time to further consider what the team had discussed.

In moving into Warm Down, John thanked the team for what they had accomplished in the meeting and asked Team B members if they would complete the Motor NCFT and develop an NCFT for the Drivetrain as well as completing the Current Design Controls and the Recommended Action Columns on the propulsion system DFMEA. Sarah said that while this might be a tall order, she was sure that Team B members would attempt to complete the tasks that had been assigned before the next Oe375 team meeting. Phil, Jennifer and Yumiko concurred with what Sarah had said. As was becoming the norm, the Warm Down feedback was positive, and on this occasion, John was particularly interested in Sam and Phil's feedback. Sam said that while he had felt up to now a bit like the poor relation in that he was the only Oxton member who was not doing a lot of work between meetings, he was pleased to see mention of the development of engineering models which was his area of expertise. Phil commented that when he started out on the Oe375 programme, he had no idea of the amount of engineering detail the team would be getting into and reiterated that he was feeling more like the engineer he was, and less like the project manager he had become, adding that he had not had to think so much in years.

As the meeting concluded and John was helping Phil return the room furniture to its more usual state, Phil reminded John about Sarah's Chat note regarding the development of specifications in Oxton. John thanked Phil and said he had not forgotten and asked if he and Sarah, who was still in the room, had the time to discuss it now. When Phil and Sarah both said they had a few minutes, John suggested that they meet in his office.

Sarah and Phil walked with John to his office, and on entering, John beckoned for them to sit at the table while shutting the door before joining them. After looking at Phil, Sarah started the conversation by saying that she was not sure how much experience John had of procuring parts in Oxton but that she, and she was sure Phil, had played the game with the Purchasing Department quite frequently. While speaking, Sarah looked at Phil and he nodded his head in agreement with what she was saying. Phil picked up where Sarah had finished and said that he had played by the rules as well. John asked Phil what he meant, and Phil explained that Purchasing's primary interest was in obtaining the lowest cost that they could so in playing the game the engineering departments always over-specified what they required knowing full well that they were likely to end up with parts at or below the specification they requested. Phil said that the mistake that he had spoken of a couple of Oe375 meetings ago when the cassette and chain were chosen from different manufacturers was down to Purchasing. Sarah reflected that the situation with Purchasing had not always been as it was now but that things seemed to have started to deteriorate several years ago. John said that, while he did not doubt what Sarah and Phil had said reflected the reality of the situation in Oxton, he was surprised. John said that his experience at Blade was that while buyers always struck a hard bargain with manufacturing suppliers, there was an unwritten agreement that they would not compromise an engineering design. John thanked Sarah and Phil for what they had told him and said that he would need to discuss the matter with Jim Eccleston. John said that he would frame any dialogue he had with Jim in a positive matter of fact way and would not reference the discussion that they had just had. John asked Sarah and Phil if they had any more advice to offer him on the matter and when they said they had not, he thanked them for their support and concluded their brief meeting.

Driving home that evening, John toyed with the idea of calling Lucy Collins to see if she had any knowledge of how Jim Eccleston's letter to Georgia Lee had been received. He thought Lucy would probably not see him calling her as too strange since she had contacted him in the first place to warn him about the impending situation. However, on thinking further, John decided that it was unrealistic to think that Lucy, who had only been involved on the periphery of the situation, would have any knowledge of how Jim's letter had been received. Any further thoughts he might have entertained to call Lucy disappeared when he arrived home to find that Jane had obviously caught the going-out bug and had booked tickets to see the latest James Bond film. Jane told John to get his skates on as she had decided that they would eat first at their favourite Japanese-Asian restaurant in the leisure complex in which the cinema was situated.

Background References

Attribution	Weiner, B. (2014) The Attribution Approach to Emotion and Motivation: History, Hypotheses, Home Runs, Headaches/Heartaches, *Emotion Review*, 6(4), 353–361, https://doi.org/10.1177%2F1754073914534502
Design Controls	Automotive Industry Action Group. (2019) AIAG-*VDA Failure Mode and Effects Analysis (FMEA) Handbook*, Automotive Industry Action Group Pages 53–56
Detection Rating Scale	McDermott, R., Mikulak, R. and Beauregard, M. (2009) *The Basics of FMEA*, Productivity Press Pages 32–33 Automotive Industry Action Group. (2019) AIAG-*VDA Failure Mode and Effects Analysis (FMEA) Handbook*, Automotive Industry Action Group Pages 61–62
P-Diagram	Phadke, M. (1989) *Quality Engineering using Robust Design*, Prentice Hall Pages 30–32
Noise Factor Types	Zhou, J. (2005) *Reliability and Robustness Mindset in Automotive Product Development for Global Markets*, SAE Technical Paper 2005-01-1212, https://doi.org/10.4271/2005-01-1212 Pages 4–6
Noise and Control Factor Table	Henshall, E., Rutter, B. and Souch, D. (2015) Extending the Role of Interface Analysis within a Systems Engineering Approach to the Design of Robust and Reliable Automotive Product, *SAE International Journal of Materials and Manufacturing*, 8(2), 322–335, https://doi.org/10.4271/2015-01-0456
FMEA Recommended Action	McDermott, R., Mikulak, R. and Beauregard, M. (2009) *The Basics of FMEA*, Productivity Press Page 38

Chapter 10

Developing Models

While driving to work the next day, John reflected on what Sarah and Phil had told him about Oxton Purchasing Department's approach to negotiations with manufacturing suppliers over part costs. John thought that it would be better to talk with Margaret Burrell, the Director of Supply, in the first instance before raising the matter with Jim Eccleston. John recalled that Margaret had been supportive of him involving suppliers in the Oe375 team and felt this was rather at odds with what appeared to be her department's myopic focus on cost. John also reflected on Sam's comment in the Warm Down about being pleased that engineering models were to be developed and thought that it would be good to develop a model that could be used to assess the effect of noise on the Oe375 propulsion system. John decided to talk to Sarah and see if he could use at least a part of a Team B meeting to discuss this and was pleased that at this stage of the Oe375 programme he had spaced the Oe375 team meetings a bit further apart to allow more time for work to be done between meetings.

On arriving at his office and opening his email with the intent of arranging the meeting with Margaret and then speaking to Sarah, John saw a meeting notice from Jim Eccleston marked with High Importance and entitled "Business Meeting" which John thought was strange. Opening up the message, John saw that the meeting was scheduled to start in half an hour with Paul Clampin and Mo Lakhani as well as himself being invitees. John wondered what the meeting might be about to require such urgency and thought that he would soon find out. In looking to set up a meeting with Margaret, John managed to find half an hour on Margaret's calendar for a chat that morning.

John reflected that in his experience hastily called meetings were seldom harbingers of good news and so John walked along to Jim's office with some apprehension and arrived in Jim's area to see Paul Clampin chatting to Vicky Harrison. John sat down on one of the chairs outside Jim's office and was joined shortly after by Mo Lakani who sat down beside him and asked how he was and whether he knew why Jim had asked to see them. John said that he was fine, and in responding to Mo's second question, he said he did not know what Jim wanted. Just then Jim came to his office door and asked them all to come in, which they did with Jim closing the door behind them. In beckoning them to sit around the table in his room, Jim joined

DOI: 10.4324/9781003286066-10

them. Jim said that he assumed that they were wondering why he had asked to see them and explained that because money was tight in the company he had decided to cancel, or at the very least postpone one of the existing or proposed programmes. Jim added that he had his own ideas as to which programme should go and added that he would like to hear suggestions from the three of them before making a final choice. After a pause in which no one spoke, Paul said that he thought they should cancel the Oe375 programme. Jim asked Paul why he thought the way he did, and Paul said that, despite having been running for some time, it seemed to him that the programme was still a long way from fruition and so was not likely to be cost-effective. Jim did not respond to Paul's suggestion but asked Mo which programme he thought should be cancelled or postponed. Mo suggested the next off-road bike programme, adding that since they still had difficulties with the launch of the current off-road programme, a postponement would give them a bit of breathing space to enable them both to fix the current problems and allow them to feed any lessons learnt back into the next programme, if, and when, it was started. Again, Jim made no comment apart from asking John for his views. John said that his recommendation would be the programme with the lowest projected rate of return. Jim said that in speaking to Joan Harris from finance, this would be the off-road bike programme and added that this was his favoured choice. Paul said that on reflection it was clearly better to cancel or postpone the off-road bike programme, adding that he supported Jim's decision.

After thanking them all for their input, Jim said that what they had discussed was confidential until a formal announcement had been made. Jim added that because of the company's financial position, he was shortly going to announce some general cost-saving measures such as only essential travel and restricted overtime, adding that he would need to sign off any essential overtime. Jim also asked Paul, Mo and John to manage their budgets carefully and said that he hoped that the measures he was taking would be short-term. As Jim brought the meeting to a close, he thanked the attendees for their time and said to John that he would see him for his one-to-one that afternoon.

When John got back to his office, he sat at his desk with a cup of coffee and reflected that the meeting with Jim was reminiscent of his time with Blade where the company seemed to go through regular cycles of cost-cutting. In comparing his Blade experience to the meeting he had just attended, he saw a difference in that this time he had some say in the form the savings would take, although he suspected Jim's mind was made up before he spoke to the three of them. John's thoughts returned to the cost-cutting initiatives in Blade which he recalled were usually introduced under a euphemism such as "competitive realignment" and formally announced top-down. He thought it interesting that he could not recollect an occasion when the ending of cost-cutting measures was formally announced in the same way, instead things seemed to slowly drift back towards where they used to be, although after any realignment things were never quite the same as they were before it. In coming back to the Oxton situation, John thought that in a strange way, he was looking forward to Jim's announcement to see how he would frame it.

Having finished his coffee, John gave Sarah a call and was pleased when she answered him rather than her Voice Mail. John explained his intention to start the development of a model for the Oe375 propulsion system and asked if he could meet with Team B for an hour or so to discuss it. Sarah said that the team were currently struggling to complete all the work that came out of the last Oe375 meeting before the next Oe375 meeting. John was not surprised to hear what Sarah was telling him since he had felt that what he had asked Team B to do in the last Oe375 meeting was a lot of work, but at the same time, he saw the development of a propulsion system model as the next step in applying the OFTEN framework. John realised, however, that in contrast to previous Oe375 meetings, the next Oe375 meeting would be unlikely to generate any assignments since these would only materialise after he had covered the subject of robustness strategies which he intended to do in a subsequent meeting. John realised that this would give Team B more time to complete their current assignments. John explained the situation to Sarah and asked her to apologise to Team B members for not telling them this when he had identified the assignments originally. Sarah said that Team B members would be relieved to hear what she had to tell them and added that she would send John's apologies.

Sarah said that the team had a meeting scheduled for that afternoon and John was welcome to come along and told John the time and venue for the meeting. John said that he could not attend the first half hour of the meeting since he had a one-to-one with Jim Eccleston but would pop along after that. Sarah said that was fine, adding that she would send a note to the other team members to explain what was happening. Sarah said that being an afternoon meeting, Yumiko would not be attending but as was usual practice, Jennifer would send an email to Yumiko after the meeting to explain what had happened. John asked Sarah if Sam was attending the meeting, and Sarah replied that she did not expect that he would since he rarely attended Team B meetings. John said that he would speak to Sam and see if he was available and asked Sarah to hold off sending her email to the other team members until he had confirmed Sam's attendance with her. John thanked Sarah saying that he was looking forward to the meeting, and after ending the call, he called Sam, although this time his call was answered by Voice Mail. Somewhat disappointed in not being able to speak directly to him, John sent Sam an Instant Message explaining the situation and was pleased to receive a quick reply from Sam in which he said that he would be very happy to attend and asked John for the meeting time and place. John told Sam that he would ask Sarah to send the meeting notice to him after which he sent Sarah an Instant Message telling her that Sam would be attending the Team B meeting and requesting her to forward the meeting notice to both Sam and him.

John then turned his attention to the People Skills aspects of the next Oe375 team meeting, looking for articles that he could take an extract from for use in the Warm Up and discovered some interesting reading. Being absorbed in what he was doing, he nearly missed the notification of his forthcoming meeting with Margaret as it pinged up on his laptop. After finishing off what he was doing, John got up from his desk and walked along to meet Margaret.

Margaret welcomed John into her office telling him that she had heard good things about what was happening on the Oe375 programme from all three of the non-Oxton representatives on the Product Design (PD) team. John said that he was pleased with what Margaret had told him and added that he thought the external members added a lot to the team, saying that he was particularly impressed with the support that Agano had shown. John found Margaret very approachable and so decided to be frank and told her that he had something he would like to discuss with her, at which Margaret smiled and said that she had guessed as much when he had requested the meeting. John said that he understood that there had been some friction between Margaret's department and the engineering areas over part costs. A frown crossed Margaret's brow, and she asked John to say more.

John explained that he expected that some of the parts on the Oe375 programme would need to be of a higher specification than the equivalent parts on the current eBike, while others would be at about the same or lower spec. Margaret said that she was not surprised by what John had said and observed that she assumed that he had not asked for the meeting to tell her this. John thought that despite his intent to be frank with Margaret, he was probably not being as straight with her as he should be. Accordingly, John told Margaret that when the Oe375 team defined a specification for a part, it was important that any negotiations that Oxton had with potential part suppliers based on cost did not compromise the engineering specification in any way. Margaret responded that she now understood where John was coming from and, said that she would like to confirm her understanding of what he had said, which was that when the specification for an Oe375 part was developed, it would reflect what was required and not be an over-estimation. John said that he could give Margaret that guarantee and added that in his experience, this was the way things should be. Margaret paused and then said that what John had described aligned with her vision of the way in which Purchasing should support Engineering in Oxton, adding that this did not necessarily reflect current practice. John did not say anything giving Margaret the chance to expand on what she had said, which she did after a moment by saying that currently her department only negotiated with suppliers on both the engineering specification and cost where it had been indicated to them that the part request had been over-specified. John asked Margaret how an over-specified purchase request was indicated, and she said that Paul Clampin designated it by using an agreed code. Margaret continued by saying that when she had discussed the process with Paul, he had explained that engineers were precious about their designs and believed their design justified the expense they had associated with it when often, in his opinion, they did not. Margaret added that Paul had said he had a keen eye for such things based on his extensive experience.

John sat back in his chair hardly believing what he was hearing, and, recalling what Phil had said, asked Margaret how many purchase requests were identified as over-specified. Margaret said that most of the requests from Paul's area are identified as over-specified these days, although when Paul had started to codify requests, it was only the odd one or two that bore Paul's code. John pondered whether to tell Margaret the other side of the story from Sarah and Phil's perspective but decided not to, firstly because he had promised not to involve them by name, and secondly because he thought that it was not his prerogative to involve himself in Paul's business in this way. Instead, he reconfirmed that all engineering specifications associated with purchase requests for Oe375 would be based on sound engineering. John added that while the Oe375 PD team would be fully aware of the cost implications of any design and would take design actions to reduce cost where they could, the design specification was not open to further discussion. Margaret responded by saying that what John had said was perfectly acceptable to her and that she would expect nothing else. In speaking in a way which to John seemed as much to herself as to him, Margaret said that she did not understand Paul's way of doing business and added that perhaps it was due to his Economics background. In picking up on what Margaret had just said, John merely repeated her last two words "Economics background", and Margaret said that Paul had a degree in Economics. While being a bit surprised, John said that Paul was not the first senior engineering manager that he had met without an engineering-related academic qualification and explained that he knew a Plant Manager in Blade who had told him in confidence that he was a psychologist by training. John added that although the person he was talking about was well respected in the Company, he did not want his colleagues to know he did not have an academic qualification in engineering, believing that this might undermine their confidence in him. Margaret said that the job of a Plant Manager was probably more to do with managing people than knowing the intricacies of the engineering processes. John and Margaret finished their meeting on good terms, and in walking back to his office, John thought to himself he was pleased he had spoken to Margaret before approaching Jim and concluded that having resolved the matter with Margaret, there was no longer any need to involve Jim.

Back in his office and behind his desk, John's thoughts turned to the Oe375 propulsion system model. He thought that he had better get his ideas into some kind of order before the Team B meeting that afternoon and began developing a sketch on his laptop to show how he saw the forces relating to eBike propulsion being used as the mathematical basis of a model. John felt quite pleased with the results of his deliberations and feeling that he deserved an early lunch in the canteen rather than a sandwich at his desk, made his way to the canteen. After getting his lunch and looking around for a place to sit, he spotted Phil and asked him if it was all right to sit at the table at which Phil was sitting along with a couple of other people that he knew by sight but not by name. Phil told John that of course it was all right for him to join them and introduced John to his colleagues. As an aside, John managed to tell Phil that he had discussed what he termed "purchasing arrangements for Oe375" with the Purchasing Department and come to a mutually acceptable agreement. In the end, John spent longer over an enjoyable lunch in animated conversation than he should have and had to rush back to his office to collect his things before making his way to Jim Eccleston's office for his one-to-one.

On getting to Jim's area, John noticed that Vicki Harrison was not at her desk but as Jim's door was open, he knocked on it and Jim invited him in. Jim sat at his usual seat at the table, and John took a seat beside him and began updating Jim on what the Oe375 team had been doing since their last one-to-one by taking him through the People Skills of Attitudes, See, Imagine, Feel and Attribution. Jim said that he had had some interesting discussions with the team's Sport Psychologist during the time he was racing, adding that the psychologist made him aware of the need to think before he took particular actions to make sure they were the best actions to take at the time. Jim added that in racing you often have to make decisions in a split second based on your experience, and on such occasions, where your decision turns out to be the wrong one, you need to update your experience databank. Jim also related the concept of attribution back to his racing days by saying that he knew individuals who tended to search around for external excuses when they performed poorly in a race and so were not able to take full advantage of improving as a sportsperson by learning from their mistakes. Jim added that one of the enjoyable aspects of being an elite athlete was that generally other people did recognise your skills and abilities.

Before moving on to look at the technical work the Oe375 team had been doing, John mentioned that team members were taking it in turns to conduct the Warm Up.

Jim asked John how this had gone and if people felt comfortable in doing this. John replied that he had been very pleased, and he had been impressed by both the manner in which the Warm Ups were managed and the effort that individuals had put into preparing for them. Jim asked how the arrangement worked, and John explained that they had set up a rota with everyone either signed-up for a Warm Up or having done one. John added that the designated person attended the Review/Preview meeting to discuss and agree on the nature of the next Warm Up. Jim asked who designed the Warm Up and what the Review/Preview meeting was. John explained that before every Oe375 team meeting, Sarah Jones and he met with whoever was conducting the Warm Up in the Oe375 meeting and looked back on the previous meeting to see if there was anything that happened in that meeting that needed to be considered in the next meeting. John added that they then considered the purpose and content of the next Oe375 meeting which included looking at the Warm Up. John added that the Warm Up was linked to the People Skill that was covered in the meeting and so he would summarise the People Skill that he wanted to look at, and Sarah, he and whoever was doing the Warm Up would design and agree how the Warm Up was to be conducted. John said that in addition to looking at the Warm Up and People Skill, the Technical Skills content of the next Oe375 meeting was considered to allow the meeting agenda to be agreed. Jim said that what John had described sounded like comprehensive meeting preparation and added that in the meetings he attended having a meeting purpose and agenda defined in addition to the meeting notice would be considered a bonus, let alone doing any additional preparation work. Jim then said that he remembered John telling him about the inclusion of preview and review meetings as part of a process for effective meetings they had discussed in one of their early one-to-ones and admitted to being remiss in not paying any further attention to it.

John moved on to showing Jim the work that Team B had done on the Oe375 propulsion system design failure modes and effects analysis (FMEA) and explained how the document had been developed, emphasising the structural and informational linkage between the interface table and the FMEA. Jim said that he was impressed with Team B's efforts, adding that he liked the logical way in which the document was developed. Jim added that following their last one-to-one he had scanned through both the Oxton FMEA Manual and a couple of FMEAs and recognised that the coherent way in which the Oe375 document was structured was a significant improvement over the more haphazard arrangement of the existing Oxton documents. Jim said he also liked John's simplification of the severity and occurrence ratings and added that he did not recall seeing a priority matrix in the Oxton manual. John explained that, while he had seen something similar in a couple of FMEA manuals, this was a simple version that he had developed. Jim contrasted the approach that John was taking with the current usage of FMEAs in Oxton, in which he said FMEAs were developed as an afterthought and, having been developed, metaphorically shut away in a drawer and forgotten about.

To John, Jim seemed intrigued with the way in which a function fault tree was developed from its associated function tree. Jim said that while he had no direct experience of using fault tree analysis (FTA), he recalled reading what he called an interesting article in a cycling magazine in which FTA was used to assess the probability of failure in terms of speed and effort based on data gathered in an experiment in which people were asked to cycle around a circuit with different road surfaces and gradients. John said that it sounded an interesting experiment which overcame the lack of real-world data on part failures by gathering its own data. Jim asked John what analysis he intended to use the function fault tree for, and John replied for the identification of system failure due to potential single and multiple failure modes. John explained that system failure could be caused either by a single failure mode acting alone or by a combination of several partial failures and added that system failure due to a combination of partial failure modes was more readily identified on a function fault tree than in an FMEA.

John showed Jim the slide depicting the P-Diagram, and after reminding Jim of the concept of noise, explained the different type of noise factors before illustrating how information from the propulsion system interface analysis was used to populate the noise and control factor table (NCFT). Jim asked John how the NCFT was used, and John directed Jim's attention to examples of recommended actions on the DFMEA that had been generated based on the noise factors, and their effects, identified in the NCFT. Jim spotted that some of the recommended actions referred to the intent to perform modelling and asked John for more information on what was intended. John said that assuming that a mathematical model exists or can be developed which shows the relationship between the parameters of a system and its inputs and outputs then, if the effect of noise on the model parameters is known, then the effect of noise on the overall system performance can be determined. Jim said while he understood what John was saying, he did not see any mention of model parameters in the NCFT, and John explained that this was because they had not included them yet. In noticing the time, Jim then said he had another meeting to sign in to and rather abruptly finished the meeting, thanked John and as usual told him to keep up the good work.

John made his way directly to the Team B meeting from Jim's office and, on arriving at the meeting room, knocked on the door. He entered the room to find the team in full flow in developing the chainset NCFT. Sarah invited John to sit down and said that the team would just finish off what they were doing before giving him their full attention. John was pleased to see Sam actively involved in team discussions, and on sitting and observing Team B at work, he was impressed by the way the team members were interacting. He was particularly pleased to see the explicit manifestations of their use of People Skills including their use of restatement and descriptive feedback. John was reassured by the usefulness of Jennifer's role as Recorder when Sam asked why the team was not developing a single NCFT for the propulsion system and Jennifer, in referring to her notes from the last Oe375 team meeting, said that John had recommended the team develop separate NCFTs for each of the Oe375 propulsion system subsystems. Sam looked at John, and in assuming that he was wanting confirmation of what Jennifer had just said, John confirmed that Jennifer had accurately recorded his recommendation. John added that his preference was to develop a series of relatively small self-contained but interrelated documents rather than one large overall table since he found that it was easier to access and use information in a smaller document. Sam seemed satisfied with John's reasoning, and the team carried on with their analysis. After about 10 minutes, Phil said that the team had reached a good point at which to stop what they were doing and handed the meeting over to John.

John thanked Phil and the team for giving him their time and explained that he would like to discuss the modelling of the effect of noise on the Oe375 propulsion system performance and to agree on an action plan for developing the model. Phil said that John's proposal sounded interesting and, in agreeing with what Phil had said, Sam added he was keen to get things moving with modelling. John asked Jennifer to allow him to share his screen, and once she had done this, John shared the sketch that he had produced earlier (Figure 10.1). John apologised to Sam, Sarah, Phil and Jennifer if the sketch was too simplistic for bike engineers. In response, Sarah said that it did no harm to remind engineers of the basics occasionally.

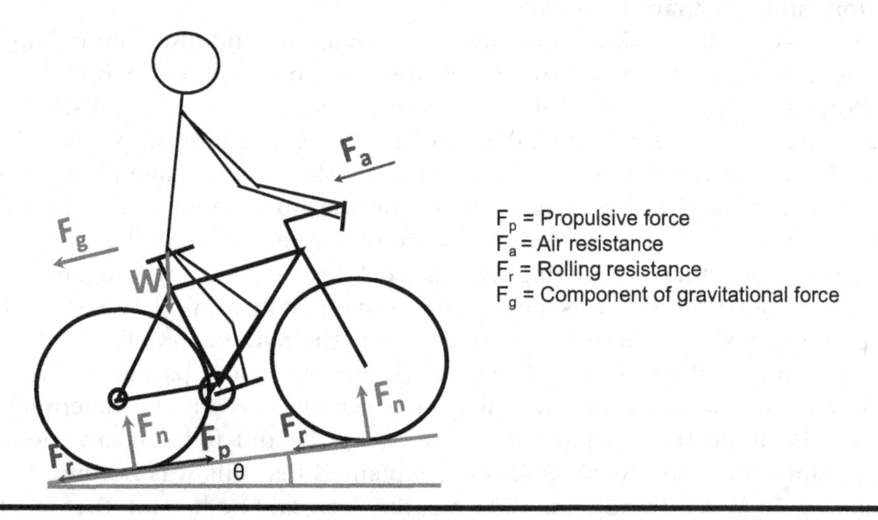

Figure 10.1 Sketch of forces acting on eBike.

John said that he assumed that the sketch of the propulsive and resistive forces acting on an eBike was self-explanatory. Sam studied the sketch and then said that it reminded him of school Physics and said he assumed that the force W was the weight of the rider and F_n was the normal reaction of the ground on the tyres. John said that while it was not explicit, W was meant to include the weight of both the rider and the eBike. Phil said he found the sketch interesting but did not see how it could be used as the basis of a model to analyse the effect of noise. John explained that if they looked further into the maths that might help them to see how the effect of noise could be included in the model and recommended that they look at the scenario of the eBike moving with constant velocity up an incline of constant gradient. John continued by saying that if the eBike was moving with a constant velocity, the propulsive force moving the eBike forward would be balanced out by the sum of the resistive forces due to aerodynamic drag, rolling resistance and the component force of gravity acting down the incline and typed this as an equation beneath the sketch.

$$F_p = F_a + F_r + F_g$$

John said that they needed to expand the equation to include the parameters that determine the resistive forces and said he would start with the one he remembered for the component of gravity parallel to the incline and typed out.

$$F_g = mg\sin(\theta)$$

John said that they need to be clear that m was the total mass of the rider and eBike, adding that of course g was the gravitational constant. Phil said that he remembered the equation for the rolling resistance and aerodynamic resistance, and John typed them out as Phil quoted them.

$$F_r = C_r mg\cos(\theta)$$

$$F_a = \frac{1}{2}C_a\rho Av^2$$

Phil said that C_r was the coefficient of rolling resistance in the rolling resistance equation, while in the equation for the air resistance force, C_a was the drag coefficient, ρ was the density of the air, A the frontal area presented by rider and eBike to the air and v the relative speed of the eBike and air. Phil added that Oxton used the product C_aA called the *Effective frontal area* which they estimate from wind tunnel testing as this was easier to estimate than A itself.

John continued by reminding team members that maintaining constant velocity was achieved when the force input by the rider and motor was just enough to overcome the resistive forces opposing the motion of the eBike and added that "just enough" meant equal to. John then explained that the effect of noise factors is to cause the resistive forces to vary and hence the force required to move the eBike to vary and said that this can be seen from the maths. Taking the example of rider mass as a noise factor, John said that a heavier rider will increase m in the equation for rolling resistance and hence increase the rolling resistance force. John checked that he had answered Phil's point about how the model was used to understand the effect of noise on eBike performance, and Phil responded that he understood and added that, like many things, it was easy when you know how.

Sam said while he was familiar with the equation for the gravitational force acting down the slope, he was not clear how the other two equations were derived. Sarah said that rolling resistance was mainly due to energy losses in the deformation of the tyre as it contacted the road which was dependent on the nature of the tyre and its loading due to the mass of the rider and eBike and added that there was also a small contribution due to friction at the surfaces in contact. Sarah added that she had never seen a theoretical derivation of the equation, but it made sense to her in that it involved the vertical component of the force of gravity which acted to deform the tyre and a constant which depended largely on the elastic properties of the tyre. Phil continued the explanation by saying that while he remembered seeing the equation for F_a derived in Fluid Dynamics at college, he hoped that Sam would not want him to derive it now and said that Sam could check that the equation was dimensionally correct and added that C_r and C_a were both dimensionless quantities. Sam said that he trusted Phil that the equation was valid.

Sarah then said that given the enormous range of different types of eBike tyres of varying composition and sizes and of course riders of different shapes, sizes and masses, that C_r and AC_a were usually determined empirically from rolling road and wind tunnel tests, respectively. Turning to John, she said that the Oxton rolling road was simpler than the rolling roads that he would be familiar with at Blade. Sarah explained that the rolling road consisted of a rotating drum driven by an electric motor on top of which a wheel and tyre are placed, with the wheel being subjected to a load which is representative of half the mass of the rider and bike given that a bike has two wheels. Jennifer added that given the expense associated with developing and running a wind tunnel, Oxton hired one on an as-and-when basis and added that such testing was mainly used for their higher-end bikes.

Sam said that he had a couple more questions and asked what the team would use as a measure of eBike performance. Phil said that power was the usual measure of performance and, after asking Jennifer to give him control of the shared screen, wrote down beneath the equations on John's sketch:

$$P = F_p v$$

Phil explained to Sam that the right-hand side of the equation represented the work done against the resistive forces per second being the force multiplied by the distance moved per second or speed and so was the power required by the rider and motor to overcome the resistive forces. Sam thanked Phil and told him that his Physics knowledge was trickling back. John then asked Sam what his second question was, and Sam asked how they would know what the relationship was between all the noise factors and the parameters and coefficients in the model. John said that some are straightforward like wind speed as this directly affected F_a while they might have to look to test results and research papers for others. Phil said that he had plenty of test data which would probably prove useful.

John said that he would like to share the model in the next Oe375 team meeting and asked Sam if that was realistic. Sam said that it should not be a problem as long as they agreed what the model would display, and which noise factor effects would be included along with their magnitudes and ranges. John said that in identifying and documenting noise factors, he recommended that they analyse the interfaces of interest first which in this case would be the interfaces that were external to the eBike through which the propulsive and resistive forces acted. John said that as he saw it, the propulsive and resistive forces acted through the two *external* interfaces of what he called *Tyre-to-Road* and *Rider and eBike-to-External Environment*. John explained that by *Road* he meant the surface upon which the eBike was ridden. Sarah said that she thought they should consider the external environment factors that acted on the tyre, such as the ambient temperature and rain, separately to those acting on the eBike and rider since they were likely to affect different model parameters. The team agreed that they needed to look at three interfaces: the two that John had mentioned, and the *Tyre-to-External Environment* interface that Sarah had identified.

Phil said that while Oxton had not done formal interface analysis for these interfaces before, they had studied them in detail over the years in rig and wind tunnel tests so that identifying the exchanges that occurred at the interfaces should not be a problem. Sam asked if they would need to develop system state flow diagrams (SSFDs) and boundary diagrams before any interface analysis was conducted. In response, John said that in situations where a system is straightforward and well-understood, developing an SSFD to derive a boundary diagram was overkill and explained that his sketch of the eBike was effectively a boundary diagram. With some discussion and a minimum of disagreement, the team then agreed on the interface exchanges for the interfaces that the team had identified (Figure 10.2).

Interface		Type	Description of Exchange	Metric	From	To	Impact
Tyre	Road	E	Normal component of weight	N	Tyre	Road	2
		E	Normal reaction to weight	N	Road	Tyre	2
		E	Propulsive force due to rider and motor	N	Tyre	Road	2
		E	Rolling resistance	N	Both	Ways	-1
		M	Dirt Debris	mm^3	Road	Tyre	-1
		M	Sharp object	mm^3	Road	Tyre	-2
		M	Water on road	ml	Both	Ways	-1
		P	Tyre-road surface contact	mm^2	Both	Ways	2
		E	Vibration transfer	Hz	Both	Ways	-1
		E	Heat transfer	J	Both	Ways	-1
Tyre	External Environ-ment	E	Heat transfer	J	Both	Ways	-1
		M	Debris	mm^3	Air	Tyre	-1
		M	Water (rain)	ml	Air	Tyre	-1
Rider & eBike	External Environ-ment	E	Drag due to air resistance	N	Air	Rider & eBike	-1
		E	Component of weight along gradient	N	Env	Rider & eBike	-1
		E	Heat transfer	J	Both	Ways	±1
		M	Dirt Debris	mm^3	Air	Rider & eBike	-1
		M	Water (rain)	ml	Air	Rider & eBike	-1
		P	Contact of rider & eBike with air	m^2	Both	Ways	-1

Figure 10.2 Interface analysis for external environment interfaces with eBike.

Having developed the interface analysis in an interface table, Sam asked if the team should have developed an interface matrix before developing the table. In responding to Sam's question, John said that while he thought that on this occasion, a matrix would be of limited value at the moment given the small number of interfaces, this was a personal choice. John said that the team could now develop an NCFT for the noise factors corresponding to the exchanges at the interfaces that they had just analysed. John reminded Team B members that while the interface analysis provided useful information in developing an NCFT, it was not simply a case of transposing information.

The team started to develop the first three columns of the NCFT by considering each exchange in turn and identifying noise factors associated with this exchange and documenting them in the appropriate noise factor category. The team had some discussion about the impact of steering and braking on propulsion as a Subsystem Interaction. Sarah said that the Physics of the forces acting on a bicycle while steering around a corner were quite complex and recommended that the team assume that the eBike was travelling in a straight line initially in developing their model of the eBike propulsive forces. Likewise, Sarah recommended the team ignore any braking force in developing the initial model as they were interested in assessing the effect of noise factors on eBike propulsion and braking would act to swamp the effect of noise factors that acted to reduce the overall propulsive force. At Phil's insistence, however, the team included braking in the NCFT as a Subsystem Interaction noise factor saying that in his opinion any unintended braking such as that caused by brake pad rub was a noise factor. The team duly completed the first three columns of the NCFT (Table 10.1).

Table 10.1 NCFT for Tyre–Road, Tyre–External Environment and Rider and Bike–External Environment Interfaces

Noise and Control Factor Table PROPULSIVE FORCES			
Interface	**Type**		**Units**
	Piece to Piece		
Tyre/Road	Tyre composition		?
	Tyre tread pattern		Pattern
	Tyre-road contact patch		mm^2
	Tyre width		ISO
	Changes over time/cycles		
Tyre/Road	Tyre wear		Tread
	Tyre pressure		bar
	Customer Usage		
Tyre/Road	Rider mass		kg
	eBike mass		kg
	Dirt on tyre		mm^3
	eBike Ride cycle		Standard
Rider & Bike/ External Environment	Effective frontal area & Rider posture	U	m^2
		D	m^2
	External Environment		
Tyre/External Environment	Road surface		Defined
	Weather (rain)		mm/hour
	Ambient temperature		°C
Rider & Bike /External Environment	Wind speed		km/h
	Wind direction		Deg wrt eBike
	Ambient temperature		°C
	Elevation		m
	Gradient		%
	Subsystem Interaction		
Propulsion/Braking	Frictional forces/torque		N, Nm
	Diverted Output		
Propulsion/Drivetrain	Drivetrain efficiency loss		%

The team had some discussion over how they were going to measure tyre composition without coming to a conclusion. John suggested that they move on and come back to this later. John said that the team needed to document the noise factor ranges and the effect of each noise factor next. Phil said that identifying the ranges over which noise factors typically varied was going to take a bit of research and said that he was happy to do this in conjunction with others outside of the meeting. John thanked Phil and moved on to look at noise factor effects. John said that because the table was being used to facilitate the development of a model, he suggested that they identify the model parameter or model parameters that each noise factor affected. John recommended that the team looke at the model parameters affected by noise factors within the noise factor type categories as he expected to find that several noise factors within each particular category would affect the same model parameters. The team started with the *Piece-to-Piece* noise factors and decided that since all noise factors were related to the choice of tyre, all piece-to piece noise factors affected the coefficient of rolling resistance, C_r (Table 10.2). John advised that each model parameter listed in the table was accompanied by the uppercase Greek letter Delta "Δ" to indicate that the effect of a noise factor was to cause variation in the parameter.

Table 10.2 NCFT Extract for Piece-to-Piece Noise Factors

Noise and Control Factor Table PROPULSIVE FORCES			
Interface	**Type**	**Units**	**Noise Factor Effect**
	Piece to Piece		
Tyre/Road	Tyre composition	Mfg-Hawk	ΔC_r
	Tyre tread pattern	Pattern	
	Tyre-road contact patch	mm^2	
	Tyre width	ISO	

Model Parameters

C_r = coefficient of rolling resistance

Sarah suggested that since Oxton sourced the majority of their tyres from the global bicycle tyre manufacturer Hawk, they select three different types of Hawk tyre to reflect the range of tyres they most often used on eBikes. The team agreed that Sarah's suggestion was a pragmatic way of incorporating what would otherwise be a myriad of tyre compositions, tread patterns and widths, and Phil said he could run some tests on the rolling road to determine C_r for the different tyre types. This also answered the problem of how tyre composition was to be measured. The team then turned their attention to the *Changes over time/cycles* noise factor group (Table 10.3).

Table 10.3 NCFT Extract for Changes Over Time/Cycles Noise Factors

Interface	**Type**	**Units**	**Noise Factor Effect**
	Changes over time/cycles		
Tyre/Road	Tyre wear	Tread	-
	Tyre pressure	bar	ΔC_r

Model Parameter

C_r = coefficient of rolling resistance

Phil said that tyre pressure had a marked influence on the coefficient of rolling resistance (C_r) and added that he would investigate this on the rolling road when he was looking at the different types of tyre. Turning to tyre wear, Sarah said that she had recently read an article where the effect of tyre wear on C_r was investigated and found to be minimal and added that admittedly this was based on only one test with one type of tyre. The team decided based both on what Sarah had said and the difficulty of associating tyre features and parameters with wear they would not incorporate this noise factor into the model initially. During their deliberations, John reassured the team that this was a reasonable strategy to take since what was important was that the effect of the noise factor was incorporated into the model and not the noise factor itself. John said that if there was any effect from tyre wear, it would cause variation in C_r and, given this effect was small, if they did want to include it in the model in the future, this could be mimicked by a minor adjustment to another noise factor that affected C_r such as tyre pressure. John added that the effect of tyre pressure was likely to swamp the effect of tyre wear anyway.

The team moved on to consider the *Customer Usage* noise factors (Table 10.4).

Table 10.4 NCFT Extract for Customer Usage Noise Factors

Interface	Type		Units	Noise Factor Effect
	Customer Usage			
Tyre/Road	Rider mass		kg	Δm
	eBike mass		kg	
	Dirt on tyre		mm^3	ΔC_r
	eBike Ride cycle		Standard	ΔC_r, $\Delta\theta$
Rider & Bike/External Environment	Effective frontal area & Rider posture	U	m^2	ΔAC_a
		D	m^2	

Model Parameters

C_a = drag coefficient
A = frontal area of rider and eBike
C_r = coefficient of rolling resistance
m = mass of eBike and rider
θ = angle of incline

In reviewing this group of noise factors, the team decided that some, such as rider and eBike mass could be easily incorporated into the model while others like the eBike Ride Cycle were much more difficult to include. John said that while the effect of Ride Cycle could be incorporated as a pattern of different types of road surface and gradients, he recommended that initially they keep things simple and consider a small number of different road surfaces with the bike travelling at constant velocity up different inclines. Sarah said that she would see what information there was on the effect of dirt on the tyre. Phil said that he had plenty of data on the effective frontal area from wind tunnel tests which he could use to identify appropriate magnitudes and ranges for this noise factor that could be used directly in the model and added that this data would include the effect of rider posture. Sam asked what *U* and *D* represented, and Phil explained that with dropped handlebars, they represented the two positions of the rider in holding the top of the handlebars in the upright position or the bottom of the handlebars in the dropped position. Sam thanked Phil and then observed that the NCFT did not include any control factors, and John said that this was a good point and explained that they would look at this in developing the model.

The team then began looking at the *External Environment* noise factors (Table 10.5).

Table 10.5 NCFT Extract for External Environment Noise Factors

Interface	Type	Units	Noise Factor Effect
	External Environment		
Tyre/Road	Road surface	Defined	ΔC_r
	Weather (rain)	mm/hour	-
	Ambient temperature	°C	ΔC_r
Rider & Bike/ External Environment	Wind speed	km/h	Δv_a
	Wind direction	Deg wrt eBike	Δv_a
	Ambient temperature	°C	$\Delta \rho$
	Elevation	m	$\Delta \rho$
	Gradient	%	$\Delta \theta$

Model Parameters
ρ = air density
v_a = relative velocity of wind and rider
C_r = coefficient of rolling resistance
θ = angle of incline

Sarah observed that while there was a significant amount of information on the effect of different types of road surface on C_r, the effect of rain was more complex. Sarah cited the fact that a thin film of water on the road surface would tend to reduce rolling resistance due to friction but as the depth of water increases, the tyre needs to displace the water and this would increase the rolling resistance. Sarah suggested that because the effect of friction between the road and tyre on C_r was significantly less than the effect of tyre deformation, they ignore rain as a noise factor. In looking at the effect of ambient temperature on rolling resistance, Jennifer said that she had read somewhere that a 10°C increase in temperature decreased rolling resistance by 5% which as far as the model was concerned was probably significantly less than the differences due to the types of tyre and tyre pressures and so again this noise factor could probably be neglected. Phil said that the team could verify what Jennifer had suggested once they had quantified the magnitudes and ranges of variation of all the noise factors.

John intervened at this point to recommend that as a simplification, the team only consider the wind direction to be head-on, and Sarah said that she agreed with this saying that if the wind was coming from a more side-on direction, it was the component of the wind acting directly against the eBike and rider direction of travel which would affect the propulsive force while any sideways component of the wind would tend to affect the stability of the eBike and rider and that was another matter. Sam added that the effect of wind speed and gradient could be incorporated directly into the model as these were represented by model parameters and added that ambient temperature and elevation would affect air density, another model parameter. Phil said he had used an online calculator to determine air density at different temperatures and altitudes and would use this to determine the effect of these two variables acting simultaneously to vary air density.

Looking next at the *System Interaction* noise factor, John reminded team members that they had decided not to include braking in the model and recommended that, at least for the moment, this included non-intended braking. John added that since braking would affect the overall propulsive force, they could add brackets to indicate that this noise factor would not be included in the model (Table 10.6).

Table 10.6 NCFT Extract for System Interaction Noise Factors

Interface	Type	Units	Noise Factor Effect
	System Interaction		
Propulsion/Braking	Frictional force/torque	N, Nm	(ΔFp)

Model Parameters

Fp = Repulsive force

John then brought the team's attention to the effect of Drivetrain efficiency loss as Diverted Output (Table 10.7).

Table 10.7 NCFT Extract for Diverted Output

Interface	Type	Units	Noise Factor Effect
	Diverted Output		
Propulsion/Drivetrain	Drivetrain efficiency loss	%	(ΔF_p)

Model Parameters

Fp = Repulsive force

Sarah said that she had read that at best the efficiency of a Drivetrain was over 98%, and at worst, around 80% depending on factors such as the size of the chainset chainring, cassette sprocket and tension of the chain. John said that he recommended that for the moment the team take the most optimistic case and ignore efficiency losses in the Drivetrain. John then asked Sam and Phil if they were OK to take the further development of the NCFT and the subsequent development of the model away and work on it with input from other members as and when required. Sam and Phil both agreed that they would be happy to do this. John said that he was very pleased with the progress that had been made and handed the meeting back to Phil.

Phil said that it was time for the Warm Down. Before getting closing comments from team members, Phil observed that while it had been an interesting and enjoyable meeting, he was concerned that looking at the Oe375 propulsion system model would mean that the team would not be able to fully complete the assignments that they had taken away from the last Oe375 team meeting. John said that if they were able to include discussion of the propulsion system modelling, in the next Oe375 meeting, then this would mean it would be unlikely that Team B would take any assignments from that meeting which would give the team more time to complete their current assignments. Phil commented on what John had just said by summarising it as "doing more work now in order to provide more time for the existing work" and said that this sounded paradoxical. John agreed that the way Phil had put it did make it sound illogical and added that he hoped that team members understood what he meant. John said that it would have been better if he had explained this at the end of the last Oe375 meeting but admitted that at that time he had not clearly sorted out in his own mind the next steps that he would like the Oe375 team to take. John added that it was a comment that Sam had made in Warm Down that had influenced his thinking. In smiling, Phil said that John appeared to be blaming Sam now, and in response, John said that Sam was only partly to blame and smiled. On moving into Warm Down and in giving their closing comments, team members all agreed that it had been an interesting meeting and looked forward to seeing the model in operation. On walking back to his office after the meeting, John thought that it had been an interesting and productive day.

On arriving at his office the next morning, John realised that he had become so absorbed in the development of the Oe375 propulsion system modelling the day before he had still not got anything to form the basis of the Warm Up for the next Oe375 meeting. Realising that he only had an hour before the Review/Preview meeting, he set to work and resumed his search for a suitable article. After a while, finding nothing suitable, John decided to write his own short article set in the context of the Oxton first-time engineering norm (OFTEN) framework and had only just finished it when Sam, who was conducting the Warm Up, knocked on his open door and was followed shortly thereafter by Sarah.

John began the Review/Preview meeting by asking Sam how things were progressing on the propulsion system modelling. Sam said that he and Phil had just met and developed a plan of action and although it was a bit of a squeeze, they should have the model ready to demonstrate at the Oe375 meeting, given they were setting aside the rest of the day to work on it. John said he was pleased and then asked Sam for his reflections on the last Oe375 meeting. Sam said that he thought that the programme was beginning to ramp up over the last few meetings and that this was exemplified in the last meeting in which he said he felt that the team had covered a lot of ground. John asked Sam if he could be a bit more specific, and Sam said that as he recollected the meeting, they had looked at a DFMEA extract, considered an extension to the propulsion system SSFD, function tree and function fault tree, begun Phase 3 of the OFTEN framework by looking at Current Design Controls including detection ratings and recommended actions on the design FMEA, along with noise factors and the P-Diagram. Sarah interjected to say that a section of the propulsion system NCFT was also developed in the meeting as well as looking at Attribution. John responded by saying that in hearing Sam and Sarah list everything, it was clear that they had covered a lot. John asked Sam if he thought that the meeting covered too much to which Sam replied that at the time it seemed a natural flow but as he was not currently working with Team B on the tasks that resulted from the meeting, John might do better to ask Sarah. John turned to Sarah and asked her for her thoughts. Sarah said that she had found the meeting interesting and agreed with Sam that they had covered a lot of ground, adding that she was all right with what they had done. Sarah said that now that Team B had a bit more time, they should be more than able to complete their assignments from the meeting.

Sitting back, John was silent for a moment before saying that at the time he did think that the assignment list was a tall order and had realised subsequently that the programme was getting to the stage where they needed to look at the way in which the team was working, adding that he would like to discuss this at the next Oe375 meeting. In turning to the Warm Up for the Oe375 meeting, John checked with Sam that he was still happy to conduct the Warm Up now that he would also be demonstrating the propulsion system model. Sam said he was fine about doing it and added that Phil and he were sharing the presentation of the model. John thanked Sam and said that as the People Skill he was going to discuss in the Oe375 team meeting was an aspect of the Communication Skill of Framing Information additional to the spoken word that the team had considered in a previous meeting. John added that this time, he was going to consider the written word by considering the structuring of reports and PowerPoint presentations to ensure that the information they contained was communicated effectively. John said that he had written a short article which lacked structure and said he would email it to Sam. John suggeste that Sam share it with all team members as a part of the *Self* phase of Warm Up and ask them to review it individually and think about how it might be structured to make it easier to read before getting together in pairs to compare notes.

John asked Sam what he thought of the exercise, and Sam said that the exercise sounded straightforward but that he was unclear if John was suggesting that team members structure the article or was just suggesting that they consider how it might be structured. John said that he was suggesting the latter since structuring the article would take too much time. Sam asked if he should print the article and give out paper copies, and Sarah recommended that he share it on his laptop as well as sending out a copy to team members immediately before the meeting. Sam said that he could document the suggestions for structuring the document by asking the pairs to give him their thoughts, and he would record these on his laptop. John said that this should work fine and added that he also had a document in which the article had been structured and told Sam he would send this to him as a separate file. John said that he would prefer it if Sam did not share the structured document as he was going to pick this up later in the meeting. John requested that Sam faithfully record the team members' input and not add any features explained on the structured document unless team members suggested them. Sam confirmed that he would ensure that he only recorded things that team members suggested. John then asked Sam if he had any more comments, and Sam replied that he was happy to conduct the Warm Up as they had agreed.

In considering the meeting agenda, John said that after looking at Framing Information, he would like to discuss the way in which the team might work together in the future and then consider the propulsion system model. John asked Sarah and Sam if there was anything else they should add to the agenda, and Sarah said that they needed to review the work that Team B had done on the assignments from the last meeting. John said that he had been thinking about this and said he would rather not include this on the agenda for this meeting but leave it until the next meeting as Team B would not have completed all the assignments before the meeting. Sarah said that this was fine by her and, smiling, added that she was sure the other Team B members would not have any objections either. Sam said that the draft agenda was fine by him and added that he was looking forward with some eagerness to continuing with the propulsion system model development. On that positive note, John thanked Sarah and Sam for their help and concluded the meeting.

John reflected that he was disappointed in himself in that he had not thought sufficiently ahead to realise the assignments from the previous Oe375 meeting could be spread out over a longer time. Although John appreciated that this was a relatively minor issue, he thought it could be damaging to his relationship with Team

B. In thinking back to his time in Blade, he realised that all too often projects and programmes had a tendency to take on a life of their own with the high intensity of workload giving the undertaking such momentum that he seldom had the time to take a meta position and reflect if things were progressing as efficiently and effectively as they might. John then reminded himself that being just another cog in the Blade wheel had meant that even if he had wanted to change things, he probably would have not been able to and compared this to his current position, albeit in a much smaller company, in Oxton.

In walking into the conference room, the next day for the Oe375 meeting, John was pleased that he had been able to sort out in his own mind the short- to mid-term workload process for the Oe375 programme and would be able to discuss it with the team in the meeting. As usual, Phil had arrived and set up the room, and in thanking him, John remarked that he must be an early riser. Phil agreed and explained that it went back to his days in production when his favourite shift was the early one. Sam arrived next which gave John the chance to check with him if he was still OK to take the Warm Up. Sam said that this was fine, and John confirmed with him that he had received the draft article extract in both unstructured and structured forms.

Once everyone had arrived or signed in, Sam, sitting at his table, said that he was taking the Warm Up and checked that those in the remote locations could hear and see him. Sam then said that he had just emailed a draft of an extract of an article on the OFTEN framework to team members (Figure 10.3).

The OFTEN Framework comprises of four phases: Function Analysis, Function Failure Analysis, Countermeasure Development and Design Verification. The initial two phases of the framework are based on the premise that in identifying how a system might fail it is important to first to know how the system functions. Having understood how a system functions the manner in which it fails to function can be more easily determined through the identification of potential function failure modes. Having identified potential function failure modes, the effects and causes of failure can be established and effective countermeasures developed in the third phase of the framework. In the final phase of the framework the effectiveness of the countermeasures are verified. Each of the four phases is supported by several engineering tools. Phase 1: System State Flow Diagram, Function Tree, Boundary Diagram, Interface Matrix & Table. Phase 2 DFMEA, Function Fault Tree, Phase 3: Vigilance, DFMEA, Robustness Strategies, Noise and Control Factor Table, Phase 4: Noise and Control Factor Table and Design Verification Testing. The flow of information between the tools in the OFTEN Framework is both iterative and distributary. For example, information in SSFD is used to determine both the Boundary Diagram and a Function Tree, while information in the Function Tree is used both in the initial population of the DFMEA and in developing the Function Fault Tree. It is not unusual for information to be identified in developing an DFMEA which indicates that data has been missed in performing Interface Analysis in the previous phase of the OFTEN Framework which will require the Interface Analysis to be updated retrospectively. The function-based approach to analysis within the OFTEN framework facilitates information flow and hence linkage between the engineering tools.

Figure 10.3 Extract of draft article on the OFTEN framework.

Sam explained that he would like team members to quickly read through the article and consider how it might be better structured and added that he was not asking people to edit the document. Sam added that having considered individually how the article structure might be improved, he would like team members to get together in pairs and compare notes after which he would collate the team findings. Sam said that he would designate the pairings in due course. After allowing what he judged as sufficient time for team members to access, read and consider the draft, Sam explained that he would ask John to set up three breakout rooms with an Oxton team member pairing with an external member and informed team members of the designated pairings of Andy and Jennifer, Wolfgang and Phil, and Yumiko and Jennifer. Sam then said that he would like each pair to agree how the structure and layout of the article might be improved. Sam stressed again that he did not want team members to edit the article but to suggest structural improvements, and by way of an example, he said that they might want to include a subheading at a particular point. Sam said that he would give the pairs 5 minutes to agree on a list of improvements and suggested that at least one of them keep a note of what they agreed. Sam then asked John to create the breakout rooms which he did.

After giving the pairs just over 5 minutes to complete the task and noticing that a couple of Oxton members were no longer looking at their laptop screens, Sam send a Chat message to all team members asking if they were ready to come back together. On receiving confirmation that each pair had completed the task, Sam asked John to close the breakout rooms. Once everyone was back in the meeting, Sam said that he expected that the teams had similar lists, and to minimise duplication, he asked each pair to call out in turn one item from their list which had not already been identified. Sam said he would type the items as they were called out and shared a document on his laptop. Sam quickly developed a list of team members' suggestions (Figure 10.4).

Suggestions for Improving Document Structure

Include:

- Paragraphs
- Title
- Bullets
- Numbering
- Subheadings
- Links
- Increased line spacing
- Changed font size

Figure 10.4 Team's suggestions for improving draft article extract.

Sam thanked the team members and asked them for their reflections on the exercise. Jennifer said that she found it straightforward and, in smiling, said that whoever wrote the article extract might know something about the OFTEN framework but clearly knows nothing about structuring documents. Yumiko said that her and Jennifer's lists were almost identical, and she found the exercise interesting given English was not her first language. Since no one else commented, Sam said that John was going to pick up their conclusions shortly. He then ran through the meeting purpose and agenda and handed the meeting facilitation to John.

John thanked Sam, and in picking up what Sam had said, he told team members that the list they had generated would prove useful in looking further at the People Skill of Framing Information. John reminded team members that they had considered the Framing of the Spoken Word in a previous meeting and added that he was going to look now at the Framing of the Written Word. John picked up his phone and held it up as if to take a photograph by moving the phone backwards and forwards while looking at the image. John said that in taking photographs, we all try to frame the photograph and in pretending to take a photograph of the Oxton team members, John said that he was trying to ensure that everyone was in shot. John added that professional photographers will often spend a significant amount of time composing a picture. John said that in framing a photograph, we are making sure that the picture contains all the relevant information that it should, and this information is structured to make the interpretation of the picture easier by composing the picture such that the viewer's eye is drawn to the key part or parts of the image. John said a report or slide presentation also benefits from thought put into its construction to ensure that it conveys the message that it is meant to convey in a way that makes it straightforward for the reader or viewer to assimilate the contents of the report or presentation. John continued by saying that writing an engineering report or developing an engineering-based slide presentation is like writing a story in that it should have a clear beginning, an end, and a relatively easy-to-follow and coherent storyline joining the two. John asked Sam if he would put the team's suggestions for improving the draft article extract on the team ShareDrive, which he did, and after accessing it, John shared it alongside what he said was his attempt at framing the OFTEN article extract (Figure 10.5).

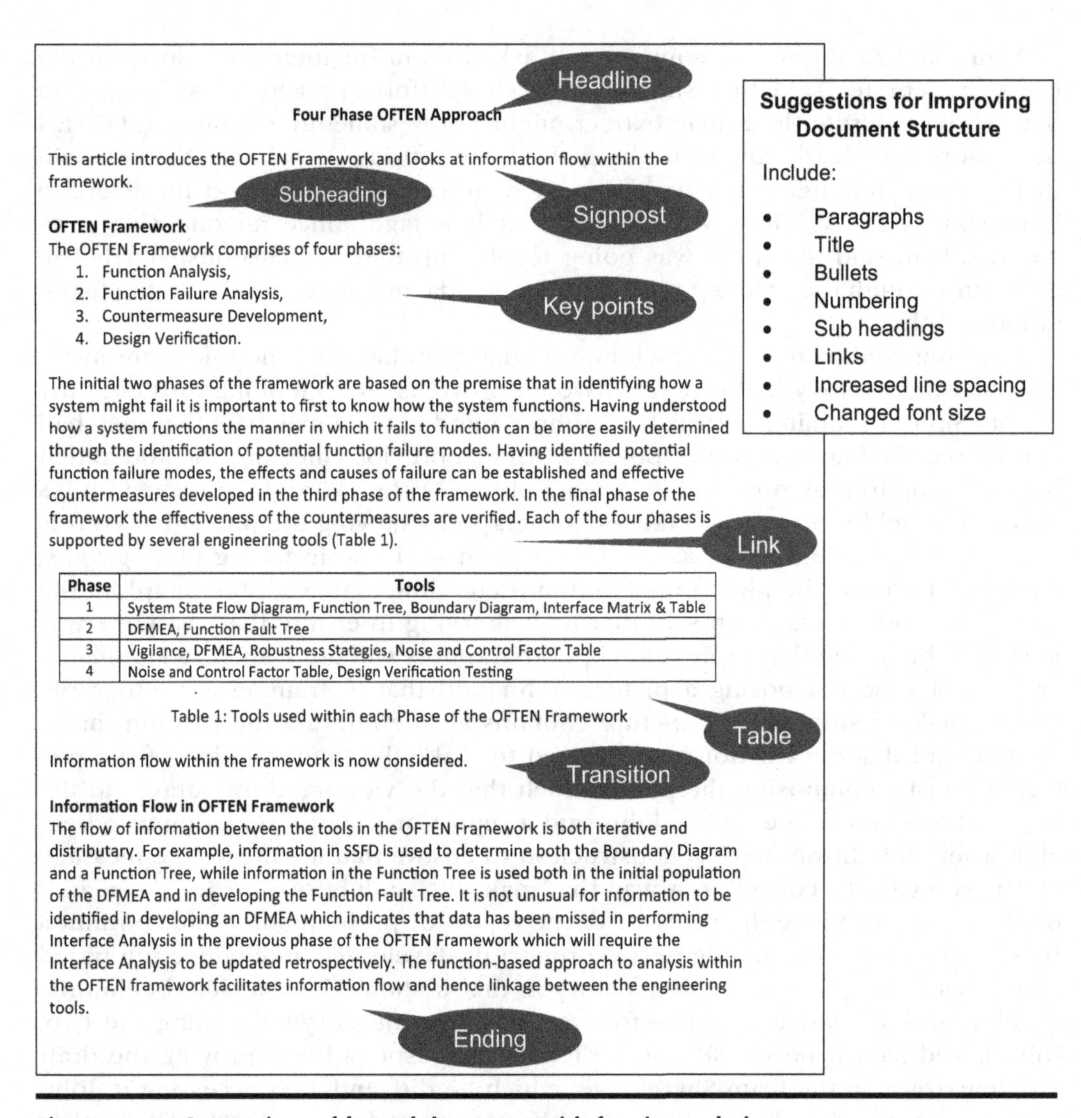

Four Phase OFTEN Approach

This article introduces the OFTEN Framework and looks at information flow within the framework.

OFTEN Framework

The OFTEN Framework comprises of four phases:

1. Function Analysis,
2. Function Failure Analysis,
3. Countermeasure Development,
4. Design Verification.

The initial two phases of the framework are based on the premise that in identifying how a system might fail it is important to first to know how the system functions. Having understood how a system functions the manner in which it fails to function can be more easily determined through the identification of potential function failure modes. Having identified potential function failure modes, the effects and causes of failure can be established and effective countermeasures developed in the third phase of the framework. In the final phase of the framework the effectiveness of the countermeasures are verified. Each of the four phases is supported by several engineering tools (Table 1).

Phase	Tools
1	System State Flow Diagram, Function Tree, Boundary Diagram, Interface Matrix & Table
2	DFMEA, Function Fault Tree
3	Vigilance, DFMEA, Robustness Stategies, Noise and Control Factor Table
4	Noise and Control Factor Table, Design Verification Testing

Table 1: Tools used within each Phase of the OFTEN Framework

Information flow within the framework is now considered.

Information Flow in OFTEN Framework

The flow of information between the tools in the OFTEN Framework is both iterative and distributary. For example, information in SSFD is used to determine both the Boundary Diagram and a Function Tree, while information in the Function Tree is used both in the initial population of the DFMEA and in developing the Function Fault Tree. It is not unusual for information to be identified in developing an DFMEA which indicates that data has been missed in performing Interface Analysis in the previous phase of the OFTEN Framework which will require the Interface Analysis to be updated retrospectively. The function-based approach to analysis within the OFTEN framework facilitates information flow and hence linkage between the engineering tools.

Figure 10.5 Comparison of framed document with framing techniques.

In comparing the two documents, John said that, while the phraseology around the document structure was not exactly the same as the team's list of suggestions, his document contained all the team's suggestions for framing apart from changing the line spacing and font size. John added that his framing included three aspects that were implicit within the original draft that he had made explicit and listed these as a beginning or *Signpost*, a *Transition* and an *Ending*. John said that there was one aspect in his framing that was not in the team's list which was the *Table*. In returning to the structural phraseology, John said that he preferred the word *Headline* to *Title* because, as he saw it, the purpose of a headline was to draw the reader's attention to the article while he saw the word title acting more as a name to distinguish one thing from another. John asked team members for their comments on the framed document, and Yumiko said the subheadings, key points and table helped her to see the overall flow of the document before she started to read the detail. Andy said that he felt the key points and tables made the document easier to read and so would help him to understand the important points without them getting lost in the detail.

In concluding the discussion, John said that, while he had spoken about the framing of documents and slide presentations, framing also applied to verbal communication in addition to the words that were spoken, adding that in his experience, few people used such framing. To illustrate the point that he was making, he said that he now wished to move on to the next agenda item and discuss with the team the way in which they might work together as they proceeded within the OFTEN framework. Before moving on to the next agenda item, John asked team members what type of framing he had just used, and Sarah said that he was signposting.

John said that the first thing that he wanted to do was to apologise again to Team B members for asking them to do a lot of work in a relatively short time when, if he had thought about it properly, there was no such a rush. Phil said that on behalf Team B he accepted John's apology and added that they did begin to think that John had caught the Oxton management bug. Out of curiosity, John asked Phil what the Oxton management bug was, and Phil replied that it affects your brain and dilates your experience of time while at the same time reducing the apparent magnitude of a task compared to its real magnitude. John smiled and said that it seemed this was a universal infection. John then explained that the Oe375 programme had reached the stage where a regular schedule of meetings started to get out of sync with what needed to be done. John said that his experience was that the relative ease of arranging meetings with meeting management software bred another type of virus which also affected the brain and makes people believe that to progress tasks all you needed to do was to organise a schedule of meetings. In explaining that while he was largely speaking tongue in cheek, John said that there was an element of truth in what he was saying and added that as he saw it there was nothing wrong with a regular schedule of meetings providing meetings were used flexibly to suit what needed to be done.

John explained that in using the Oe375 meeting schedule flexibly, he suggested that on some occasions, Oe375 team meetings could be handed over to Team B to carry on working on assignments. John said he thought it was important that, whatever task a scheduled meeting was used for, sufficient notice of a meeting purpose was given to all Oe375 team members so that those not attending a particular meeting were able to rearrange their work schedule accordingly. John asked team members for their views. Andy said that while he did not want to give the impression that he found the Oe375 meetings an unwanted chore, which was far from the truth, he had no objections to missing some meetings if he had prior notice. Wolfgang agreed with Andy and said that he had more than enough work to do and would welcome what he described as some holes appearing in his calendar when Oe375 meetings became Team B meetings. Sarah said she thought that using the scheduled meetings flexibly was a good idea and suggested that at the end of each Oe375 team meeting immediately before Warm Down, when they were looking at Team B assignments, they agree when the next Oe375 team meeting would be held. Wolfgang said that he would keep the complete schedule of meetings in his calendar and would appreciate knowing as soon as possible of any meetings that were to be Team B meetings. In looking at Jennifer, Sarah said that she would ensure that this was the case. The other team members agreed with Sarah's suggestion, and in concluding this discussion, John said that he would like to attend all meetings currently designated as Oe375 team meetings and would act as a coach in any of these meetings that were reallocated to Team B. Phil said that John was of course welcome at any of Team B meetings whenever they were scheduled. John thanked Phil and suggested that the team move on to the next agenda item.

Phil took over the meeting and said that he wanted to discuss the development of a model for the Oe375 propulsion system and added that he was signposting which drew smiles. Phil shared the sketch similar to the one that John had drawn, although John thought that it was somewhat clearer and more detailed than his original version (Figure 10.6).

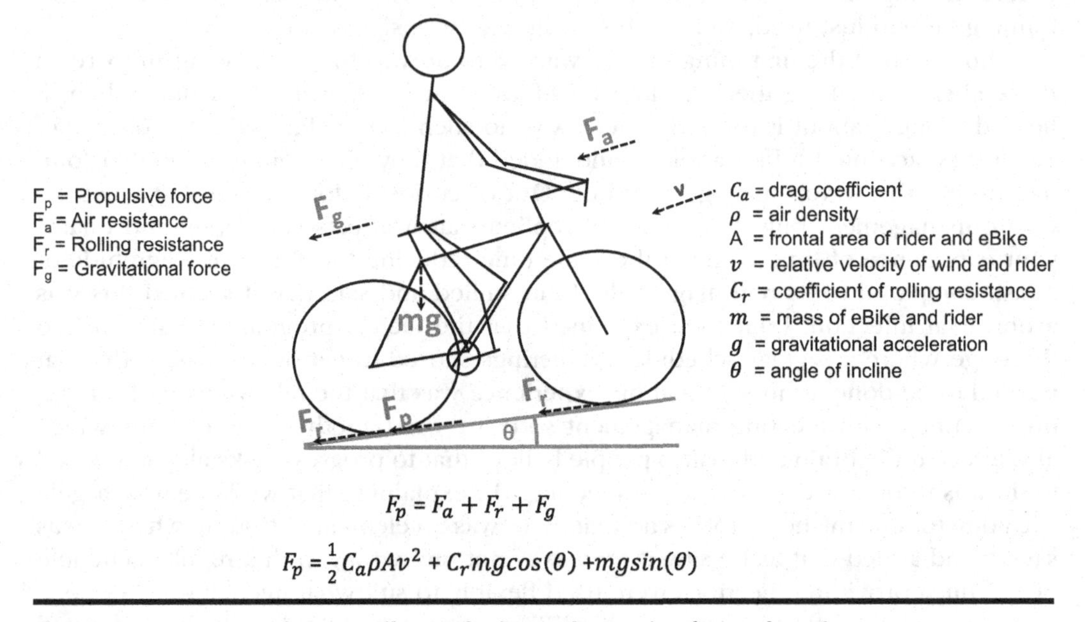

Figure 10.6 Forces acting on eBike producing and opposing forward motion.

Phil explained what the forces were and said that when the propulsive force was equal to the sum of the forces opposing motion, the eBike would travel at constant velocity and pointed out the equation beneath the sketch saying that it was a standard equation based on the Physics of the situation. In drawing team member's attention to the right of the sketch, Phil defined the parameters in the equation and added that this was a mathematical model of the eBike forward motion. Phil then said that if any or all of the parameters on the right of the sketch vary, this would cause variation in the forces opposing forward motion of the eBike which would result in the eBike going faster or slower for a given power input from the motor and rider depending on whether the sum of the opposing forces decreased or increased. In reminding team members that noise factors tend to disrupt the performance of a system, Phil explained that the effect of noise factors on eBike performance could be quantified if their effect on the parameters of the eBike model was known and reiterated that the eBike performance they were considering was its forward motion. Phil said that team members would recall that noise factors affect a system through its interfaces and for this reason Team B performed interface analysis on the interfaces of interest and listed these as the tyre/road, tyre/external environment and rider and bike/external environment and added by "Road" the Team B meant the surface on which the eBike was ridden. John was impressed by Phil's description and remembered that it was only a couple of days ago that he had said that he did not see how a model could be used to assess the effect of noise on eBike performance.

Phil paused and asked if the team members had any questions or comments on what he had said. Wolfgang effectively asked the same question that Sam had asked John in the Team B meeting in saying that as he understood it, within the OFTEN framework approach, a system SSFD and boundary diagram should be developed before conducting interface analysis. In response, Phil said that John had recommended that for an existing system which is well-understood you can develop a boundary diagram directly and added that the sketch he was sharing was in effect a boundary diagram. Phil looked towards John, and John assuming that Phil was wanting him to agree with what he had just said explained that while an SSFD could be developed in this case, it was unlikely to yield any new knowledge. As no one else spoke, Phil continued by saying that having conducted the interface analysis, Team B then identified the noise factors affecting the eBike motion along with their ranges and control factors and documented these in an NCFT which he shared (Table 10.8).

Table 10.8 NCFT of Noises Acting to Affect the eBike Propulsion System

Noise and Control Factor Table PROPULSIVE FORCES								Control Factor		
Interface	Type		Units	Range			Noise Factor Effect	Type of tyre	Tyre pressure	Propulsion system model
				Min	Max	Normal				
	Piece to Piece									
Tyre/Road	Tyre composition		Mfg-Hawk	Hawk HS	Hawk RS	Hawk PA				
	Tyre tread pattern		Pattern	As above			ΔC_r	X	X	X
	Tyre-road contact patch		mm^2	As above						
	Tyre width		ISO	28-622	47-662	37-662		X	X	
	Changes over time/cycles									
Tyre/Road	Tyre wear		Tread				-			
	Tyre pressure		bar	3.0	5.0	4.0	ΔC_r	X	X	X
	Customer Usage									
Tyre/Road	Rider mass		kg	50	90	75	Δm	X	X	X
	eBike mass		kg	20	25	22.5		X	X	X
	Dirt on tyre		mm^3	None	High	Low	ΔC_r			
	eBike Ride cycle		Standard				ΔC_r, $\Delta\theta$			X
Rider & Bike/ External Environment	Effective frontal area & Rider posture	U	m^2	0.250	0.347	0.285	ΔAC_a			X
		D	m^2	0.228	0.317	0.260				X
	External Environment									
Tyre/Road	Road surface		Defined	F gravel	C gravel	Asphalt	ΔC_r	X	X	X
	Weather (rain)		mm/hour				-			
	Ambient temperature		°C	0	35	15	ΔC_r	X	X	
Rider & Bike/ External Environment	Wind speed		km/h	0	25	15	Δv_a			X
	Wind direction		Deg wrt eBike	0	180	All	Δv_a			X
	Ambient temperature		°C	0	35	15	$\Delta\rho$			X
	Elevation		m	0	1500	200	$\Delta\rho$			X
	Gradient		%	0	6	3	$\Delta\theta$			X
	System Interaction									
Propulsion/Braking	Frictional force/torque		N, Nm				(ΔF_p)			
	Diverted Output									
Propulsion/Drivetrain	Drivetrain efficiency loss		%	2	20	5	(ΔF_p)			

Phil explained that the *Noise Factor Effect* column in conjunction with the Control Factor of ***Propulsion System model*** indicated which model parameter each noise factor affected and hence how the effect of each noise factor would be incorporated into the model. Phil added that the propulsion system model was seen as a control factor as it would enable the team to ascertain the effect of each noise factor on eBike propulsion and by adjustment of parameters within the model reduce the effect of noise. Phil cited the example of finding the optimum tyre pressure or reducing the contribution of the frame to the effective frontal area. Wolfgang asked what "F gravel" and "C gravel" were as the minimum and maximum road surface, and Phil explained that these terms stood for fine gravel and coarse gravel, respectively.

Phil brought team members' attention to the fact that some of the noise factors were highlighted by the light shading and slightly larger font. Phil explained that while some noise factors affected a model parameter directly, for others, the relationship between the noise factor and a model parameter was more indirect and added that he had highlighted those noise factors where the relationship was a bit more complex. In continuing, Phil said that he would not insult the intelligence of team members by going through the straightforward examples save to illustrate what he meant and cited ***eBike mass*** as directly quantifiable. In looking at ***Tyre composition, Tyre tread pattern*** and ***Tyre–road contact patch***, Phil said that Sarah had suggested that instead of attempting to quantify these directly that they chose three tyres that would represent the range of tyres that Oxton would typically use on an eBike. Phil added that the composition of tyres and tread patterns varied greatly both within and across tyre manufacturers and said that any attempt to include this directly as a noise factor would prove extremely difficult because manufacturers did not publish the specifics of the composition of their tyres. Phil added that it was also the case that the relationship between a particular composition and the tyre's properties was closely guarded by tyre manufacturers. At this point, Andy reflected that his experience of attempting to get such information from a particular tyre manufacturer was like trying to get blood out of a stone and added that he wondered at times whether tyre design was more of a black art than a science, which raised smiles.

Phil continued by saying that he had selected two tyres of each type documented on the NCFT and determined C_r for each type through performing tests on the Oxton rolling road. Phil explained for the external team members' benefit that the Oxton rolling road was a drum which rotates at constant speed with a wheel and tyre sitting on top subjected to a load to simulate the effect of the weight of the rider and eBike. Phil said that he had performed the tests at the different tyre pressures documented on the NCFT so enabling the effect of tyre pressure to be included in the model and shared the results of the testing of one type of tyre (Table 10.9).

Table 10.9 Rolling Road Test Results

Tyre	*Hawk PA*		
Tyre Pressure/bar	3.0	4.0	5.0
Rolling Resistance/W	25.00	22.00	20.10
Cr	0.0075	0.0066	0.0060

Wolfgang asked how the test took account of the potential rise in temperature of the tyre during the test, and Phil replied that the test equipment was in a temperature-controlled area and each tyre was allowed to warm up for 20 minutes before the test results were taken. Phil then drew team members' attention to the *eBike Ride Cycle* noise factor on the NCF table and said that while the model allowed for different road surfaces and gradients to be inputted, he admitted that this was a simplification in modelling a typical Ride Cycle. Phil added that there were other simplifications in the model and cited the example that the model did not take account of the effect of rider mass on tyre pressure. In considering the noise factors of *Effective Frontal Area* and *Rider Posture*, Phil said that team members would know that these two factors are often considered together as *Effective Frontal Area (EFA)* because of the difficulty of measuring them independently and added that EFA was the product C_aA. Phil said that he had determined the values for EFA shown in the NCFT from previous wind tunnel data that Oxton had and added that as members will have probably realised "U" in the table meant the Upright riding position with dropped handlebars while "D" stood for the Dropped position.

Phil then turned to the last two highlighted noise factors in the NCFT of *Ambient Temperature* and *Elevation*. Phil said that since the principal effect of increasing these noise factors was to decrease air density, they had included the effect of both of them in the model by developing a look-up table (Table 10.10) which was used by the model to determine air density for particular values of altitude and ambient temperature.

Table 10.10 Look-up Table for Air Density in kgm^{-3}

	Temperature/°C		
Altitude/m	*5*	*15*	*10*
0	1.269	1.224	1.146
200	1.245	1.203	1.122
1500	1.097	1.058	0.986

Phil explained that he had developed the look-up table from an online calculator which assumed that atmospheric pressure was standard. In looking towards John, Phil then said that John had told Team B that what was important was to include the effect of each significant noise factor in the model which did not mean that each noise factor had to be included independently. Phil asked if there were any questions or comments before he handed over to Sam whom he said was going to show the team the model. Yumiko asked why some of the noise factors in the NCFT had no relationship to a control factor. In thanking Yumiko for her question, Phil said that this was something that he should have mentioned and in looking at the three noise factors of *Tyre Wear, Dirt on tyre* and *Rain*, he explained that while Team B members could not find anything definitive about the effect of these noise factors what they could find seemed to indicate that their effect was minimal.

Wolfgang asked Phil if he would explain his statement that it was not necessary to include the effect of every noise factor in the model. Phil said that what he believed he said was that the effect of each noise factor did not need to be included independently in the model. Phil shared the NCFT again (Table 10.8) and said that the effect of tyre pressure on the coefficient of rolling resistance was significant and cited the results that he had given in Table 10.9. In doing some quick mental arithmetic, Phil said that C_r decreased by approximately 30% for a doubling of tyre pressure, and added that on the other hand if ambient temperature doubled C_r would typically decrease by 5% so that the effect of temperature increase could be mimicked in the model by increasing the tyre pressure by a further small amount. Phil explained that the mimicking of one noise factor with another noise factor with the same effect had to be done within careful limits and added that if all noise factors that decreased C_r were mimicked by tyre pressure, then the tyre might burst which generated smiles.

Andy said that he assumed that the effect of ambient temperature on C_r was not included as a noise factor in the model for the reason that Phil had just given. In telling Andy that he was correct in his observation, Phil added that Team B were also aware that tyre–road contact patch varied with tyre diameter but, as they believed that this effect was small compared to tyre pressure, they had not taken it into account. Wolfgang asked why there were no control factors indicated for Braking Resistive force and Drivetrain efficiency loss, and Phil replied that Team B had decided not to include these in the propulsion system model at this stage. Since there were no more questions, and on looking at the time, Phil suggested the team take a break. As Sam was leaving the room, John managed to catch him and told him how much he enjoyed his Warm Up and added that he was impressed with his facilitation. Sam thanked John for his feedback and then continued on his way out of the room.

When all team members had returned, Sam started his presentation of what he called "The eBike Noise Effect Simulator" by thanking Phil for explaining how the effects of noise had been modelled. Sam then explained that with Phil's help, he had used proprietary software for model-based design to develop a user interface and model to simulate the effect of noise on the propulsion system. Sam then shared the user interface which he described as the simulator dashboard (Figure 10.7).

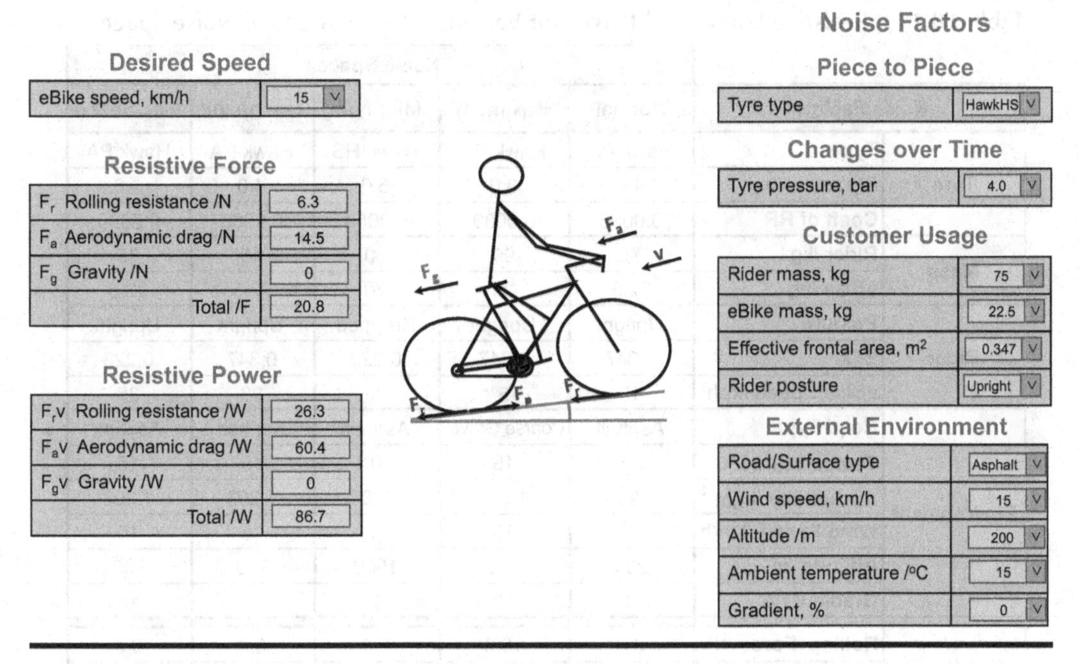

Figure 10.7 eBike propulsion system noise effect simulator dashboard.

Sam explained that the magnitudes of the noise factors on the right-hand side of the simulator were selected from drop-down menus which listed the three levels of the noise factors shown on the eBike propulsion system NCFT (Table 10.8). Sam continued by saying that once the desired speed of the eBike had also been selected from another drop-down menu on the left-hand side of the simulator, the resistive forces and resistive powers were calculated by the model and displayed beneath the desired speed. Sam then shared a table of the results that he had obtained by using the simulator for various combinations of noise factor magnitudes (Table 10.11).

Table 10.11 Resistive Forces and Power for Various Combinations of Noise Space

Noise Factors		Normal	"Maximum"	Minimum	Reasonable	Max Speed
		Noise Space				
Tyre	Type	Hawk PA	Hawk RS	Hawk HS	Hawk PA	Hawk PA
	Pressure /bar	4.0	3.0	5.0	4.0	4.0
	Coeff of RR	0.0066	0.0089	0.0050	0.0066	0.0066
Mass	Rider /kg	75	90	50	75	75
	eBike /kg	22.5	25	20	22.5	22.5
Rider	Posture	Upright	Upright	Dropped	Upright	Upright
	EFA	0.347	0.347	0.228	0.347	0.228
	eBike Speed /kph	15	5	5	20	25
Environment	Road Type	Asphalt	Coarse Gravel	Asphalt	Asphalt	Asphalt
	Temperature / °C	15	15	0	15	15
	Air Density /kgm^{-3}	1.203	1.293	0.986	1.203	1.203
	Wind Speed /km/h	15	15	0	15	15
	Altitude /m	200	200	1500	200	200
	Gradient /%	0	10	0	3	0
Resistance	Rolling Force /N	9.5	15.0	3.4	6.3	6.3
	Drag Force /N	15.6	6.9	0.2	19.7	25.8
	Gravity Force /N	0.0	112.3	0.0	28.7	0.0
	Total Forces /N	25.1	134.2	3.6	54.7	32.1
	Rolling Power /W	39.5	20.8	4.8	35.1	43.8
	Drag Power /W	64.9	9.6	0.3	109.6	178.9
	Gravity Power /W	0.0	155.9	0.0	159.3	0.0
	Total Power /W	104.4	186.3	5.1	304.0	222.7

In reviewing the table of results, Sam drew team members' attention to the fact that the column of noise factor levels headed "Normal" corresponded to the data shown on the simulator dashboard (Figure 10.7). Sam then showed the simulator in operation by inputting the noise factor levels from the column headed "Maximum" and demonstrated that the simulator outputs of resistive forces and power corresponded to the values given at the bottom of this column.

Sam then asked if there were any questions or comments, and Yumiko said the simulator was impressive and looked easy to use, which prompted a murmuring of agreement from others. Jennifer asked Sam why the column labelled "Maximum" did not show the maximum of all the total forces and power given on the table, and Sam pointed out that the eBike speed was 5 km/h for this column. Jennifer said she understood and observed that riding at the maximum speed of 25 km/h up a 10% incline against a wind of 15 km/h on a coarse gravel path was hardly realistic. In response, Sam changed the input eBike speed to 25 km/h on the simulator which showed that the total resistive power was 1076W. Sam said that out of interest, he had looked up the fastest time for a professional cyclist riding up Mount Ventoux, the famous Tour de France climb, which corresponded to an average speed of around

20 km/h and added that Mont Ventoux has an average gradient of approximately 7.5%. Sam said that he put typical pro-cycling data into the simulator with this gradient and bike speed along with a wind speed of 20 km/h which he said was modest for Mont Ventoux, adding as an aside "as the name implied". Sam continued by saying that with this input data, the resulting total resisting power was about 400W which if compared to the 1076W that they had just calculated demonstrated that the simulator had to be used with a dose of common sense. Sam said that if he were riding up this mountain, he would want an eBike even if he shaved his legs, which prompted laughter in the Oe375 team. In explaining, what he called, the other *Noise Space* conditions, Sam pointed out that the column labelled *Max Speed* corresponded to a zero gradient, with *Reasonable* being a more realistic challenge and *Minimum* being the least challenging combinations of noise factor magnitudes. Sam asked for questions and comments. Andy asked Sam what he meant by the term *Noise Space*, and Sam said it was a term that he had heard John use that he understood to mean the noises that a design would likely to be subjected to. John intervened to say that he liked Sam's definition and added with a smile on his face that he used the term because it sounded more technical than it was.

Andy then asked Sam how he and Phil knew that the noise factor levels represented the maximum and minimum levels that an eBike rider would meet, and in response, Sam said that they were not meant to be definitive but were their best estimates based on experience and some research. Andy thanked Sam for his reply and followed up by asking if he and Phil had any data to support the combinations of noise factor levels they had inputted to the simulator. This time Phil answered the question by saying that the combinations of levels were again their best estimates. Wolfgang said that as he saw it, the table of results only confirmed what every bicycle manufacturer already knew that aerodynamic resistance becomes the dominant resistive force at higher speeds with the gravitation force becoming dominant over rolling resistance for steeper inclines. Phil said while the table of results did confirm what bicycle manufacturers already knew, he saw this as a good thing and added if the results did not confirm what was already known, the Oe375 team would not have a lot of confidence in the simulator. Phil then explained that possibly what every bicycle manufacturer did not know was the relative influence that individual noise factors had on eBike performance in a particular set of circumstances. Wolfgang conceded the point that Phil was making, and after a pause, in which no one said anything, Sam handed the meeting back to John.

John said that he was very impressed with the work that Sam, Phil and Team B had done and thought it deserved a round of applause which was duly forthcoming. In picking up Andy's question, John said that it was unlikely with any system that the worst combination of noise factors would occur simultaneously and, in many cases, as Sam had ably demonstrated, the theoretical worst-case combination was not necessarily realistic. John continued by saying that one thing that had struck him in listening to Phil and Sam was that the Oe375 team probably did not have a good enough understanding of the various conditions under which the Oe375 eBike would be used and asked if other team members thought the same. Sarah said that she was sure that John was correct in his assumption and added that although PD teams in Oxton learnt from the experience of the things that had gone wrong with previous models, the company did not do a lot of upfront market research to understand how customers used their lower-end products. Jennifer joined the discussion by saying that she thought that the Oe375 team need to do some market research, adding that they probably should have done this earlier. In response to Jennifer's suggestion, John said that, although he thought it was a good idea, it would probably be difficult to do at the moment, thinking to himself as he spoke about the current financial situation in Oxton.

Wolfgang intervened to say that he thought he might be able to help and explained that Teutoburg had a web-based survey which was used by potential customers to help them select the most appropriate type of bike or eBike for their needs. Wolfgang continued by saying that the survey asked potential customers a series of largely multiple-choice questions that probed the ways in which they saw themselves using a bike or eBike before an algorithm pointed them to a small number of alternative options based on their answers. As he was speaking, Wolfgang accessed and shared the survey, pointing out the link on the Teutoburg home page and giving an overview of the way the survey was used. Wolfgang said that while he was not sure how long the company kept the survey data, he was sure it was kept for several months to allow customers to return to a partially complete or completed survey. John said that what Wolfgang had described sounded interesting and asked Wolfgang firstly if Oxton might have sight of the data and secondly what form the data took. In answering John's first question, Wolfgang said that Teutoburg would clearly not share any personal customer details, but he would enquire what might be done. Wolfgang answered John's second question by saying that as far as he knew, the data was stored as raw answers to individual questions and added that, if the Oe375 team wanted to add additional questions, he felt sure that this would be considered, as such questions would be likely to improve the survey. Wolfgang said that he would get back to John within the next few days. John thanked Wolfgang for his suggestion and said that more objective data would only improve the use of the noise effect simulator and said that he was sure it would prove an invaluable tool.

John then said that he needed to bring the meeting to a close through Warm Down. John reminded people about what he had said earlier about this meeting not resulting in any assignments and the next meeting looking at robustness strategies. The team then went into 'pass the pen' as a matter of course with the feedback being very positive and members saying that they had found the meeting very interesting and were looking forward to the next meeting.

In walking back to his office, John was pleased with how the meeting had gone and felt that he had at least repaired some of the damage his overloading of Team B might have caused. On getting back to his desk and reviewing his email inbox, he saw an email from Jim Eccleston entitled Company Finances. On opening the email, John saw it was written in typical Jim Eccleston style being brief and to the point. Jim said that as company employees might have heard or read in the press, Oxton Cycles was not currently in a very healthy financial state. Jim explained that to retain job security, he was taking several actions including cutting back significantly on overtime and travel. Jim said that, in consultation with Oxton management, he had also decided to postpone the start of the design and development programme for the next off-road bike code named O442. Jim continued by saying that he was implementing a plan to improve the company financial position but that this would not bear fruit until the medium term, and in the meantime, he apologised for any hardship his measures might cause and hoped that employees would bear with him and the company. Although bringing bad news, John was impressed by the forthright nature of Jim's message and his recognition that a reduction of overtime was likely to have a detrimental impact on some employees.

On returning home that evening after what he thought had been a productive day, John was greeted by Jane who asked him if he had spoken to Lucy Collins. John said that he had not received a call from Lucy, and Jane explained that Lucy had called on the landline and had said that she had something interesting to tell him. John said he would call Lucy back shortly, and Jane advised him Lucy had said she was going out and that she was off to Belgium for a couple of days and would ring him on Friday after work which left John wondering what Lucy's news might be.

Background References

Framing Information	Syer, J. and Connolly, C. (1996) *How Teamwork Works: The Dynamics of Effective Teamwork*, McGraw-Hill Pages 243–244
Bicycle propulsion model	Martin, J., Milliken, D., Cobb, J., McFadden, K. and Coggan, A. (1998) Validation of a Mathematical Model for Road Cycling Power, *Journal of Applied Biomechanics*, 14(3), 276–291, https://doi.org/10.1123/jab.14.3.276 Pages 278–281

Chapter 11

Preventing Failure

John arrived in his office the next day to see that Jim Eccleston had set up another urgent meeting in the space of a few days although this time he saw that he was the only invitee. As John had an hour before the meeting, he busied himself by working on the technical content of the next Oe375 meeting. John arrived at Jim's office a couple of minutes early for Vicky Harrison to tell him that Jim had popped down to see Numa McGovern in HR and would be back shortly. Jim arrived some 10 minutes later and in apologising to John for keeping him waiting ushered him into his office shutting the door behind him. Unlike their one-to-ones, where Jim always sat down at his table with John, this time Jim sat at his desk and beckoned John to sit on the seat facing him.

In typical Jim style, he got straight to the point by telling him that Paul Clampin was leaving Oxton and explained that he had been offered another job as CEO of a furniture manufacturer in, what Jim described as, "the dark side" and John took to mean the other side of the Pennines. Jim said that, given the current financial position of the company, he was not intending to replace Paul, at least for the moment. Jim explained that he was proposing to cover Paul's workload between himself, John and Mo Lakhani and asked John for his thoughts. John, in giving himself space to think through his response, asked Jim if Paul's intended departure was a surprise to him. Jim said he would like to get John's thoughts on taking on more workload and responsibility and John replied that he would like to know what extra workload he had in mind before answering his question directly. Jim said that he was thinking of John taking over one of Paul's current programmes and managing some of his people. John asked Jim which programme and people he had in mind and Jim responded that he was considering handing the O254 hybrid commuter budget bike to him and those of Paul's people working on the Oe375 programme which would be Sarah Jones and Jennifer Coulter. Jim added that the O254 team included some good people that he thought John would get on well with. John said that he thought that Jim's suggestion was a good fit given the similarities between the Oe375 and O254 programmes and the fact that he knew Sarah and Jennifer well. Jim said that he was hoping that John would say that, although he pointed out that being a person's manager puts a different perspective on things.

DOI: 10.4324/9781003286066-11

John asked Jim when Paul was leaving, and Jim responded by saying that he was leaving at the end of the week. John's reaction was one of surprise and remarked that this seemed rather sudden. Jim said that Paul had only told him yesterday and added that while Paul was due to work two months' notice, he had mentioned during their conversation that his new company were looking for him to start as soon as practicable. Jim added that while Paul's new appointment was dependent on a satisfactory reference from himself this would not be a problem. Jim said that while he would normally have expected a handover period, in Paul's case this might do more harm than good. John was not sure what Jim meant by this last remark but decided not to pursue it. Jim then puzzled John a bit more when he asked him not to make contact with the O254 team or say anything to Sarah and Jennifer, adding that he would like to introduce John to the O254 team and meet with Sarah, Jennifer and him first thing on Monday morning. Jim added that he would ask Vicky to set up a meeting with John, Sarah and Jennifer to that effect. In looking at Paul's calendar Jim saw that the O254 team had a meeting on Monday afternoon and said they should use that to introduce John. John said that Jim's proposal sounded fine to him, and, after a pause, Jim said that having an afternoon meeting with O254 was leaving things late and he was sure that what he called the Oxton rumour mill, would be in full swing long before then. Jim said that he would send out a brief companywide email over the weekend announcing Paul's departure for pastures new and saying that he would meet with those directly affected by Paul's move on Monday. Jim then suggested to John that he should speak to Numa McGovern before the end of the week to discuss his new responsibilities in managing Sarah and Jennifer and added that Numa was aware that Paul was leaving. In closing their meeting Jim asked John not to say anything about Paul's departure to anyone for now. Jim thanked John for his time and got up and opened the door for him and asked Vicky if she would come into his office.

Yet again John found himself walking away from a meeting with Jim feeling as if he had been hit by a whirlwind. Despite the suddenness and unexpected nature of what Jim had said John felt excited. On getting back to his office with a cup of coffee and sitting at his desk John decided that he was more than ready for the new challenges Jim had put his way. He had been very pleasantly surprised both by the way that the Oxton members of the Oe375 team had taken to the Oxton first-time engineering norm (OFTEN) framework and the way they had got on with programme work and challenges largely on their own without him providing any major support between team meetings. While this had left him with the unexpected benefit of space to think and read more about the bicycle industry, the People Skills and right first-time design, he looked forward with some relish to working at the pace that he was accustomed to at Blade. He was also looking forward to using the People Skills in the context of managing people, where in his experience they came into their own. John was puzzled, however, by Jim's apparent reluctance to acknowledge Paul's departure in the normal open way and his comment about a handover with Paul doing more harm than good. In thinking that Jim was hinting at Paul's character and way of working he could only conclude Jim had thought that in any extended handover period Paul might attempt to sabotage some of the work he was leaving behind. John concluded that at a minimum Paul would not want to make things easy for his successor.

John had arranged a meeting with Sarah and Phil that afternoon to let them know of the outcome of his meeting with Margaret Burrell over design specifications. Given Jim's wishes, John decided that he would not mention Paul Clampin's imminent departure but see if Sarah knew anything about it. John then opened Numa's calendar and, finding that she did not have any free time for a meeting this week, he gave her a call to see if there was a possibility of fitting in a meeting with her in the next few days. Numa answered the phone in her usual cheery manner and said that she had been expecting John's call and said that she assumed that John would like a chat with her. John said that he would like to meet with her, and Numa suggested, since she was quite busy over the next few days, that they had a working lunch that day. Numa told John not to worry about bringing food as she would organise some sandwiches and asked if he would prefer coffee or tea. John said he preferred tea at lunch time and Numa asked him if he could spare half an hour at 12:30. After John confirmed that he could make that time, Numa said she was looking forward to seeing him and finished the call. As he was putting his phone down John realised that he had not mentioned the reason for the meeting but assumed that since Numa had said that she was expecting him to call she knew what he wanted to talk about.

John then turned his attention to working on a plan for taking Oe375 forward, now knowing that he would have less time available to guide the team on a step-by-step basis. John arrived at Numa's office spot on 12:30 and knocked on her open door. She got up from her desk and invited him in and directed him to sit on one of her two comfortable chairs and closed the door behind him. John noticed the plate of sandwich rounds cut into quarters covered in cling film along with a small bowl of fruit, a teapot, cups and a jug of milk on the low table between the chairs. After sitting down with John, Numa took the cling film off the sandwiches and offered John one of the two paper plates which were underneath the sandwich plate. Numa said that she was not sure whether John preferred brown or white bread, so she had ordered a mix. John spotted his favourite brown bread and cheese and opted for that initially. Numa told him to take a few sandwiches since they were of, what she described as, a size where they disappeared in a couple of bites. Having taken several sandwich sections herself, Numa asked John how she could help him.

John said that he wanted to discuss the mechanics of taking responsibility for the supervision of Sarah Jones and Jennifer Coulter. Numa replied that she knew Jim was proposing that John took over a couple of Paul Clampin's people, but she did not know who until John had just told her. Numa said that in terms of the mechanics she would make the necessary changes to her database and added that was of course the easy part. Numa said that she would like to discuss the personal development of Sarah and Jennifer with him, and John asked Numa to explain further. Numa said that it was important to her that Oxton managed its employees in a way which furthered their personal development to make sure that the company both made the best use an employee's talents and helped the employee to realise their personal goals. Numa said that a critical part of this process was the 360-degree appraisal process that she was trying to introduce in the company and asked John if he was familiar with this process. John said that he knew something about it since its introduction in Blade had been mooted at one time but never happened. Numa reminded John that, as the name implied, it meant an individual getting feedback on a yearly basis not only from their management but also from their peers and subordinates.

Numa said that while the 360-degree appraisal process was working well in some areas of the company, in other areas it was not used, adding that Paul was vehemently opposed to implementing it in his area. Numa said that she felt it unfortunate that Jim had not fully backed it and had left it to individual areas to decide for themselves whether to use it. John said that he was very happy to use the system and asked Numa if she had any more information on it. Numa responded that the company that supplied the questionnaire software had a reasonable online training course and she said that she would send John the link. John asked if he could see Paul's latest appraisals for Sarah and Jennifer, and Numa replied that while he would have to ask Sarah and Jennifer for their agreement, the appraisals would not help him much adding that he would be better off starting from a clean slate. John asked Numa what she meant, and she replied that Paul only paid lip service to the appraisal process. Numa continued by saying that Paul seemed to use a standard template for all his direct reports and added that Paul's appraisals were completed in a way which was not sufficiently negative for any appraisee to have grounds to challenge it but at the same time it did nothing to advance their personal development. Numa, in changing tack to the new situation, said that she was pleased that John was taking over some of Paul's responsibilities and while she was aware that John was working closely with Sarah and Jennifer on Oe375 she recommended that he leave it a few months before triggering the appraisal process to allow the change in their supervision to settle down.

John asked Numa what was happening about a card and a leaving present for Paul. Numa said that she had discussed this with Jim, and he had told her that Paul had said that he preferred to leave without any fuss. Numa continued by saying that given the imminence of Paul's departure, she thought that it would be difficult for the company to arrange anything and added that, while most people will be pleased to see Paul on his way, she thought it unlikely that many would want to contribute to a leaving present. John was rather surprised at Numa's candour but thought that she was only reflecting the reality of the situation. Numa then changed the subject and she and John spent the rest of their meeting finishing off the sandwiches, drinking their tea and talking more generally of John's experiences so far of working at Oxton. Numa concluded the meeting by thanking John for their discussion and wishing him good luck in his widened role and invited him to take a piece of fruit with him. John selected a red apple, and in thanking Numa, left.

On returning to his office John saw that he had a voicemail message from Mo Lakhani in which Mo asked John to call him back at his earliest convenience. John called Mo who picked up his phone and thanked John for responding quickly. Mo began by saying that, while Jim had asked him not to speak about it until next week, he assumed that Jim had also told John about Paul Clampin's situation. John confirmed that this was the case and said he was under the same verbal embargo as Mo. Mo explained that he would need to cancel their update meeting on Monday to discuss Oe375 progress for, what he said was, obvious reasons. John said that he understood Mo's position and he expected to be equally busy next week. Without any prompting from John, Mo said that he had been expecting this to happen for some time given that most people in Oxton had worked out Paul's modus operandi and Paul's track record of not staying with a company for any length of time. Not wanting to prolong the discussion John just responded to Mo with the one word "interesting" and said that he had to go.

John needed to clear his head before meeting with Sarah and Phil. Since he had worked over lunch and the sun was shining, he decided to take a brief walk in the Oxton site grounds. Duly refreshed after his walk John returned to his office to find Sarah and Phil waiting outside. John apologised for keeping them waiting and Phil replied that Sarah and he were a couple of minutes early. Opening his office door John beckoned Sarah and Phil inside and after entering they seated themselves at John's table. After shutting his door John joined them. John gave Sarah and Phil a brief summary of the outcome of his meeting with Margaret Burrell which he said he found to be very constructive. He confirmed that as far as the Oe375 programme was concerned in procuring parts for the eBike Margaret's department would take any engineering specification as read. The discussion then strayed to the work that Phil had been doing with Sam on the Oe375 propulsion system model and the general perception that the current eBike was underpowered. John said that if Oe375 performance was to meet customer expectation then, amongst other things, the propulsion system as a whole would need to be optimised and added that this could only be achieved if the team had a clear understanding of customer expectations. Sarah said that while in the design of premium road bikes performance was assessed against particular UK and European Pro race stages, the lower-cost bikes and eBikes were designed based on technical experience. John asked Sarah what she meant by technical experience and Sarah explained that it largely meant attempting to improve on an existing model without incurring significant cost increase. Phil said that, as he saw it, the term underpowered was not quantified and typically in this situation a Product Design team would simply opt for a more powerful motor, providing the programme budget would allow it.

Sarah said that the fundamental problem was that the design process used on budget bikes was bottom-up rather than top-design. John realised that the discussion had moved some distance from the original intent of the meeting, but since that had been met, he decided that he would like to pursue the topic further since it might help in his future dealings with the O254 team. Given where their discussion was taking them John felt it unlikely that Sarah had heard anything about Paul's departure. John asked Sarah why she described the design process for budget bikes as bottom-up. Sarah responded by saying that the design team started by reviewing the current model and identifying where things needed to be improved, or it might be nice to improve if the programme budget allowed, on a component-by-component basis. Sarah added that the team only really considered the larger systems and the total bicycle during Design Verification, or later after launch, when they found that they had to make adaptations to the overall design to properly accommodate some of the revised components. John asked Sarah if she would give him an example of what she had just described. Sarah said that in one instance, which was the subject of a recall, the positioning and design of the rear mudguard was such that when wider tyres were fitted objects could get jammed between the mudguard stay and the tyre inhibiting wheel rotation. Sarah added that the geometry of the stay was changed with recalled bikes having to have stays replaced. Phil added that this was a classic example of where the Supply Department opted for cheaper mudguards without fully understanding the consequences of their actions. Phil added that the Supply Department was not entirely to blame and explained that the Product Design team had to take some of the responsibility for passing component specifications to

Supply in total isolation of the system of which they were to be a part. At this point, John thought the meeting had gone full circle and thanked Sarah and Phil for their time and said that he was very hopeful that such issues would not happen on Oe375.

After Sarah and Phil had left, John thought about what they had said and its relevance to his involvement with O254. In taking on more responsibility John wondered if he would still have the luxury of extended meeting preparation time in the future. John decided that before discussing technical issues with the O254 team he would focus initially on the team itself and explore the hopes and fears that team members had with his taking over from Paul. Driving home that evening John reflected on what had been another eventful day at Oxton and in comparing the recent happenings to his time at Blade he decided that perhaps the smaller size and hence lower inertial mass of Oxton compared to Blade made it more susceptible to more dramatic changes. He reflected that, while things never stood still at Blade with the company constantly evolving, not necessarily always for the better, the pace of change in the larger company did seem more measured.

John's first meeting the next day was the Oe375 Review/Preview meeting which he had scheduled for first thing in the morning as Yumiko was attending since she was taking the Warm Up in the following Oe375 team meeting. Yumiko signed in before Sarah arrived and John took the chance to ask her a question that had puzzled him from their first meeting and asked Yumiko how she had got into the bicycle industry having studied aerospace engineering for her PhD. Yumiko said that for family reasons she wanted to live near her parents and Agano seemed an obvious choice when a position was advertised that she was lucky enough to be given. Since Sarah had still to arrive John then asked Yumiko how she was finding attending the Oe375 team meetings. Yumiko said that she found the meetings interesting and, although she was familiar with most of the technical methods that they had looked at, she had not seen them used within an explicitly defined Framework such as OFTEN. John asked Yumiko what she thought of the People Skills, and she replied that she thought they were useful and helped Team B in their work. Yumiko then said that while she enjoyed working with Team B, the time difference made things difficult as she could only attend, what was for the UK, the early morning meetings. Yumiko added that Jennifer was very helpful in letting her know what the team had been doing in her absence and giving her some tasks that she could work on for the team although she observed that working alone was never the same as working with others. Just then Sarah arrived and, having heard the last part of what Yumiko had said, said that she and the other Team B members saw Yumiko as a valued member of the team. Yumiko thanked Sarah for her comment and John thanked Yumiko for her feedback. Turning his attention to the meeting in hand John said that he was pleased to have Yumiko join them and summarised the format of the Review/Preview meeting for her benefit.

John asked Yumiko for her thoughts on how the last Oe375 meeting had gone and Yumiko said that she was impressed with the work that Sam and Phil had done in developing the eBike propulsion system noise simulator. Yumiko added that Jennifer had sent her a note explaining what had happened in John's meeting with the Team B and, while she had looked through the documentation that had been developed in that meeting, the first time that she had seen the simulator was in the Oe375 meeting. In agreeing with Yumiko Sarah said she was also impressed with what Sam and Phil had done and added that she thought the simulator was a good way of explaining to someone what Noise was, and how it affected system performance. On reflecting further on the meeting Sarah said that she found the Warm Up and John's presentation of Framing useful and Yumiko agreed with her. John said that in the last Review/Preview meeting, it was agreed that a lot, perhaps too much, technical information was included in the preceding Oe375 meeting. John asked Sarah and Yumiko for their reflections on, what he called, the technical information density in the last Oe375 meeting. Yumiko said that she did not think that too much was packed into the last meeting and Sarah agreed. John thanked Yumiko and Sarah for their feedback and said that in the next Oe375 meeting, having now quantified the effect of Noise on the performance of the Oe375 propulsion system, he wanted to look at ways of improving this performance through increasing system robustness. John added that he was getting ahead of himself and before looking at the Oe375 meeting agenda he wanted to consider the meeting Warm Up which he recommended was based on the People Skill that was going to be considered in the meeting. Yumiko said that she had assumed that the Warm Up and People Skills presentation in the meeting would be closely linked since this had become the norm.

John said that the People Skill he wanted to look at in the Oe375 meeting was another aspect of the communication skill of questioning. John reminded Yumiko and Sarah that the team had looked at Questioning for Clarification in one of their early Oe375 team meetings and said that this time he wanted to discuss Questioning for Information. John continued by saying that there were different types of questions that could be asked when gathering information with two high-level types being open and closed questions. John said that, as he was sure Sarah and Yumiko were aware, in closed questions the person being questioned is limited to a number of answers to choose from with the most limiting being Yes/No and added that this compared to open questions in which people are free to respond to the question in any way they see fit.

John said that he had thought of basing the Warm Up on the use of closed questions and said that when he was a boy his family used to play a game called Twenty Questions at Christmas. John said that the game was the basis of a radio programme that was broadcast in the UK. John explained that in playing the game one person choses an object and the other players have to guess what it is through a series of up to twenty closed questions which the person who has chosen the object answers with a Yes or No as appropriate. John said that if the players do not guess what the object is after asking all of their 20 questions, then the person who chose the object wins. John suggested that Oe375 team members play the game in pairs with each person choosing an object in turn and asked Yumiko what she thought of his idea. Yumiko said that it sounded fun. Sarah then suggested that to make it relevant the object could be something that people use at work. Yumiko said that she liked Sarah's idea and that by restricting the object choice in this way they could reduce the number of questions asked to ten. John said that he liked Sarah's and Yumiko's suggestions and added that a shrewd questioner would ask fairly broad questions initially like "*Do you use this object every day?*" and then ask more focused questions. John recommended to Yumiko that she did not tell Oe375 team members this, and Yumiko agreed that this might spoil the fun. John asked Yumiko if she was happy to take the Warm Up based on what they had discussed, and she said that she was looking forward to it. Yumiko added that she did not think it would take very long and in agreeing, John suggested that she allow each pair to think of a second object and repeat the process. John then suggested that after the pairs had finished asking and answering questions Yumiko should facilitate a team discussion in which team members could talk about their experiences and explain how hard or easy they found the exercise along with any tactics they had used. Yumiko said that she would be happy to do this, and in concluding the discussion on the Warm Up, John recommended to Yumiko that she stress to the pairs that they could only ask and answer Yes/No questions.

John said that he had already mentioned two agenda items that he intended to include in the next Oe375 meeting of Questioning for Information and Robustness Strategies and asked Sarah and Yumiko what else should be included. Sarah said that Team B had finished the tasks that they took away from the Oe375 meeting before last and added that she would ask Phil if he was OK to present a sample of what Team B had done. Having agreed on the agenda, John thanked Yumiko and Sarah for their help and concluded the meeting. After Yumiko had signed out of the meeting Sarah told John that she had heard a rumour that Paul Clampin was leaving Oxton and asked John if he knew anything about it. John replied that Paul had not said anything to him about leaving, and Sarah left John's office with a smile on her face which left John wondering whether Sarah had seen through the way that he had been economical with the truth in answering her question.

John drove into work the next day reflecting on what Sarah had asked him at the end of the Review/Preview meeting and wondering if he would be asked about Paul's departure in that morning's Oe375 team meeting. He decided that, since it would be difficult to deflect any questions on the subject without lying, he would be open in his reply although, as it turned out, he need not have worried as Paul's name was not mentioned. As usual, Phil had set up the room and was sitting at his usual table staring at his laptop when John walked into the conference room. Phil looked up and exchanged greetings with John. No sooner had John booted up his laptop and signed into the meeting than Yumiko joined and wished Phil and him a good morning. John asked Yumiko if she was all set for the Warm Up and she replied that she was looking forward to it. After the other team members had arrived or signed in Yumiko said that she was taking the Warm Up and shared a slide (Figure 11.1).

- **Self**: Identify two objects that you use at work
- **Others**: Form a pair (A & B)
  - Andy & Phil, Wolfgang and Jennifer, Sarah and Sam
- Person A: Select one of your objects
- Person B: You have to guess what A's object is by questioning them*
- Switch over people, and repeat (*A asks B, B asks A, then repeat*)
- **Team**: Discuss your experience

Rule

**You can ask up to 10 questions and these must have Yes/No answers*

Figure 11.1 Warm Up slide – using closed questions.

Yumiko gave team members a short while to read the slide and then said that she would like each of them to think of two objects that they use at work. Yumiko said that team members should get together in the pairs shown on the slide with each team member trying to guess the other's object through questioning by asking up to ten questions. Yumiko added that it would be better if the object was not something that was used every day like a computer but something that would be more difficult to guess. Yumiko asked John if he would set up two breakout rooms for the first two pairs, and John replied that he would do this. Yumiko stressed that only questions with a Yes/No answer should be asked and gave the example of a question such as "*Is the object green?*" Yumiko then asked team members to think of their objects, if they had not already done so, and join their pair with the person "B" beginning the questioning. Yumiko said that pairs should swap over once they have either guessed the object or asked all ten of their questions and added that the whole process should be conducted twice. Yumiko said that she would keep an eye on how things were going and call everyone together for a discussion when everyone had finished the questioning. John then created the two breakout rooms.

In watching what was happening at Oxton John saw that the four Oxton team members seemed engaged either speaking, staring at their laptop screens or out into space. John noticed some smiles, a few laughs and a couple of groans. After about 10 minutes, seeing Sarah and Phil talking to each other, Yumiko assumed that they had finished and after a couple more minutes she sent a Chat message around the team asking if they were all finished and quickly received confirmation that everyone had completed the exercise. Yumiko then asked John to close the breakout rooms and after he had done this, she asked team members for their feedback on the exercise.

Andy said that Phil had chosen two rather obscure objects, a paper clip and calculator, and said that while he had been able to guess the first object, he had not guessed the second. Andy then asked, "Who uses a calculator these days?" Sam responded to Andy's question by saying that obviously Phil did and added that he assumed this was because he could not find his slide rule, which caused laughter. Sam said that he was also able to guess one of Sarah's objects which he said was a stapler but not the correcting fluid and in mimicking Andy he asked, "Who used correcting fluid these days?" Andy quickly picked up on Sam's statement and said that obviously, Sarah did as her quill pen tended to splotch ink, which caused more laughter. There was general agreement that the exercise was made harder because team members were restricted to Yes/No questions. Sarah said that Yes/No questions were not very helpful when the object was obscure like Sam's tie. Jennifer asked Sam when he wore a tie and Sam explained he kept a tie in a desk drawer for the occasions when he need to look smart such as when outside visitors were coming although adding that it was not very useful when he wore a T-shirt to work. As no one else volunteered feedback Yumiko asked what tactics team members used in asking questions and Wolfgang said that he did not guess Jennifer's first object of a bag of coffee and so for her second object he used quite broad categories in his first couple of questions like "*Can you eat it?*" and narrowed things down after that. Yumiko asked Wolfgang what Jennifer's second object was, and he said it was a document file. Yumiko then thanked everyone for taking part and said that John was going to talk more about questioning shortly. In concluding the Warm Up, Yumiko summarised the meeting agenda as Questioning, Feedback from Team B, and Robustness Strategies and handed the meeting over to John for the People Skills input.

John thanked Yumiko for what he said was a fun Warm Up and said that he wanted to look further at the important communication skill of questioning. John reminded the team that they had considered *Questioning for Clarification* in an earlier team meeting and said that he wanted to look now at another aspect of questioning which he called *Questioning for Information*. John said that, as team members were aware, questioning was a key communication skill within an engineering team which facilitated the sharing of technical information. John said that one question that a team member might ask of another is "*How large is the temperature range for the Noise Factor of ambient temperature?*" He then asked team members to consider that the question was asked in a subtly different way of "*How small is the temperature range for the Noise Factor of ambient temperature?*" John asked the team how the different forms of the question might influence the answer that was given. Sam said that he thought the question which used the word *large* would get an answer of a larger range than the question with the word *small*. John asked if any other team members agreed with Sam, and there was general agreement in the team that what Sam had said was reasonable.

John said that what Sam had suggested was born out in a seminal study conducted about fifty years ago in which groups of university students were asked to estimate a quantity using a series of questions which used quantitative adjectives or adverbs. John continued by saying that the students were divided into two groups with one group given what were almost the same questions as the other group but where the quantitative adjectives or adverbs were the antonyms of those in the questions given to the other group. John gave examples of *large* and *small*, *high* and *low*, and *hot* and *cold* as antonyms. John said that the result of the study showed that almost without exception there was a significant difference in the average estimation of each quantity between the two sets of students. John added that there was an overriding tendency for the adjective or adverb that implied a larger magnitude such as *large* or *high* or *hot* to be associated with the greater average magnitude of the estimated quantity.

Andy said that he found the study interesting, and he was aware that he sometimes chose his words carefully when asking work colleagues and friends a question. There were murmurs of agreement with what Andy had said. In picking up the point that Andy had made, John said that there were several features of a question which could influence the response that a question elicits. John shared a slide (Figure 11.2).

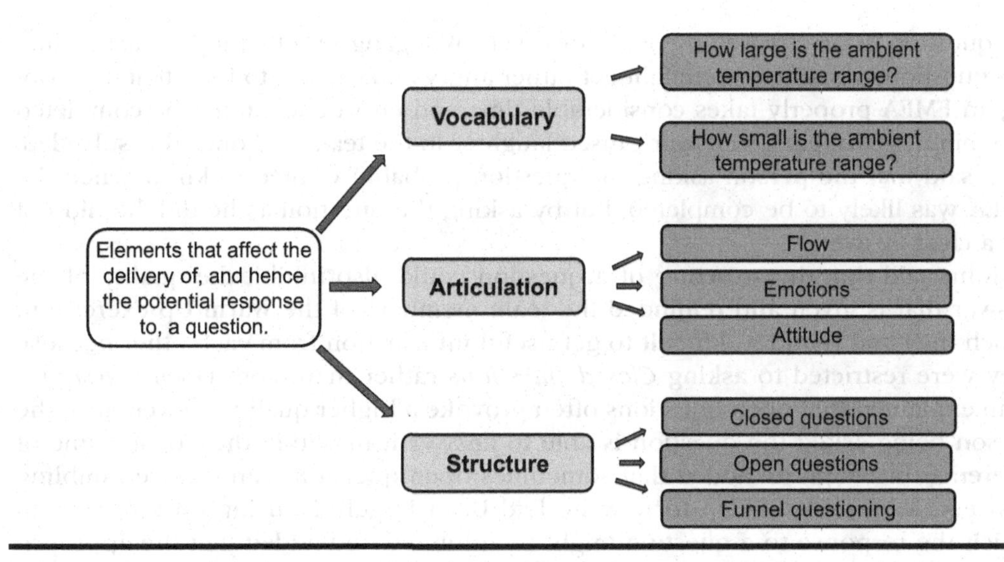

Figure 11.2 Features that can be adjusted to influence the delivery of a question.

John explained that, in addition to the words that are used in asking a question, the way in which a question is articulated can also influence the response that is given. Directing team members to the item *Flow* associated with *Articulation* on the slide John said that questions asked in quick succession can confuse the person being questioned. John continued by saying that this technique was used on people who are being interrogated when the interrogator was seeking a particular answer with the rapid flow of questions being used to maintain control of the situation. John said that he hoped that none of the team members had been interrogated at work, but they may have met a manager who had used this technique in a meeting. John noticed some wry smiles and took this to mean that some team members shared his experience in this matter. Turning to the second item on the slide under Articulation of *Emotions*, John said that a question asked in an emotionally charged manner will often provoke an answer which responds more to the emotional content of the question than the question itself. John explained that questions asked in an angry manner will tend to be answered defensively while questions asked in a flippant manner may well be answered flippantly. Phil said that he had attended Design Reviews where the manager leading the review had got quite angry, which Phil explained was self-defeating since the angrier the manager got, the less inclined those being questioned became to share information.

John thanked Phil for his example and continued by saying that another way in which the articulation of a question can influence the answer it elicits is questions which are biased by the questioner's *Attitude*. John asked Wolfgang how he might respond if he asked him "*Haven't you finished developing that FMEA, yet?*" In asking the question, John stressed the last word "yet". Wolfgang said that if John asked him this question in that way, he might get rather angry and explain to John that developing an FMEA properly takes considerable time and effort and cannot be completed in 5 minutes. Wolfgang's answer caused laughter in the team and once this subsided, John said that the person asking the question probably wanted to know when the FMEA was likely to be completed, but by asking the question as he did, he did not get a clear answer.

John said that the structure of a question could also inhibit the quality of the answer that is given and reminded the team members of the Warm Up exercise in which they had found it difficult to get useful information from each other because they were restricted to asking *Closed Questions* rather than more *Open Questions*. John explained that open questions often provoke a higher quality answer since the person being asked the question is able to answer it in within their own frame of reference, although he added that sometimes open questions can receive rambling answers. John said that up to now he had been largely focusing on the ways in which the response to a question might be inhibited and added that the appropriate use of structure in questioning can be highly effective in eliciting information. John explained that he might begin an analysis by asking the open question "*What exchanges occur at the interface between the tyre and road?*" and subsequently narrow the discussion down by asking "*What exchanges of energy occur?*" John added that later he might look to conclude the analysis of this interface by asking "*Have we identified all the exchanges at the tyre-road interface?*" John explained that the technique of starting with a general open question and moving to more specific questions and potentially ending up with closed questions was termed *Funnel Questioning*.

John asked if there were any questions or comments about what they had discussed. Jennifer said that she sometimes framed questions in the hope of getting the answer she wanted to hear and now realised that in technical discussions this was probably not the best thing to do. Yumiko said that she sometimes gave the answers to questions that she thought people wanted to hear and John said that he had found that this was sometimes useful in speaking to family members but often not very helpful in engineering team meetings. John added that sometimes in responding to a family member's question his assumption about what he thought someone wanted to hear was wrong and this could lead to a difficult situation which raised smiles from team members. John concluded the discussion on the People Skill by saying that in the Oe375 team, questions asked to obtain information are best framed in a neutral manner, using neutral vocabulary, to allow any reply not to be unduly influenced by the way in which the question was asked. John then handed the meeting over to Phil for his report back on recent Team B assignments.

Phil thanked John and said Team B had done quite a bit of work lately and he wanted to go quickly through the Oe375 cassette noise and control factor table (NCFT) and corresponding extracts from the propulsion system design FMEA as examples of this work. Phil shared a combination worksheet (Figure 11.3).

Internal Interface Table: Propulsion System

Interface Ref	Interface		Type	Description of Exchange	Impact
3-D	Chain	Cassette	E	Chain force transmitted to cassette	2
			P	Physical contact between chain and cassette	2
			M	Transfer of lubrication between chain and cassette	1
			M	Dirt/debris transferred between chain and cassette	–1

Oe375 Propulsion System		D Cassette	E3 Rear Wheel	E6 Rider	E7 Environment
1	Pedal				
2	Chainset				
3	Chain	X			
4	Cassette		X	X	X

External Interface Table: Propulsion System

Interface Ref	Interface		Type	Description of Exchange	Impact
4-E2	Cassette	Rear Wheel	E	Fix cassette to wheel hub	2
			P	Locate cassette on wheel hub	2
			E	Cassette locked to rear wheel	–1
			E	Vibration transmitted from wheel to cassette	–1
4-E6	Cassette	Rider	E	Rider cleans cassette	1
			M	Rider uses cleaning fluids	1
			E	Noise generated at cassette	–1
			I	Rider checks cleanliness of cassette	1
4-E7	Cassette	Environm't	M	Dust/Dirt transferred to cassette	–1
			M	Water transferred to cassette	–1
			M	Loss of lubricant to environment	–1
			E	Road saline solution causing corrosion	–1

Noise and Control Factor Table
CASSETTE

Interface Ref	Type	Units	Range Min	Range Max	Noise Factor Effect	Control Factor: Manufact'g controls	Regular maintenance	Cassette specification	User manual	Sealed lubrication
	Piece to Piece									
	Variation in cassette teeth profile	mm/deg	–	Visual Std	Chain slips Chain fails to engage	X				
4-E2	Variation in cassette location on wheel hub	mm/deg	–	Visual Std		X				
3-D	Variation in cassette clamp force	Nm	34Nm	38Nm		X				
	Changes over time/cycles									
4-E6, 4-E7	Lubricat'n loss from cassette teeth & chain	ml	0	–	Excessive cassette wear		X		X	
4-E6, 4-E7	Lubrication loss from cassette bearings	Nm	zero	0.1	Increased bearing friction					X
3-D	Cassette teeth wear	mm	0.1	0.2	Lower chain engagement		X	X	X	
4-E2	Cassette comes loose on wheel hub	Nm	10Nm	20Nm	Vibration; frictional loss	X				
	Customer Usage									
4-E6	Poor/no maintenance, cleaning of cassette	Hz	Monthly	Weekly	Excessive cassette wear		X		X	
3-D	Duty cycle; Demand on cassette	W/Rev	50	150	Variation in rider power				X	
	External Environment									
4-E6, 4-E7	Debris ingress	mm^3	–	Visual Std	Loss of lubrication		X		X	
4-E6, 4-E7	Dirt ingress	mm^3	–	Visual Std	Chain slips		X		X	
4-E6, 4-E7	Water ingress	ml	–	Visual Std	Chain fails to engage		X		X	
4-E6, 4-E7	Saline corrosion	mm^2	–	Visual Stdl	Increased frictional loss		X		X	
	Subsystem Interaction									
3-D	Dirt/Debris transferred from chain	mm^3	–	Visual Std	Increased frictional loss		X		X	
4-E2	Vibration from wheel/tyre	Hz							X	
	Diverted Output									
3-D, 4-E2	Frictional loss	%	0	2%	Increased rider effort			X		

Figure 11.3 Propulsion system interface matrix and table extracts and cassette NCFT.

Phil explained that the interface matrix extract only showed interfaces with the cassette and asked if there were any questions or comments on the detail of what he was sharing. Andy asked why the noise factors which were associated with the control factor of *Regular Maintenance* were also linked to the Oe375 User Manual. Phil responded to Andy's question by saying that Team B intended that a maintenance schedule be detailed in the eBike Instructional Manual and added that he would pick this up when he shared the appropriate propulsion system design failure mode and effects analysis (DFMEA) extract shortly. Wolfgang asked why two interfaces were referenced for some noise factors, and Phil said this was where the nature of the exchanges involved two interfaces. Phil cited the example of lubrication loss from the cassette bearings and explained that even though the bearings were sealed such loss might occur if the person cleaning the cassette sprayed a solvent-based cleaner on the side of the cassette on a regular basis. Phil added that the cleaning involved the cassette–rider interface and the loss of lubricant to the environment involved the cassette–environment interface. Wolfgang thanked Phil for his clarification. Sam then enquired why there was not a minimum magnitude for the noise factor ranges with a maximum being set by a visual standard. Phil said that these were attribute factors with the magnitude of the noise factor either being acceptable or not.

As no one else spoke John asked what interface the first noise factor on the NCFT "*Variation in cassette teeth profile*" corresponded to. Phil asked Sarah to answer John's question and Sarah said that while it did not correspond to any of the interfaces with the cassette that the team had reviewed in developing the NCFT, this type of noise factor was common on the P-Diagrams that she had seen. John then asked Sarah a follow-up question about the nature of the manufacturing controls listed as the control factor for this noise factor. At this point, John assumed that Sarah had realised why he was asking the questions since she replied that the controls would be associated with the manufacture of the cassette. Sarah quickly followed up her response by saying that this type of control would be the responsibility of the company manufacturing the cassette and as such this noise factor was the responsibility of the cassette supplier and not Oxton and so the noise factor was not appropriate at the system level at which the current analysis was being conducted. Yumiko said that the variation of the cassette teeth profile on the product manufactured by Agano was tightly controlled and was insignificant when compared to the subsequent wear on the cassette in customer usage. Jennifer said that she had deleted the noise factor from the table. John thanked Sarah and Yumiko for their input and said that it was good to see that the discipline of developing the NCFT based on information documented during interface analysis had no longer been breached. Sam asked John what he meant by his remark, and John said that he assumed that Team B had not been able to designate an interface from the propulsion system analysis that corresponded to the noise factor because the noise factor was not derived from the interface analysis, and on this basis, Jennifer had deleted the noise factor.

Since there were no more questions, Phil said that Team B had updated the Oe375 propulsion system design FMEA to include current controls and the detection ratings after which they used the information in the cassette NCFT to add the recommended actions. Phil said that Andy and Wolfgang might recall the extract of the propulsion system DFMEA extract that he was sharing being developed up to the Occurrence column in the Oe375 team meeting before last (Figure 11.4).

Function	Potential Failure Mode	Potential Effect(s) of Failure	Severity	Potential Cause(s)/ Mechanism(s) of Failure	Occurrence	Current Design Controls Prevention	Current Design Controls Detection	Detection	Recommended Action
Convert force in chain to torque at rear wheel	Temporary loss of conversion of force in chain to torque at rear wheel	Rider experiences jolt in pedalling Increased effort required by rider	6	Poor contact between cassette and chain due to cassette teeth and chain wear - intermittent chain slippage	4	None	Hardware in loop test Bicycle DV testing	7	Select cassette based on modelling
				Loss of lubrication on cassette teeth causing premature wear - chain slip	6	Handbook recommendation	Bicycle DV testing	7	Update user manual recommendation
				Excess lubrication attracting dirt and dust causing premature chain and cassette wear and slipping	4	Handbook recommendation	Bicycle DV testing	7	Update user manual recommendation
				Dirt and dust causing premature chain and cassette wear and chain slip	4	Handbook recommendation	Bicycle DV testing	7	Update user manual recommendation
				Corrosion causing poor contact between chain and cassette - chain slip	3	Handbook recommendation	Bicycle DV testing	9	Update user manual recommendation
				Cassette not coupled fully to wheel hub causing cassette to slip	1	No Oxton control	Bicycle DV testing	7	None
				Cassette not located properly on wheel hub causing chain to slip over cassette teeth	1	No Oxton control	Bicycle DV testing	7	None
				Vibration from wheel hub causing loss of contact between cassette and chain	3	Clamp torque specification	Bicycle DV testing	8	Review clamp torque specification
				Incompatibility of higher speed chain and cassette across groupset manufacturers causing chain slippage	2	None	Bicycle DV testing	8	Update Oxton component selection process

Figure 11.4 Extract from propulsion system DFMEA including current controls and recommended actions.

Phil explained that Team B used information in the cassette NCFT to update the Recommended Actions column of the design FMEA and paused to allow the non-Team B members to read and absorb the content of this column. Phil said that Team B had decided that the User Handbook recommendations on maintenance for four causes should be significantly improved. Phil explained that Team B felt that Oxton not having any prevention control for the two causes which they had rated as not likely to occur was understandable given the reliable nature of the design of the interface between the cassette and wheel hub. Phil added that, on the other hand, not having a prevention control to ensure that compatible Drivetrain components were selected even where any failure cause associated with this had a low occurrence rating was inexcusable. John appreciated the neutral wording of the Recommended Action against the mistake-based cause in the bottom row of the DFMEA extract and was subsequently pleased that this wording did not raise any questions. Phil asked if anyone had any questions or comments and Andy asked why so many items had relatively high detection ratings. Phil replied that bicycle Design Verification testing was not designed to identify the failure causes associated with these ratings and cited the example of corrosion which he said would probably not occur over the duration of a typical bicycle test. Since there were no more questions or comments Phil handed the facilitation of the meeting over to John.

John thanked Phil and said that although Phil had only shown a small sample of the work that Team B had been doing it was clear from what he had illustrated that Team B had done a lot of work in developing the propulsion system design FMEA and NCFTs in a symbiotic manner. Yumiko asked John what he meant in using the word symbiotic. John explained that in the context in which he was using the word it referred to a relationship between three entities that each benefitted from at least one of the other two. John said that the Design FMEA benefitted from the interface analysis and NCFT, with the NCFT also benefitting from interface analysis while the NCFT could be used to validate interface analysis content.

John said that although it was not explicitly apparent in the recommended actions listed in the DFMEA extract that Phil had just taken the team through, a key strategy for design improvement was to make the design robust to noise. John added that he would like to look at strategies for achieving design robustness and explained that robustness in a product design could be achieved in several ways and shared a slide (Figure 11.5).

A. Select the best design concept

B. Reduce the sensitivity of the design to the noise factor

C. Remove or reduce the noise factor

D. Insert a compensation device

E. Disguise the effect

F. Boost the design

Figure 11.5 Strategic ways to achieve robustness.

John said that the robustness strategy *A. Select the best design concept* meant choosing the design concept that was going to be the most robust given the Noise Space in which the product was intended to be used and added that the design concept had to be cost-effective and feasible. John cited the use of a chain drive versus a belt drive as an example of alternative design concepts and recalled the presentation that Yumiko had given in discussing the relative advantages of the two design concepts.

In directing team members' attention to the second robustness strategy on his list of *B. Reduce the sensitivity of the design to the noise factor* John said that, as he had mentioned previously, the famous Japanese engineer Genichi Taguchi promoted the technique of what he called "Parameter Design" in which the magnitudes of design parameters are varied to find the combination which corresponded to the most robust design. In illustrating the approach John explained that at the system level the Oe375 team will have to select the most robust combination of chain and cassette speeds to best suit the projected duty cycles of the eBike. Yumiko said that Agano used parameter design techniques in optimising chain and cassette performance and added that using such techniques often allowed performance to be increased at no cost increase or even reduced cost.

After thanking Yumiko John moved on to the next robustness strategy *C. Remove or reduce the noise factor* and said that this was often difficult to do for two reasons. John said that many noise factors such as those associated with the external environment cannot be removed while it is difficult to remove others without compromising the overall eBike performance. To illustrate the point, John said that while the heat generated in the windings of an eBike motor detrimentally affects motor performance it would be rather foolish to remove the windings as the source of the Noise, which caused laughter among team members. John continued by saying that a better strategy in this case was to reduce the noise. John said that the effect of heat in the windings could be reduced by encouraging this heat to dissipate to the environment through the cooling effect of air flowing over the motor casing. Yumiko asked John if he would give her an example of a noise factor that had been removed in bicycle design. John said that he could not think of one and asked the team for help. After a pause, Phil said that when tyres, inner tubes and valves were not as well sealed as they are today, each bicycle came with a pump that attached to the frame and said that this was even the case for racing bikes. Phil said that he did not believe the pump was removed to improve robustness against drag and gravity after analysis of an NCFT, but it made a good story, which generated laughter. After thanking Phil, John said that while Phil was speaking, he had thought of the use of a chainguard which protected the chain from the noise factor dirt and debris ingress. John quickly added that while he was thinking of the chainguard he was still listening to what Phil was saying and restated Phil's example of the bicycle pump which prompted smiles.

John turned to the next robustness strategy *D. Insert a compensation device.* John said that he saw the inclusion of an electric motor on a bicycle to form an eBike as an example of a highly successful implementation of this strategy in which the added propulsion source helped to reduce the effect experienced by the rider of key noise factors like drag, gravity and rolling resistance. In looking at the robustness strategy of *E. Disguise the effect* John said that he could not think of a propulsion system example where the effect of a noise factor was disguised and asked team members for their help. After a pause, Phil said that he could not think of a propulsion system example other than colouring the chain brown to hide any dust on it, which drawing smiles, and then cited the non-propulsion system example of covering up welding imperfections on a frame by applying extra paint.

John thanked Phil and then said that if all else fails and the effect of a noise factor still needs to be addressed, the final robustness strategy on the list of *F. Boost the design* can be used and explained that this meant taking actions like using higher grade materials or tightening manufacturing tolerances. In illustrating this robustness strategy John referred to the use of upgraded materials in Drivetrain components and added that adopting this robustness strategy often added cost. John asked if team members had any questions or comments and Andy said that he assumed that upgraded materials were used to reduce the sensitivity of the Drivetrain to noise and added that this appeared to him to be strategy B by another name. John explained that the difference between strategies B and F was that the intent with B was to increase system robustness at no or reduced cost while strategy F was invariably achieved at increased cost. John thanked Andy for his comment and said that he was correct in the sense that it was not always clear which robustness strategy a design action fell into. John added that, like the noise factor types, categorising a particular design action by robustness strategy is of lesser importance than identifying the action in the first place. John said that the list of robustness strategies was only meant for guidance to facilitate engineering thinking on a broader front. Since there were no more comments John said that it was time for the break.

John took the opportunity during the break to send a Chat message to Yumiko saying how much he enjoyed her Warm Up. John said he liked the way that she had clearly set out the exercise instructions with the Warm Up steps on the slide that she had developed and added that he thought that she had facilitated the discussion at the end of the exercise well by, for example, encouraging all team members to speak. Yumiko responded immediately thanking John for his note and saying that she had enjoyed the experience.

Once the break was over John said that having looked at the different robustness strategies, he wanted to return to the NCFT that Team B used in developing the model for assessing the effect of Noise on the Oe375 propulsion system. John explained that he would like the team to decide which robustness strategy to adopt for each noise factor listed in the table. Sam intervened and asked John if he could show the team some analysis that he had done which he thought was relevant to the discussion. John said that of course he could and gave control of the shared screen to Sam who displayed a graph (Figure 11.6).

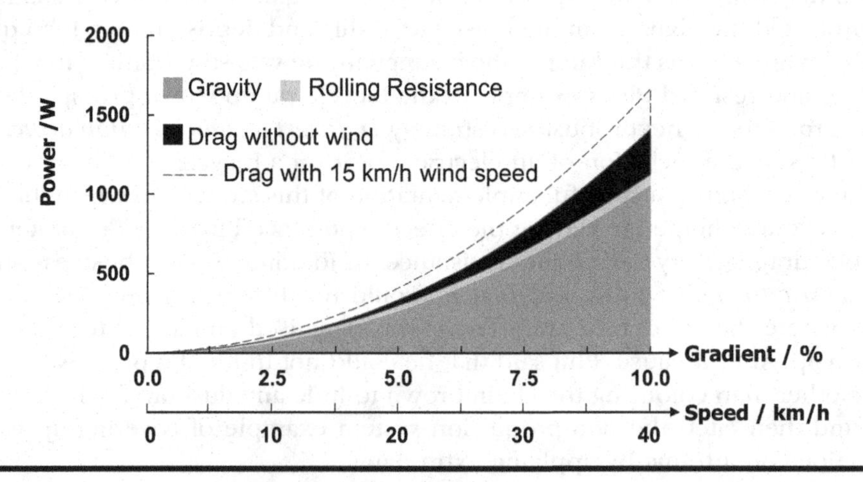

Figure 11.6 Resistive power versus gradient/speed.

Sam explained that the graphic was based on data from the eBike propulsion system model and was a stacked area plot showing the power required to overcome the resistive forces of gravity, rolling resistance and aerodynamic drag. Sam asked if all team members were familiar with a stacked area plot and Jennifer said that she would appreciate an explanation. Sam explained that the plot he was sharing was effectively four plots in one, stacked one on top of the other with the lowest plot being for the power required to overcome the force of gravity plotted against gradient. Sam said the lighter grey plot was for the power required to overcome rolling resistance plotted against gradient while the darkest grey plot was for the power required to overcome drag with zero wind speed this time plotted against the eBike speed. Sam added that the final dotted line plot was for drag with an opposing wind speed of 15 km/h plotted against eBike speed.

Sarah said the plot looked interesting and observed that the maximum speed of 40 km/h was in excess of the maximum legal speed for eBikes in the UK and Europe of 25 km/h. Sam agreed with Sarah and said that since the model could be used for any theoretical speed, he thought he would include the higher value. Sam said that although the plot was unrealistic in showing a constantly increasing gradient it showed the importance of paying attention to the resistance due to gravity for Oe375. Andy asked Sam if he could say more. Sam explained that cycling literature tended to focus on aerodynamic drag as the dominant resistive power and added that drag was the overriding resistive power when cycling at higher speeds since drag increased with square of the speed while the gravitational power only increased in direct proportion with the speed. Sam said that Wolfgang had pointed this out in the last Oe375 meeting when looking at the table of results from the propulsion system model. Sam said that he saw the graphic that he was sharing as a clear indication that as far as Oe375 was concerned the gravitational resistance was a significant noise factor because of the maximum speed of 25 km/h placed on eBikes. Sam asked if anyone had any questions and John said that he thought that what Sam had just shown the team had important implications for what they were going to look at next.

John said that that he wanted to consider which robustness strategy might be best applied to each noise factor in the Oe375 propulsion system NCFT that Team B had developed, and Phil had shown at the last Oe375 team meeting. John suggested that the team look at each of the noise factor groups in turn and shared the piece-to-piece noise factor section of the table (Table 11.1).

Table 11.1 NCFT Extract; Piece-to-Piece Noise Factors with Robustness Strategy

Noise and Control Factor Table PROPULSIVE FORCES				Control Factors			
Interface	Type	Units	Noise Factor Effect	Type of tyre	Tyre pressure	Propulsion system model	Robustness Strategy
	Piece to Piece						
Tyre/Road	Tyre composition	Mfg-Hawk	ΔC_r	X	X	X	A. Select most robust tyres
	Tyre tread pattern	Pattern					
	Tyre-road contact patch	mm^2					
	Tyre width	ISO		X	X		

In referring to the NCFT extract, John said that he had hidden the interface reference and noise factor range information to make the extract more legible. After a pause, Sarah said that she thought that they should select the most robust tyre options and following a short discussion which confirmed that the team agreed with Sarah's recommendation John typed a short comment to this effect in the *Robustness Strategy* column that he had included in the table. John brought team members' attention to the coding he had used at the beginning of the comment which this time was "*A*" for *Select the best design concept* and referred them back to the list of robustness strategies that he had shared earlier (Figure 11.5). Wolfgang observed that bicycle tyres were of a fixed design concept so there were no alternative concepts available. While agreeing generally with what Wolfgang had said Phil said that the design of tyres varied considerably in terms of parameters like width, profile, tread, composition, tubed or tubeless, puncture proofing and so on and added that, as he saw it, they would be selecting the most robust design. Sam asked how they would know what the most robust tyres were. Andy said that HJK had a good relationship with Hawk and explained that he might be able to get some inside information on what developments were happening for eBike tyres. John thanked Andy and said that his connections would prove useful.

Sarah then asked John if he had seen Oxton's hardware-in-loop (HIL) rig, and John said that he had. Sarah said that they could use the rig to simulate the resistive power indicated by Sam's plot to investigate the relative robustness between tyres. Wolfgang said that he was familiar with the HIL technique and added that he would be interested in seeing the Oxton arrangement. Sarah said that she thought she had a schematic diagram of the original Oxton setup on her laptop and started to search for the image. After searching for a short time, Sarah shared the graphic (Figure 11.7) and explained that it was based on the previous Oxton setup which had now been updated but it would give the external team members an idea of what it looked like.

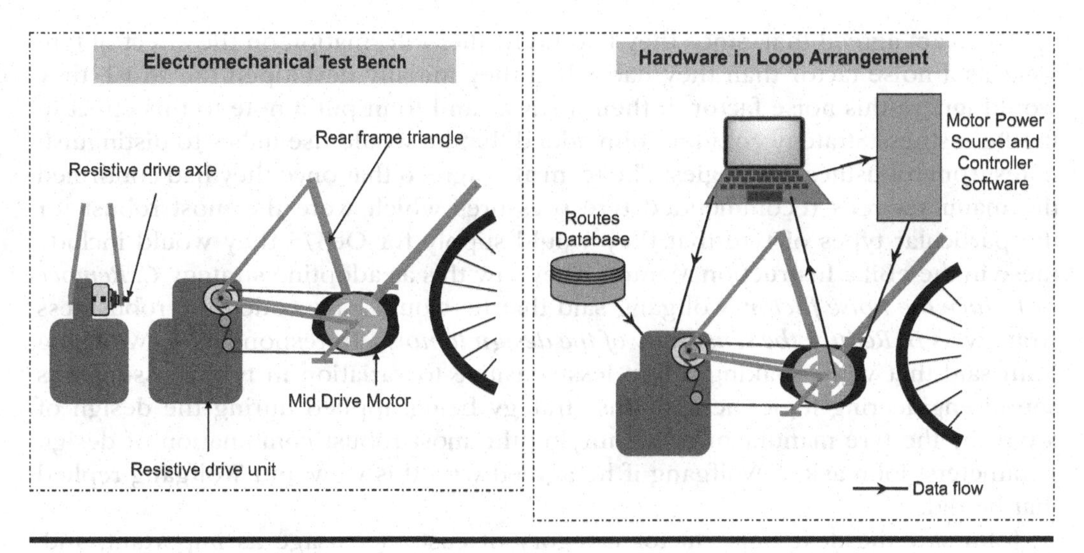

Figure 11.7 Previous Oxton hardware-in-loop rig setup.

Sarah said that the electromechanical testbench shown on the left of the graphic was essentially a high-end off-the-shelf bicycle trainer in which the load on the Drivetrain was provided by an electric motor working against the Drivetrain. Sarah said that the resistive motor was software-controlled as indicated by the schematic on the right-hand side of the graphic allowing the resistive force to simulate the forces that would be experienced on several well-known Grand Tours and other professional cycling routes as well as what she described as more everyday routes. Sarah added that the rig provided speed, distance, gradient, and power data output measured by various sensors. Sarah explained that the test bench had now been replaced with a more sophisticated arrangement with a more responsive electromagnetic resistance. Sarah added that the angle of inclination of the bike could also be varied in the new rig in line with changes in gradient in the Ride Cycle to give a more realistic ride experience and the input from the rider could also be simulated by the resistive unit compensating for rider input so dispensing with the need for rider input if required.

John thanked Sarah for her explanation, which he said was very useful, and added that he would like to continue looking at the noise factors in turn and shared the changes over time/cycles noise factors (Table 11.2).

Table 11.2 NCFT Extract; Changes over Time Noise Factors with Robustness Strategy

Noise and Control Factor Table PROPULSIVE FORCES				Control Factor			
Interface	Type	Units	Noise Factor Effect	Type of tyre	Tyre pressure	Propulsion system model	Robustness Strategy
	Changes over time/cycles						
Tyre/Road	Tyre wear	Tread	-				*Minimal effect*
	Tyre pressure	bar	ΔC_r	X	X	X	C. Specify most robust pressure

The team agreed that, since they had no further information on the effect of tyre wear as a noise factor than they had when they initially developed the NCFT, they would ignore this noise factor in their analysis and John put a note to this effect in the Robustness Strategy column. John added that he would use italics to distinguish notes from robustness strategies. The team also agreed that once they had confirmed the manufacturer's recommended tyre pressures which were the most robust for the particular types of tyre that they would supply for Oe375 they would include these in the eBike Instruction Manual. They saw this as adopting strategy *C. Remove or reduce the noise factor*. Wolfgang said that he would have coded the robustness strategy as *B. Reduce the sensitivity of the design to noise*. In responding to Wolfgang John said that while making a tyre less sensitive to variation in tyre pressure was sound engineering logic, he saw this strategy being applied during the design of a tyre by the tyre manufacturer identifying the most robust combination of design parameters. John asked Wolfgang if he agreed with this view and Wolfgang replied that he did.

John saw the next noise factor category of customer usage as important, and, because of the importance of the Oe375 eBike to Oxton's long-term survival, stated his view that the Oe375 team should ensure that eBike reflected the company's historical reputation for high quality, value-for-money product. John explained that this had implications for the performance targets and hence the robustness strategies that the team adopted across the whole eBike and the part that that the propulsion system, that they were now considering, played in this. John then shared the customer usage extract from the NCFT (Table 11.3).

Table 11.3 NCFT Extract; Customer Usage Noise Factors with Robustness Strategies

Noise and Control Factor Table PROPULSIVE FORCES					Control Factor			
Interface	Type		Units	Noise Factor Effect	Type of tyre	Tyre pressure	Propulsion system model	Robustness Strategy
	Customer Usage							
Tyre/Road	Rider mass		kg	Δm	X	X	X	B. Increase robustness to rider mass
	eBike mass		kg		X	X	X	B. Reduce mass
	Dirt on tyre		mm^3	ΔC_r				*Neglect*
	eBike Ride cycle		Standard	ΔC_r, $\Delta\theta$			X	B. Optimise propulsion system
Rider & Bike/ External Environment	Effective frontal area & Rider posture	U	m^2	ΔAC_a			X	B. Streamline frame
		D	m^2				X	

In line with what John had just said, the team decided that the propulsion system should be robust to a relatively wide customer base in terms of rider mass and adopted robustness strategy B for this noise factor. The team moved on to consider the eBike mass and Phil said that as a result of helping Sam develop the Propulsion Resistive Power model, he had been looking at the masses of competitor bikes and found that the allocated mass of 22.5 kg for Oe375, which was based largely on the current model, was not competitive. Phil explained that he had come across several eBikes where the mass was significantly less than the team's estimation. As a result of what Phil had said the team decided they need to reduce the eBike mass as the robustness strategy for this noise factor. Following further discussion, the team agreed that since this was not just a case of specifying a limit on the maximum mass and they would need to look at reducing the eBike mass by considering the masses of the individual components of which the bike was composed, and this aligned to robustness strategy B.

Turning to the eBike Ride Cycle the team saw this as an important noise factor that would heavily influence potential customers' perception of Oe375. The recent discussion with Sarah and Phil about the somewhat serendipitous Oxton design process for the company's lower-end bikes flashed through John's mind and he was determined that the design of the Oe375 would be grounded on a more objective process. John told team members that, as he saw it, for Oe375 to be robust to the noise factor of the Ride Cycle the first thing the team would have to do was to define the Ride Cycle objectively. Sarah recommended that this be done on the basis of appropriate cycling routes and added that Oxton had a library of these which could be used in association with the HIL rig. The team agreed with Sarah's recommendation and John suggested that before they used the HIL equipment they should define the propulsion system power requirements through analysis of the properties like the length, frequency, and steepness of inclines. Phil added that they would need to agree and develop the control strategy for the motor as a part of any analysis. John said that based on their discussion he recommended that the team adopt robustness strategy B to improve the robustness of the Oe375 propulsion system to the noise factor of Ride Cycle.

In considering the final noise factor in the customer usage category on the NCFT of effective frontal area and rider posture, the team agreed with Jennifer when she said that the Oe375 team had little control over this noise factor. Jennifer said that aligned to John's vision for Oe375 she hoped that customers of all shapes and sizes would buy the eBike and said that these customers, having bought the eBike, were more likely to adopt a riding posture that was based more on what they found comfortable than on minimising aerodynamic drag. In agreeing with what Jennifer had said Yumiko said that the Oe375 team did have control of the aerodynamics of the eBike itself. Phil agreed with Yumiko and added that since Oe375 was not likely to be used in time trials they would have to balance the aerodynamics of the eBike against the practicality of its use as a commuter and/or recreational eBike. The team agreed that they should adopt the robustness strategy B for this noise factor as far as the design of the bike frame was concerned and transferred their attention to the external environment noise factors (Table 11.4).

Table 11.4 NCFT Extract; External Environment Noise Factors with Robustness Strategies

Noise and Control Factor Table PROPULSIVE FORCES				Control Factor			
Interface	Type	Units	Noise Factor Effect	Type of tyre	Tyre pressure	Propulsion system model	Robustness Strategy
	External Environment						
Tyre/Road	Road surface	Defined	ΔC_r	X	X	X	B. Tyre pressure to suit surface
	Weather (rain)	mm/hour	-				*Ignore*
	Ambient temperature	°C	ΔC_r	X	X		A. Select most robust tyres
Rider & Bike/ External Environment	Wind speed	km/h	Δv_a			X	*See projected Frontal Area & Rider posture*
	Wind direction	Deg wrt eBike	Δv_a			X	
	Ambient temperature	°C	$\Delta \rho$			X	
	Elevation	m	$\Delta \rho$			X	*See Ride cycle*
	Gradient	%	$\Delta \theta$			X	

There was agreement in the team that they should specify the appropriate tyre pressure for different road surfaces in Oe375 User Manual with the recommended pressures being dependent upon which tyres they selected to use, and the predominant surfaces on which the eBike was likely to be used. The team designated this action as robustness strategy B. Following discussion of the remaining noise factors in this group, the team decided that they would all be addressed by actions proposed for other noise factors and documented their conclusions accordingly in the NCFT. John then shared the last group of noise factors of System Interaction along with the Diverted Output category (Table 11.5).

Table 11.5 NCFT Extract; System Interaction Noise Factor and Diverted Output with Robustness Strategies

Noise and Control Factor Table PROPULSIVE FORCES				Control Factor			
Interface	Type	Units	Noise Factor Effect	Type of tyre	Tyre pressure	Propulsion system model	Robustness Strategy
	System Interaction						
Propulsion/Braking	Frictional force/torque	N, Nm	(ΔF_p)				C. Eliminate non intended braking
	Diverted Output						
Propulsion/Drivetrain	Drivetrain efficiency loss	%	(ΔF_p)				C. Reduce efficiency losses

Sarah said although Team B had decided not to include any unintended braking force or Drivetrain efficiency loss in the propulsion system model initially, she thought that they should still define a robustness strategy aimed at reducing these two factors. John said that, as far as Oxton was concerned, he saw this as robustness strategy C and added a comment to this effect to complete the team's update of the NCFT (Table 11.6).

Table 11.6 NCFT for Propulsive Forces Including Robustness Strategies

Noise and Control Factor Table PROPULSIVE FORCES				Control Factors			
Interface	Type	Units	Noise Factor Effect	Type of tyre	Tyre pressure	Propulsion system model	Robustness Strategy
	Piece to Piece						
Tyre/Road	Tyre composition	Mfg-Hawk	ΔC_r	X	X	X	A. Select most robust tyres
	Tyre tread pattern	Pattern					
	Tyre-road contact patch	mm²					
	Tyre width	ISO		X	X		
	Changes over time/cycles						
Tyre/Road	Tyre wear	Tread	-				*Minimal effect*
	Tyre pressure	bar	ΔC_r	X	X	X	C. Specify most robust pressure
	Customer Usage						
Tyre/Road	Rider mass	kg	Δm	X	X	X	B. Increase robustness to rider mass
	eBike mass	kg		X	X	X	B. Reduce mass
	Dirt on tyre	mm³	ΔC_r				*Neglect*
	eBike Ride cycle	Standard	ΔC_r, $\Delta\theta$			X	B. Optimise propulsion system
Rider & Bike/ External Environment	Effective frontal area & Rider posture — U	m²	ΔAC_a			X	B. Streamline frame
	Effective frontal area & Rider posture — D	m²				X	
	External Environment						
Tyre/Road	Road surface	Defined	ΔC_r	X	X	X	B. Tyre pressure to suit surface
	Weather (rain)	mm/hour	-				*Ignore*
	Ambient temperature	°C	ΔC_r	X	X		A. Select most robust tyres
Rider & Bike/ External Environment	Wind speed	km/h	Δv_a			X	*See projected Frontal Area & Rider posture*
	Wind direction	Deg wrt eBike	Δv_a			X	
	Ambient temperature	°C	$\Delta\rho$			X	
	Elevation	m	$\Delta\rho$			X	*See Ride cycle*
	Gradient	%	$\Delta\theta$			X	
	System Interaction						
Propulsion/Braking	Frictional force/torque	N, Nm	(ΔF_p)				C. Eliminate non intended braking
	Diverted Output						
Propulsion/Drivetrain	Drivetrain efficiency loss	%	(ΔF_p)				C. Reduce efficiency losses

John said that it was important to look at the robustness strategies in the round and not make the system more robust to one noise factor in a way that made the system less robust to other noise factors resulting in less overall system robustness. In picking up John's point Sarah said that an example of this would be to use road tyres on a Mountain Bike to reduce rolling resistance and explained that road tyres have little tread which would give them little grip in off-road muddy conditions. John thanked Sarah and said that the propulsion system model allowed the team to see the effect of applying different robustness strategies across the propulsion system.

Wolfgang intervened and said that it was not necessary to develop an NCFT to identify the example that Sarah had just given and added that he wondered why the team needed to go to all the bother of developing an NCFT at all in identifying robustness strategies for the different noise factors. John asked Wolfgang to say more. Wolfgang cited the examples of streamlining the frame to reduce drag and changing tyre pressures to suit different road surfaces and said that these were statements of the obvious. John agreed with Wolfgang that the two examples that he had mentioned were strategies that the team could easily have identified without the NCFT. John quickly accessed the eBike propulsion system noise effect simulator, and without sharing it, drew team members' attention to the robustness strategy of increasing the eBike robustness to the noise factor of rider mass. He selected noise factor magnitudes in the simulator and at the same time, seemingly without drawing breath asked Wolfgang if he could say off the top of his head what the difference in the Total Resistive Force would be between a rider of 60 kg riding at 5 km/h on asphalt slope of 3%, on Hawk RS tyres, with a tyre pressure of 3.0 bar, on an eBike of mass 22.5 kg in an upright position, with an Effective Frontal Area of 0.250 and windspeed of 15 km/h, at an altitude of 200m and at an ambient temperature of 15°C, compared to a rider of 90 kg with an Effective Frontal Area of 0.347 riding on Hawk HS tyres, with a tyre pressure of 5.0 bar under otherwise identical conditions. Wolfgang replied that he did not know the difference off the top of his head but could calculate it in a few minutes. John apologised to Wolfgang and admitted he had asked an unfair question adding that in case anyone was interested the answer, according to the propulsion system noise effect simulator, was 9 N. John then explained that the strength of the NCFT was that it included all key noise factors and their effects allowing combinations of noise factor effects to be investigated. Phil added that Sam and he had made extensive use of the NCFT in developing the propulsion system model. Wolfgang conceded that the NCFT might have its uses.

John said that he had found the last discussion interesting and then explained that Oxton could only take responsibility for implementing some of the robustness strategies such as *Select the most robust tyres* and *Reduce mass* of the eBike. John added that other robustness strategies such as *Reduce efficiency loss* in the Drivetrain would largely be the responsibility of the parts supplier. John continued by saying that, yet other robustness strategies such as *Streamline frame* would be more of a joint responsibility with the part supplier. Andy asked John if responsibility for implementing the robustness strategies was shared between Oxton and part suppliers how could all the noise factors listed on the NCFT be investigated in combination. John responded by saying that it was Oxton's responsibility to explain to a part supplier the Noise Space in which the supplier's part would be operating in order to confirm the part's robustness.

In following on from what John had said, Yumiko explained that Agano were working to adapt one of their current road eBike Drivetrains for use in other types of eBike with lighter weight frames such as hybrid road/off-road. John said that what Yumiko had outlined sounded interesting and asked her how they were adapting the Drivetrain. Yumiko said that they were looking to tune it for use on lighter-weight eBikes with a mid-drive motor by using Parameter Design considering things like the impact of motor torque on the chain. Yumiko added that the Drivetrain already had low frictional losses and Agano were aiming to maintain this in any new eBike application. Yumiko said that she would set up a meeting for John with her manager and herself if he was interested to find out more and

John said that he was. Yumiko asked John if he would email her some times when such a meeting would be convenient, and John said that he would do so. John then thanked team members for their contributions in the meeting and said that it was time for Warm Down.

The Warm Down comments focused on the Warm Up and the robustness strategies. The Warm Up was generally seen as a fun exercise while several team members said they found the different approaches to achieving design robustness interesting and thought that the documenting of robustness strategies on the NCFT would prove useful. After the meeting had closed John helped Phil set the room furniture back to its normal state. In walking back to his office John felt pleased with how the meeting had gone and was looking forward to the meeting with Agano that Yumiko had offered. John thought that this meeting might lead to furthering the relationship between Oxton and Agano and potentially lead to significant, tangible, and mutually beneficial results from the involvement of suppliers in the Oe375 team. John also remembered Andy's offer of help in getting information from Hawk and Wolfgang's earlier offer to give Oxton access to their customer purchasing survey and saw them in the same context. In thinking of Wolfgang's offer, John felt guilty in failing to follow up on this so far and made a mental note to rectify his inaction by arranging a meeting with Wolfgang as soon as he could. In reflecting further on his future meetings with Agano and Wolfgang, John decided to ask Margaret Burrell if she would like to attend both meetings.

On getting back to his office John gave Margaret a call only to get her voicemail. John left a message saying that he had something interesting to discuss with her and asked her to give him a call when convenient. John then gave Wolfgang a call and on him answering his phone John apologised to him for not having contacted him sooner about reviewing the Teutoburg customer data. Wolfgang said that this was not a problem and explained that in the interim he had contacted the IT and Legal Departments in Teutoburg both of whom had now given the go-ahead for discussions between their two companies. John asked Wolfgang how the Teutoburg customer data was protected, and Wolfgang explained that the company's Terms and Conditions, which customers had to agree to before taking the survey, allowed for data to be shared anonymously within the Teutoburg group and associated companies. Wolfgang added that he had been told that Oxton could be classified as an associate company particularly given his involvement in the Oe375 programme. Wolfgang observed that he imagined that few people actually read the terms and conditions before agreeing to them. John thanked Wolfgang and asked him how the data might be accessed, and Wolfgang replied that the survey software allowed for a number of formats of report to be generated and shared. John said that he would like to set up a meeting between Wolfgang and anyone else from Teutoburg that Wolfgang felt should attend along with Margaret Burrell the Oxton Director of Supply and himself to discuss things further. Wolfgang said that he knew Margaret and gave John a few time slots when he was available in the next week to hold the meeting. John said that he would send Wolfgang a meeting notice once he had found out Margaret's availability.

John then took the opportunity to ask Wolfgang what he said was a very important question that he had been mulling over for some time. Wolfgang said that he was intrigued and asked John what his question was. In response, John asked Wolfgang how he had become a Manchester United supporter. Wolfgang said that was an easy one and explained that his grandmother had met his grandfather, who was in the British military at the time, immediately following the Second World War and added that his grandfather came from Manchester. Wolfgang explained that his grandparents settled in Germany and his grandfather used to take him to Old Trafford as a boy at least once a year when he went back to see his family. Wolfgang said that although his grandfather had died some time ago his father and he still maintained the family tradition. John thanked Wolfgang and said that he had answered two questions in one with the second question being how he had acquired his fluency in English. John thanked Wolfgang for his help and said that he was looking forward to the meeting next week.

Margaret rang John later in the day and seemed very interested in what John had to tell her and said that she would be very happy to join him in any meetings with Agano and Teutoburg and advised John to drop Jim Eccleston a note to make him aware of the proposed meetings. On reviewing their calendars, they found a couple of slots in the next week when they were both available for the Agano meeting and a time which matched Wolfgang's availability. John sent a note to Yumiko and told her that he would like to invite Margaret to the meeting and told her of Margaret's and his availability. John then sent a meeting notice to Wolfgang and a note to Jim telling him of the proposed meetings and their purpose.

Driving home that evening John thought while every week at Oxton seemed eventful this week had been particularly significant and, despite some trepidation about starting his expanded role on Monday, he was looking forward to seeing what the next week brought. As he was nearing home John remembered that Lucy Collins had said that she would call him. On reaching his house and opening the front door John was greeted by Jane and the first thing that he said to her was to ask if Lucy Collins had rung. No sooner had Jane said that she had not, the landline phone rang. John rushed to answer the phone and was pleased to hear Lucy who, after exchanging pleasantries, told John that it had happened again and this time closer to home. John asked Lucy to explain, and she said that Blade HR had received an anonymous voicemail overnight accusing someone, whom John would probably know, of inappropriate behaviour which the person in question had denied vehemently and dismissed as nonsense. John asked Lucy who the accused person was and what they had been accused of and Lucy said that, since the matter was being kept internal to Blade, she did not feel able to reveal this information. As with her previous phone call to John, Lucy asked him to keep the fact that she had called him confidential, and John assured her that he would.

John took the opportunity to ask Lucy if she knew what had transpired in Blade after the receipt of Jim Eccleston's letter. Lucy said that she did not know who Jim Eccleston was or have knowledge of any letter that he had sent. John apologised to Lucy and said he should have explained to her that Jim Eccleston was the CEO of Oxton Cycles. John added that after conducting an investigation in Oxton about the accusation involving John, Jim had written a letter to Georgia Lee in the Legal Department in Blade saying that he had not found a scrap of evidence to support the allegation and asking her to think twice before sending him on another fool's errand. Lucy said that she had not heard anything more about the accusations against John since she handed in her report and added that, because she never gave any credence to the accusation, she thought John might be interested to hear about the second similar incident. John said that he was very grateful to Lucy for sharing this information with him and asked her how she came to hear about it. Lucy said that while Blade had successfully kept the incident in which John was named under wraps on this occasion information had leaked onto the Blade grapevine and she speculated that this was because the accused person was internal to Blade. With that Lucy said she had to go, and John thanked her again for calling him and Lucy said that she would let John know if anything relevant to him resulted from this recent incident.

Jane listened patiently as John related to her what Lucy had said, and then said that she thought that it was unfortunate that Lucy had called since they had managed to put the Blade allegations against John behind them. While agreeing with Jane, John said that awful as it must be for this second accused person, it was somewhat reassuring to know that it appeared that he was not the only one being targeted by, what he called, this lunatic. Try as he might, his thoughts kept returning that evening to what Lucy had said and he wondered if what had happened recently might in the end help to find the person who had so wrongly accused him.

Background References

Question delivery	Harris, R. (1973) Answering Questions Containing Marked and Unmarked Adjectives and Adverbs, *Journal of Experimental Psychology*, 97(3), 399–401. Stapleton, K. (2018), Chapter 4: Questioning, In Hargie, O. (editor) *The Handbook of Communication Skills*, Routledge, 121–145.
Robustness Strategies	Zhou, J. (2005) ***Reliability and Robustness Mindset in Automotive Product Development for Global Markets***, SAE Technical Paper 2005-01-1212, https://doi.org/10.4271/2005-01-1212. Pages 6–7

Chapter 12

Moving Levels

John returned to work after spending a surprisingly relaxing weekend with Jane, despite his phone conversation with Lucy on the Friday evening and the upheavals to come in Oxton. He had read the email that Jim Eccleston had sent out to all employees over the weekend and while by now he was getting used to Jim's messages being brief and to the point, he thought that this note was particularly brusque in its content and tone. Jim had merely mentioned that Paul had left the company to take up the offer of a job elsewhere and made no mention of any achievements that he might have had while with Oxton or good wishes for his new venture. John had also seen that Jim had set up a meeting with Sarah, Jennifer and himself in his office first thing on Monday morning. Vicky had followed-up Jim's meeting notice by sending a text to each of the meeting participants telling them about the meeting, and John assumed that Vicky had done this in case any of them did not check their emails over the weekend. In thinking things over after seeing Jim's meeting notice, John felt that Sarah and Jennifer would be anxious after Paul's departure from Oxton had been confirmed, and they were told that they would now be reporting to him. In seeing that their calendars were free for the first hour of the day, he decided to have a chat with them immediately after the meeting with Jim. He thought about sending them a meeting notice but decided against this since they would not know that they would be reporting to him until the meeting with Jim and might wonder about the apparent urgency of any meeting with himself. John also decided that the meeting might be more relaxed if it seemed like a spontaneous chat. John then spent a while mulling over what he would say to them.

On arriving at Oxton on Monday morning John left the bag containing his laptop in his office and hung up his coat before making his way to Jim's office to find Sarah and Jennifer chatting with Vicky about yesterday's late afternoon Premier League match. Jim was in his office and called them all in and sat down at the table with them. To John, the meeting seemed more like a celebration than a wake with the tone being set by Sarah's remark that Oxton's gain was Paul's new company's loss after Jim had mentioned Paul's leaving as a reason for calling the meeting. John could not help

DOI: 10.4324/9781003286066-12

smiling at Sarah's tweaking of the saying and was relieved to see the others, including Jim smiling, as well. As John expected Jim kept the meeting short by explaining that Sarah and Jennifer would now work for John and John would take over the running of the O254 budget hybrid commuter programme in addition to Oe375. Sarah asked Jim what the implications would be for Oe375 given John was taking on an additional programme and Jim responded that this was something that John would talk to them about. Jim then called the meeting to a close and wished the three of them good luck in working in the changed situation.

As they left Jim's office, and knowing he was likely to be bombarded with questions, John asked Sarah and Jennifer if they had a few minutes to discuss things in his office. John managed to keep the conversation to generalities until they reached his office where Sarah and Jennifer sat down at his table, and he shut his office door before joining them. John began the discussion by asking Sarah and Jennifer what they thought, and they both replied that they were pleased. John asked them if they had known that Paul was going before Jim had told them, and Jennifer said that there were rumours swirling around in Paul's department on Friday and in the end, Paul had told them in confidence that he was leaving at the end of the day. Sarah said that Paul had stressed that although the suddenness of his departure might make appear that he was not leaving of his own accord, he was leaving by mutual agreement with Jim Eccleston for a more senior job with a significantly better remuneration package in another company. Jennifer said that she had asked Paul which company he was going to but all that he would say was that it had nothing to do with bicycles. John told Sarah and Jennifer that Jim had told him that Paul was going to become the CEO of a furniture manufacturer. Sarah then asked John if his new responsibilities meant that he would have less time to spend on Oe375 and John replied that this was inevitably the case and because of this he had drafted a plan for taking the Oe375 programme forward which he would like to share with the team.

John continued by saying he was excited that Sarah, Jennifer and he would be forming their own department and was looking forward to working with them more closely. Sarah asked John how he saw them all working together, and John said that he wanted them to work that out together rather than him saying how it was going to be. John added that he intended to set up a series of weekly department meetings which he described as three-to-three meetings. John said that one thing that he was insistent on was that Sarah and Jennifer considered themselves as working with him and not for him. John explained that, of course, there were times when John's seniority would need to be recognised such as signing off expense claims and added that one thing his seniority did not mean was that he was always right or that Sarah and Jennifer had to agree with him by default. John said that in his experience, creativity often resulted from constructive disagreement. In thinking about his future working relationship with Sarah and Jennifer over the weekend, John had realised that their relationship would require time to mature.

John asked Sarah and Jennifer if they thought it would be possible for him to meet the Oxton members of the Oe375 team that morning to talk about the new situation and Sarah replied that Jennifer and she were going to the Team B meeting that started shortly. Sarah suggested that John join Team B for the second hour of the meeting and added Yumiko might still be in the meeting as sometimes she left after the first hour and other times she stayed for the whole meeting. John said that he was fine with Yumiko being at the meeting and asked if Sam would be attending. Sarah said that Sam probably would not be at the meeting and said that she would talk to him and see if he could attend the second hour. John then asked Sarah and Jennifer if they had any more questions and Jennifer asked if they would need to move their desks. John said that he did not think that was necessary since they could use his office whenever the three of them needed to get together and his and Paul's offices were equidistant to Sarah and Jennifer's desks. John said that he assumed that they would like to sit in the same area and alongside the same people as they did now. Sarah and Jennifer confirmed that they would like to stay where they were, and Jennifer then corrected John by telling him that he should have said Paul's old office. John smiled and asked if they had any more questions and Sarah replied that they had lots but that they needed to get ready for the Team B meeting, and with that Sarah and Jennifer left telling John that they would see him later.

When John knocked at the door and entered the Team B meeting, those present immediately stopped what they were doing and looked at him as he sat down at the table with them. Phil broke the silence by wishing John good luck in his expanded role and this was followed by the other team members expressing similar sentiments. John was pleased to see that Sam had joined the meeting. Being aware that Yumiko was signed into the meeting, John briefly explained his new situation without any reference to the financial reasons behind his change of responsibility and then turned his attention to the Oe375 team in the guise of Team B. John said that he would like to explore the team's feelings about his sudden change of role and asked for comments and questions. As John expected the recurring theme in team members comments was the point raised by Sarah and Jennifer earlier, that John would be unlikely to be able to devote as much time to the Oe375 programme in the future as he had in the past.

John had decided in the interim period between seeing Sarah and Jennifer and him meeting with Team B that the best way of addressing team members' anxieties was to take the team through the draft plan that he had developed for taking the Oe375 programme forward. Before sharing his plan, however, John summarised the Oe375 team's accomplishments to date by reminding the team that they had nearly taken all of the analysis of the Oe375 propulsion system through Phases 1 to 3 of the Oxton first-time engineering norm (OFTEN) framework and were working on the design failure modes and effects analysis (DFMEA) for the propulsion system. John said that the team had also developed a model to simulate the effect of noise on propulsion system performance and had identified robustness strategies and other countermeasures to address the noises that affected the propulsion system.

John told the Oe375 team that he felt that Paul's departure occurred at a favourable time in the Oe375 programme in that the team had applied the OFTEN framework Phase 1 to 3 tools to a key system of the eBike and so were perfectly capable of repeating this analysis for the other eBike systems. John then said he wanted to outline the key steps that he saw the Oe375 team taking in moving forward and shared a slide (Figure 12.1).

1. Complete Design FMEA for Propulsion System
   - and NCFTs including Control Factor/Robustness Strategy
2. Conduct OFTEN Phases 1* through 3 analysis for other eBike Systems
   - Frame
   - Wheel & Tyre Set
   - Braking
   - Steering
   - Seating
3. Action Design FMEA Recommended Actions
   - Robustness strategies
   - Other countermeasures for Failure Modes
4. Conduct Design Verification

** Decide for which Systems SSFD is developed*

Figure 12.1 Draft plan for Oe375 programme.

John said that he was sharing a draft version of the plan since he wanted the team to take ownership of it. John pointed out that in Steps 1 to 3 of the plan the team would be using the same tools in the same order that they had used in analysing the propulsion system. John explained that he thought it was important that the team completed OFTEN framework Phases 1 to 3 for all the eBike systems before looking at Phase 4 Design Verification and reassured the team that he would discuss in detail how Design Verification might be conducted before they started this phase. In addressing the use of the system state flow diagram (SSFD) John reminded the team of his view that developing an SSFD for those systems which were well understood from the engineering perspective rarely added a great deal to the existing knowledge. John suggested that the Braking and Steering systems would probably benefit from the development of an SSFD and said that this was only a suggestion as he wanted the team to decide this for themselves when they came to start the analysis of these systems.

Sarah asked John how the functional and architectural decomposition of a system should be achieved if an SSFD was not developed. John asked Sarah to say more, and Sarah explained that the development of a DFMEA relied on the identification of the Affected System Function during interface analysis which was dependent on the functional decomposition of the system being analysed and has been completed in the form of a function tree. John told Sarah that she had raised a good point, and in taking the example of the frame, he said that Oxton did considerable analytical analysis on frames for their high-end models and this knowledge and experience should be sufficient to develop a boundary diagram and function tree for the frame without developing an SSFD first. John said that he was happy to discuss this point further when the Oe375 team started to look in detail at the other individual eBike systems.

With this point in mind, John agreed with the team that they would keep their schedule of Oe375 team meetings, and he said he would attend as many of these as he could acting in an advisory and coaching capacity. John said that he would prefer not to lead the meetings and explained that this was not only because he now had less time available to support the Oe375 programme but because he felt that, given the Oe375 team's performance up to now, the team was perfectly capable of proceeding without his leadership. John recommended that the team elect their own leader from within the team. John said that because of the change of circumstances he would no longer be introducing a different People Skill in each meeting but urged the team to maintain the Warm Up and Warm Down sessions in their meetings. John added that if anyone was interested in giving an occasional short presentation on a People Skill he had reference material that he would share. John said that in stepping down from the leadership role he would also no longer be leading the Review/Preview meetings and added that as he felt that these were important, he hoped that the new leader would continue running them. John also said that having seen Phases 1 to 3 of the OFTEN framework conducted once, he felt that the external members in the team should decide for themselves which, if any, future, Oe375 meetings they would attend and said that he would speak to Andy and Wolfgang individually about this. John added that in any event, he would notify all the current Oe375 team members when the Oe375 programme was about to enter Phase 4 of the OFTEN framework. John said that he would prefer that the Oe375 team decided for themselves without him being present if they wanted to adopt his plan and in doing so identify what changes they wished to make to the plan. John added that before leaving the team to consider the draft plan he was happy to answer any questions for clarification that the team might have. Phil said that the thought that John had explained the plan clearly and added that the plan seemed straightforward to him. Yumiko asked John if he would still be involved in the meeting with her manager and her. John replied that he would and thanked Yumiko for setting up the meeting saying that he was looking forward to it. Since there were no more comments John said that he would leave the team to their discussions and left the room to a round of applause.

John subsequently learnt from Sarah and Jennifer that the Oe375 team had adopted his plan with minor revisions and agreed to start the next stage of their analysis with the Braking system. They had also decided to take each of the different systems in turn through Steps 1 to 3 of the plan before starting the analysis of the next system. As John hoped, Sarah told him that she had been elected to lead the team and she reassured John that she would continue with Phil's assistance to run the Review/Preview meetings. As things turned out the two series of Oe375 and Team B meetings tended to merge although they were still distinguished by name and the inclusion of a weekly programme status review in the Oe375 team meetings which John attended. The Team B meetings focused almost exclusively on furthering the OFTEN framework analysis across the eBike systems. The status review was facilitated by Sarah's introduction of a programme task log. Ostensibly John led the status reviews although since Sarah and Phil chose what they would use as examples of the analysis that Team B had been conducting, this was a moot point. John was happy with this arrangement since on the few occasions on which he had asked to see further analysis to confirm the integrity of what Sarah and Phil were showing him they had largely been able to furnish this without difficulty. John was able to make minor adjustments to both the programme direction and the detail of the analysis by what quickly became an expected feature of the meeting known as "John's Tip of the Week".

John spoke to Andy and Wolfgang about the new situation. Andy decided to dip in and out of the Oe375 meetings depending on what system was being analysed, and Wolfgang said that while he would be available for a discussion as and when required he would wait until Oe375 moved on to Phase 4 of the OFTEN framework before re-joining the team meetings. Yumiko continued as a working member of the Oe375 team while they were considering Groupset design including discussion of the interactions that other bicycle systems might have with the Groupset.

John had his first meeting with the O254 team on the Monday afternoon immediately following Paul Clampin's departure and, as he had said he would, Jim attended the initial part of the meeting to introduce John to those that had not met him before. Jim then gave the team what John thought was a highly abridged version of why the management changes were taking place and concluded by saying that he was sure that things would progress well under John's guidance. Jim then left the meeting without taking any questions. In thinking at the weekend about his first meeting with the O254 team John had concluded that the team would be likely to have been unsettled by Paul's departure and he had decided to start the meeting with an extended Warm Up in which the team members identified their hopes and fears for the situation with which they were suddenly faced. Judging from their demeanour, John deduced that team members were rather puzzled by not diving straight into the meeting's core business although, because they appeared to accept what he was asking them to do, John suspected that his reputation might have gone before him.

On reflecting after the meeting John thought that the Warm Up achieved its purpose in two key areas. Firstly, he got team members to discuss their concerns which largely centred on their fears that the change of management would disrupt the O254 product development process and would cause them to miss their timing and budget targets. Secondly, John felt that conducting a Warm Up clearly signalled to the team that he intended to take a different approach in guiding the programme to the one that Paul had taken. John reinforced the difference in approach while focusing on the team's concerns over programme timing and budget when he told team members that he expected that the way in which he was going to ask them to work would initially take significantly longer than was planned for the product design process that Paul had been actioning. Seeing what he took to be a look of disbelief on the faces of a couple of team members John continued by saying that the more rigorous product design process he was asking the team to follow would result in far fewer failures surfacing during Design Verification at, or prior to, launch. John added that as a consequence these two parts of the product development process should take significantly less time than would have happened with the team's current approach. John said that, given this, he was expecting to meet the programme timing and budget, although he metaphorically crossed his fingers as he gave the team this assurance. John spent the rest of the meeting gaining an understanding on the status of the O254 programme before introducing the team to Warm Down. On reflecting after the meeting, John was generally pleased by the way it had gone, although he was disappointed to find that the scale of the task that he faced in turning round the O254 programme appeared to be a lot closer to what had been his worst-case scenario prior to the meeting than it was to his vision of the best case.

John and Margaret's initial meetings with Teutoburg and Agano both took place during the first week of John's changed role in Oxton with both meetings proving to be productive and subsequently resulting in a stronger relationship between Oxton and the two companies. As a result of reviewing the Teutoburg customer data the

Oe375 team decided to focus the eBike more towards a road/gravel hybrid eBike than a road/mountain hybrid. In electing to shift the Oe375 core customer group subtly there were some in the Oe375 team who saw this as nothing more than a marketing ploy. It took Wolfgang to point out that to include what he called Gravel DNA, Oe375 would need to have the visible gravel features of wider tyres, dropped handlebars as well as the less visible features of comfort and ease of riding in off-road mode that wider tyres, a wider range of gearing and appropriate frame geometry offered. Mainly in deference to Phil, the team also agreed to offer a straight handlebar option. The team subsequently decided that Oe375 would have four primary build variants with the cheapest variant having mechanically operated disc brakes and the other three variants hydraulic brakes with all four variants being largely distinguished by having different levels of Groupset. The team also decided to include several of the excitement features that they had identified early in the programme as standard and others as optional extras. Standard features included an antitheft alarm and easily detachable wheels with an ultrafast charger and fingerprint unlock linked to the rider's smart phone being offered as options.

The meeting with Agano proved to be equally beneficial as, and synergistic to, the strengthened relationship with Teutoburg. Yumiko's manager Kusuko Tanaka had explained to Margaret and John that Agano had developed a new efficient low-loss Groupset along with a complementary mid-drive motor for electric road bikes and were looking to extend the use of these systems with electric hybrid and gravel bikes. With Jim Eccleston's blessing, Team B agreed to share documentation with Agano on the Oe375 interfaces that involved Groupset and Electric Drive components to allow Agano to tune their systems to the Oe375 eBike requirements without compromising the Drivetrain and Electric Drive efficiency. Both companies agreed that Yumiko should remain a member of Team B while OFTEN Groupset documentation, which included interface tables, NCFTs and DFMEAs, was being developed. Oxton and Agano also collaborated on the development of the Electric Drive control strategy so enabling Agano to tune the existing road eBike Electric Drive Control Unit and smart control rider interface they had developed to Oe375 requirements. Towards the start of their enhanced partnership, Margaret Durrell and Jim Eccleston initially signed a letter of intent with Agano, later followed by a formal contract, for them to supply all Groupset and Electric Drive components for Oe375 including those for the different build variants.

In looking back after several months had elapsed, John was surprised how what seemed a very different way of doing things in the early weeks after Paul Clampin's departure had now become normal. After the initial disruption things had settled down with both the Oe375 and O254 teams getting into their stride. In considering his experiences with the O254 programme John recalled how surprised he had been to learn in the initial meetings with the O254 PD Team how closely the programme decision-making reflected what Sarah and Phil had told him about the budget bike product design process in that fateful week in which Paul Clampin left Oxton. John had gathered that Paul's main interest in the O254 programme was ensuring that it kept to time and on or below budget, with any engineering details having a lower priority, particularly if they conflicted with his top two priorities. John had found that he had needed to temper his questioning of the team in his first few meetings for fear of embarrassing them when he had asked them to explain the engineering logic behind some of the key decisions that had been taken. Nevertheless, John felt that he had been able to have a beneficial effect on the O254 programme even though the team had progressed some way into the programme timing when he took it over. On Jim's advice, John invited the O254 Technical Expert, Peter Andersson to join him in running Review/Preview meetings for O254. Peter sat alongside Sarah Jones and Jennifer Coulter in what was Paul Clampin's office area and was familiar with what was happening on the Oe375 programme. Jim had taken Peter under his wing in what became known as "The Oxton Reshuffle". Peter became invaluable to John in both preparing for, and running, the O254 meetings and was instrumental in persuading O254 team members to participate in running the meeting Warm Ups.

While John was not initially expecting O254 team members to adopt the People Skills as such, one aspect that he did insist on was that the team topped and tailed their team meetings with him with Warm Up and Warm Down. John was glad of the preparatory work he had done on Warm Ups with the Oe375 team and was able to reuse individual Warm Ups with the O254 team and given the relationship of each Warm Up to a particular People Skill he seemed to drift naturally into summarising each People Skill in the O254 meetings. Much to the team's surprise John also insisted on taking a formal break around the middle of their meetings. A couple of team members came up to him later and said how much they appreciated the break and explained that Paul had always insisted on the team using every minute of each meeting, for what he called, productive work. John had also concluded that it was not appropriate to take the team through the totality of the OFTEN framework although he spent an hour in his second meeting with them summarising the framework after several team members told him that they were curious as to what was happening on the Oe375 programme. John also spent an hour early on with the O254 team summarising what he saw as the key People Skills, as team members had said they had also heard about their use on the Oe375 programme and wondered what this was about. John learnt from several team members that they had found the change of management style between Paul and himself to be stark and it had taken them a few meetings to adapt to his way of doing things with one telling him that for the first month or so she fully expected things to revert at any time to, what she described as, the previous regime's way of working.

With Team B's permission, John spent time with the O254 team reviewing relevant self-contained sections of the Oe375 DFMEAs as Team B developed them, having first taken the O254 team through the functional approach to FMEA. John was pleasantly surprised to see how O254 team members were able to adapt the DFMEA to the hybrid commuter bike that they were working on. John thought it paradoxical that he was recommending that the O254 team use the conventional Oxton approach to DFMEA of adapting an existing document although this time he recognised that the base documentation was developed on solid foundations. John also advised the O254 team to place significantly more emphasis on Mistake Identification and Current Controls Detection in developing their DFMEAs than was the case with Oe375.

Under John's guidance and active help Peter Andersson and Isla Muir, another member of the O254 team, reviewed Quality Data such as warranty repairs for those parts that were carried over from the bicycle that O254 was replacing. This was done to understand any system-level customer-related problems with these parts in trying to ensure that such problems were not carried over with the parts to O254. Wolfgang Hurst also provided customer feedback that Teutoburg had collected for the outgoing model that O254 was replacing. John was struck by the most frequent issue in the Teutoburg data which related to the fitting of aftermarket mudguards onto a frame and front forks without eyelets. Having clip-on mudguards on his bike John sympathised with this situation in that his experience was that by generally being shorter than eyelet-mounted full-length mudguards they provided less protection from the wet. John felt that not providing mudguard eyelets on a bike designed for the commuter market was a mistake and the O254 team agreed with him.

The O254 team reviewed the O375 NCFTs to help them identify the noise factors and noise factor magnitudes that should be excited in each detection test that they conducted. One problem that was revealed from detection testing was associated with poor shift performance of the rear derailleur of O254. The decision had been made to partially route the brake and gear cables on O254 internally within the frame and an investigation revealed that at the lower limit of manufacturing piece-to-piece variation at two places in the rear derailleur cable routing cable curvature was below the Groupset manufacturer's minimum bend radius. Again, John saw this failure mode as being due to a mistake. John was not quite sure why a budget commuting bike would have internal cable routing knowing that this feature was usually reserved for more upmarket bicycles and assumed that it was seen as a marketing feature.

John was pleased that in adapting the DFMEAs to their own product the O254 team found several critical failure modes and numerous other less severe failure modes. John realised that if it were not for the DFMEAs these would probably not have been discovered until Design Verification, or worse at, or after, product launch. John felt that any initial scepticism that O254 team members might have had about John's recommendation to customise the Oe375 DFMEAs to their own product quickly evaporated when they began to discover the failure modes and develop countermeasures to them. John found that the repurposing of the Oe375 DFMEAs acted as a natural conduit for the introduction of other key tools and techniques from the OFTEN framework to O254 team members, so avoiding the perception that the team was spending significant amounts of time in training as had happened with the Oe375 team.

John took care to limit the areas of the O254 bike that he recommended the team apply OFTEN techniques to, seeing his work with the O254 programme as being more the facilitation of selective remediation than top-down design. The first use the O254 team made of interface analysis was as a result of reviewing the unintended function failure mode associated with disc brake rub on the Oe375 Braking DFMEA. The O254 team considered increasing the low occurrence rating of 3 given by the Oe375 team for this failure mode to a moderate rating of 4 for the O254 bike which, along with the severity rating of 5, caused the priority rating to move from moderate to high risk (Table 8.3). John cautioned against treating the DFMEA rating scales as a science, nevertheless the O254 team felt this failure mode was worthy of further investigation. John suggested that the team review and amend as required the interface analysis that the Oe375 team had conducted on the parts associated with the brake calliper and rotor mounting to get a better understanding of the potential causes of the failure mode. Having done this the O254 team were able to replicate the intermittent failure mode on a rig under specific conditions using a prototype frame and front forks along with the type of wheels and braking system they were expecting to use on O254. The team was subsequently able to reduce the probability of occurrence of the failure mode significantly by simple changes to the geometry of the mounting of the brake callipers.

It seemed to John that his relationship with Jim changed after Paul Clampin's departure. His one-to-one meetings with Jim continued but now on a two-weekly basis rather than one. The meetings had broadened out to discussing general company issues in addition to John updating Jim on the progress of the Oe375 and O254 programmes with this reporting being briefer because, although he was reporting on two programmes, once he had gone through the different robustness strategies, he had no more new tools to discuss until the programmes reached Phase 4 of the OFTEN framework. John felt somewhat awkward initially in reporting to Jim on the Oe375 programme now that Sarah was leading the team, but Jim had said that since Sarah reported to John rather than him, he would rather John kept him abreast of the programme status. Sarah was perfectly happy with Jim's wishes in this respect when John had told her what Jim had said. In one of their one-to-one meetings, Jim had explained one aspect of Paul's departure that had puzzled John. Jim told John that the reason that he wanted to keep Paul's departure low key was because he did not wish to be put in a position in any formal gathering to have to say a few kind words about Paul's time at Oxton. Jim confided in John that he had found Paul's overtly acquiescent behaviour to himself offensive when he knew he presented a very different persona to others in the company. Jim said he was pleased when Paul had told him about his intent to leave and added that, although felt bad about it afterwards, he had written him a good reference to ensure that this was no impediment to his departure.

The Oe375 team worked in a disciplined manner through the plan they had adapted from John's initial draft (Figure 12.1) and, although he was intimately involved in what the team was doing, John was surprised when he realised that the team had nearly completed the analysis of all the eBike systems up to, and including, Phase 3 of the OFTEN framework. Sarah had mentioned to John in their last three-to-three meeting that Yumiko had asked whether she might show Team B some of the work that Agano had been doing on the motor that was intended for Oe375. John said that he thought that this was a great idea and asked when she intended to do it. Sarah responded by saying that she had agreed with Yumiko that she would make a presentation at the next scheduled Oe375 team meeting and added that she was going to finalise the arrangements with Yumiko at the Review/Preview meeting first thing in the morning. Sarah said that she realised the Oe375 meeting was a programme status meeting but as she knew John would be available, she intended to repurpose the meeting. Sarah asked John if he would also be willing and able to attend the Oe375 Review/Preview meeting scheduled for first thing on the following day as she would like to ensure that they made best use of Yumiko's experience and expertise in the Oe375 meeting. John told Sarah he thought that this was a good idea and confirmed that he would attend both meetings.

As he strolled over to the meeting room the next morning John found himself looking forward to hearing what Yumiko had been doing in Agano. John was not disappointed by what he heard as Yumiko summarised the analysis work that she and two colleagues had been doing on the eBike motor and ran quickly through the slide presentation she intended to give to the Oe375 team. John was impressed by how Agano's analysis related to the Oe375 propulsion system's requirements that were identified through the modelling work that Sam and Phil had instigated. John immediately saw an opportunity to reinforce the link between systems engineering levels in the OFTEN framework and asked Yumiko if she would send him a copy of one of her slides, explaining that he would like to use the graphic in developing a slide that put her work into the context of the Oe375 programme. Yumiko said that she would send John a copy of her presentation saying he was free to use whichever slide he liked. John thanked Yumiko and made a mental note to send a copy of one of Yumiko's slides to Sam along with a request for Sam to generate a plot which aligned to the slide using the Oe375 propulsion system model.

Yumiko then said that rather than simply being a one-way presentation to the Oe375 team she would like to facilitate some discussion and asked the others how they thought she might best do that. Phil suggested that Yumiko do what he called "a John" and explained that what he meant by this was that Yumiko should ask team members a question here and there to get both their input, and gauge how well they had understood the technical content of the presentation. John was struck by Phil's description and, although he had not consciously thought about his approach with the Oe375 team, he thought Phil's description was apt. Yumiko thanked Phil for his advice and said that she hoped that she could do John's style of presentation justice. John said that he would like to understand how Agano had partitioned their DFMEAs and interface tables and would send an email to Yumiko requesting some specific information.

Sarah concluded the discussion of Yumiko's presentation by saying that she was looking forward to seeing more detail of the work that Agano had been doing and then turned her attention to the rest of the agenda for the Oe375 team meeting. Phil said that he was happy to facilitate the Warm Up as a lead into Yumiko's presentation. Sarah said that she thought that Yumiko's presentation would probably take up most of the meeting after Warm Up and said that she would leave it up to Yumiko to call for the break. Sarah added that, should there be any time at the end of the meeting, the team could pick up where they had left off at the end of their last meeting in their analysis of the Oe375 Steering System. Having agreed the agenda, Sarah concluded the meeting by thanking everyone for their input. After Yumiko had signed out John asked if Andy and Wolfgang would be attending the Oe375 meeting and Sarah said that they would not, since she had an arrangement with them of informing them a few days before any meeting where she thought the attendance of either of them would be mutually beneficial. John said that he thought this was just as well since there might be some sensitivity in Yumiko sharing detailed technical information with representatives of external companies.

As he walked into the conference room the next day, John found himself again looking forward to what he assumed would be an interesting meeting. Not unexpectedly he found Phil already there and the room set up. John was pleased to see Sam arrive and once Sarah and Jennifer had also arrived and Yumiko had signed in, Phil said that he was conducting the Warm Up. Phil asked team members to identify what they thought was the most important feature of the Oe375 eBike that would define its appeal to customers. John thought that Phil's question was incisive and struggled at first to answer it before deciding on what he called the eBike "Wow factor", which, he explained when his turn came to speak was the impression that the eBike should give a customer who walked into a bicycle shop and saw it for the first time. Phil asked John if he would be a bit more explicit and John said that the lines and paintwork of the eBike should make it stand out from the crowd. John was interested to hear the other team members' thoughts which ranged from the eBike's performance to its terrain flexibility. Phil thanked team members for their comments and after introducing the agenda turned the meeting over to Yumiko.

Yumiko thanked Phil and said that she wanted to illustrate some of the work she had been doing with a couple of colleagues on the A539 motor that was intended for Oe375 which built on some of the analysis that Team B had done. Yumiko said that because there were several types and configurations of eBike motor she would outline the design of A539 and shared a slide (Figure 12.2).

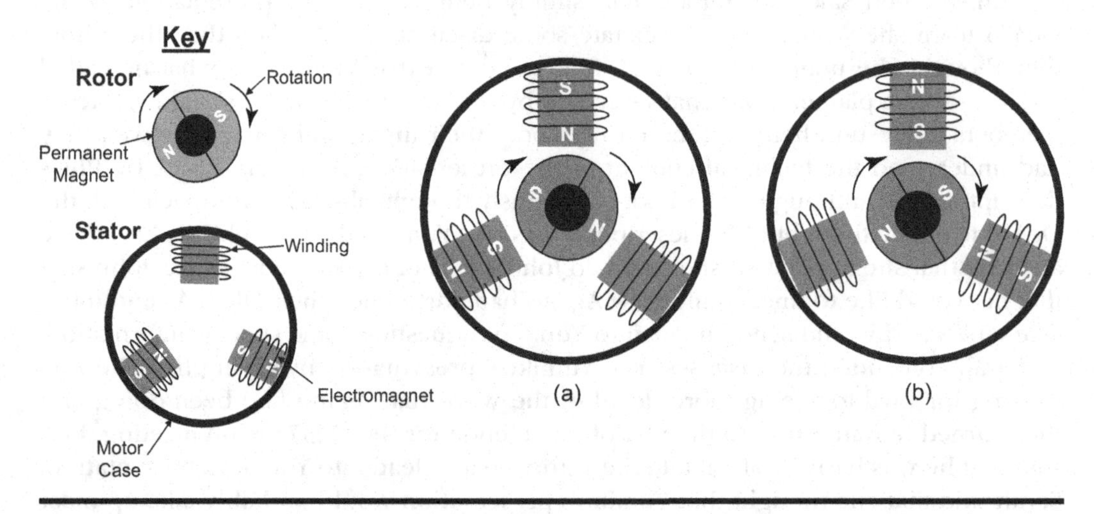

Figure 12.2 Sketch of single-phase BLDC motor.

In explaining the content of the slide Yumiko said that the sketch showed a single-phase brushless DC motor and while A539 was three-phase motor the general principles were the same with the permanent magnet rotor being central to the motor and the electromagnetic stator surrounding the rotor. Yumiko said that she was sure that team members knew how the motor operated with the stator magnetic poles being switched as the rotor turned by changing the current direction through the coils to ensure that the electromagnetic force between stator and rotor kept the rotor rotating in the same direction. Yumiko said that the shaft of the rotor was attached directly or indirectly to the chainset.

Yumiko explained that despite the architecture of A539 being pre-defined, on John's advice, she had used the same tools that Team B had used within the OFTEN framework and added that she had used the forerunner to the OFTEN framework which was FTED and spelt this out as first-time engineering design which John had described in a published paper. Yumiko then thanked John for his help. John said that she was welcome and added that his help did not amount to much more than a few emails and short telephone conversations. Yumiko said that nevertheless John's advice had been very helpful. Yumiko then said that she would show the analysis that Agano had conducted on the A539 in the context of the Oe375 Electric Drive system and referred the team members to a slide that John had shown in one of the Oe375 team's early meetings of the eBike partitioning and decomposition (Figure 12.3).

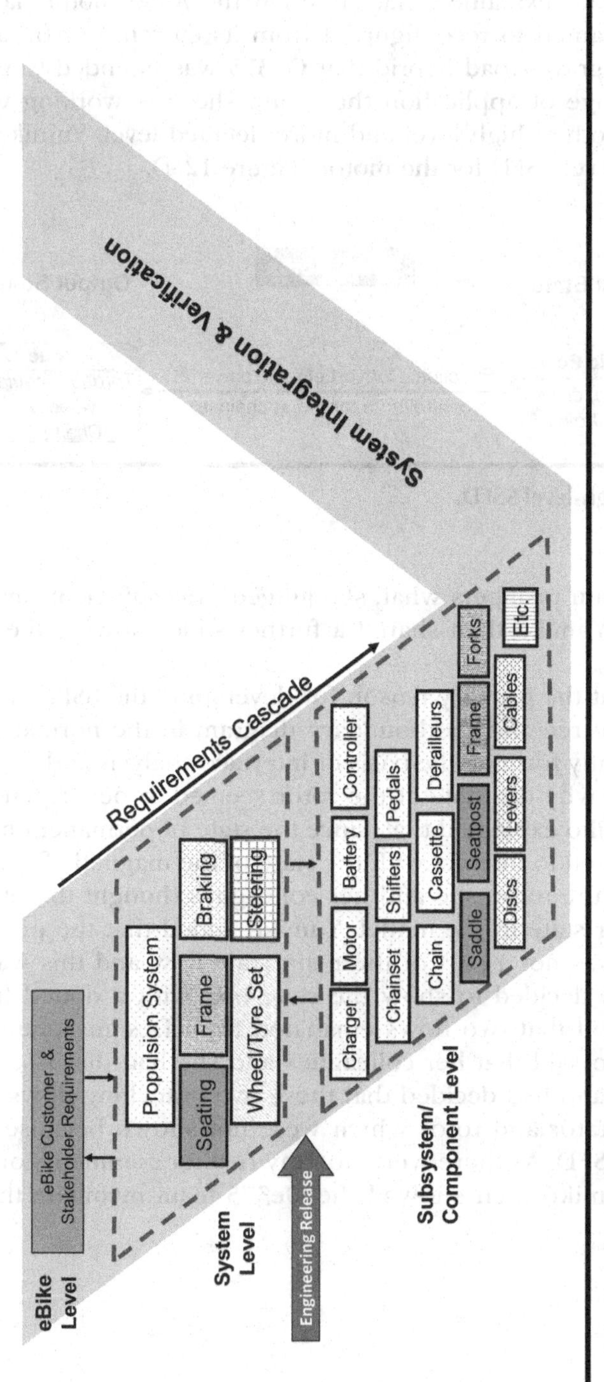

Figure 12.3 eBike partitioning and decomposition (copy of Figure 4.3).

Yumiko said that as team members knew the motor was a subsystem of the propulsion system with the requirements for the motor being cascaded from the higher systems level. Yumiko explained that although the A539 motor had already been designed, Agano wanted to reconfigure it from its current use on an electric road bike to use on the gravel/road hybrid that Oe375 was intended to be. Yumiko said that given the change of application the group she was working with decided to develop an SSFD both at high level and more detailed level. Yumiko shared a slide showing the high-level SSFD for the motor (Figure 12.4).

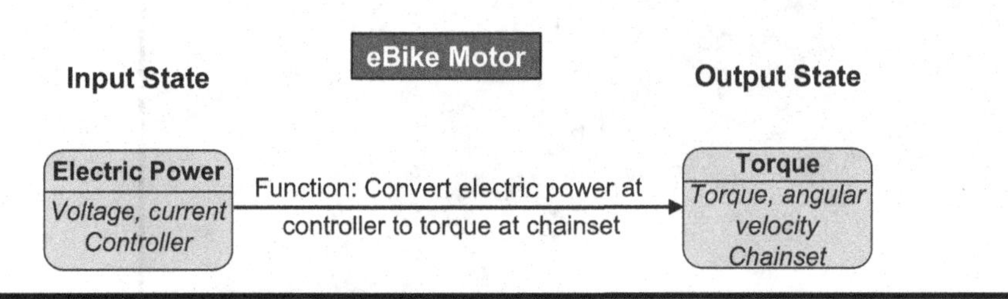

Figure 12.4 eBike high-level SSFD.

Having given team members what, she judged, was sufficient time to absorb the detail of the slide Yumiko then shared a further slide showing the detailed motor SSFD (Figure 12.5).

Yumiko said that the primary reason for developing the SSFD was to be able to derive the function tree and the boundary diagram in the normal way with these graphics subsequently forming the basis of interface analysis and hence the DFMEA. Phil asked Yumiko why the state at the rotor seemed to be suspended in what he called mid-air. Yumiko explained that, since the state of permanent magnetic field at the rotor was required to interact with the state of the magnetic field at the stator to generate torque at the rotor, she and her colleagues thought that it was important to include the rotor state in the model. Yumiko added that the permanent magnet state of the rotor was not a part of the main state flow and this was why she and her colleagues had decided to show the state box with a dotted line. Sarah then said that she noticed that two flows emanated from Design Solutions rather than from states. Yumiko said that her colleagues and she had discussed the point that Sarah was making and had decided that these two branching flows emanated from states within the stator and rotor which were not shown because of the level of resolution of the SSFD. As there were not any further comments on the point that she had made, Yumiko then showed the Oe375 team members the function tree

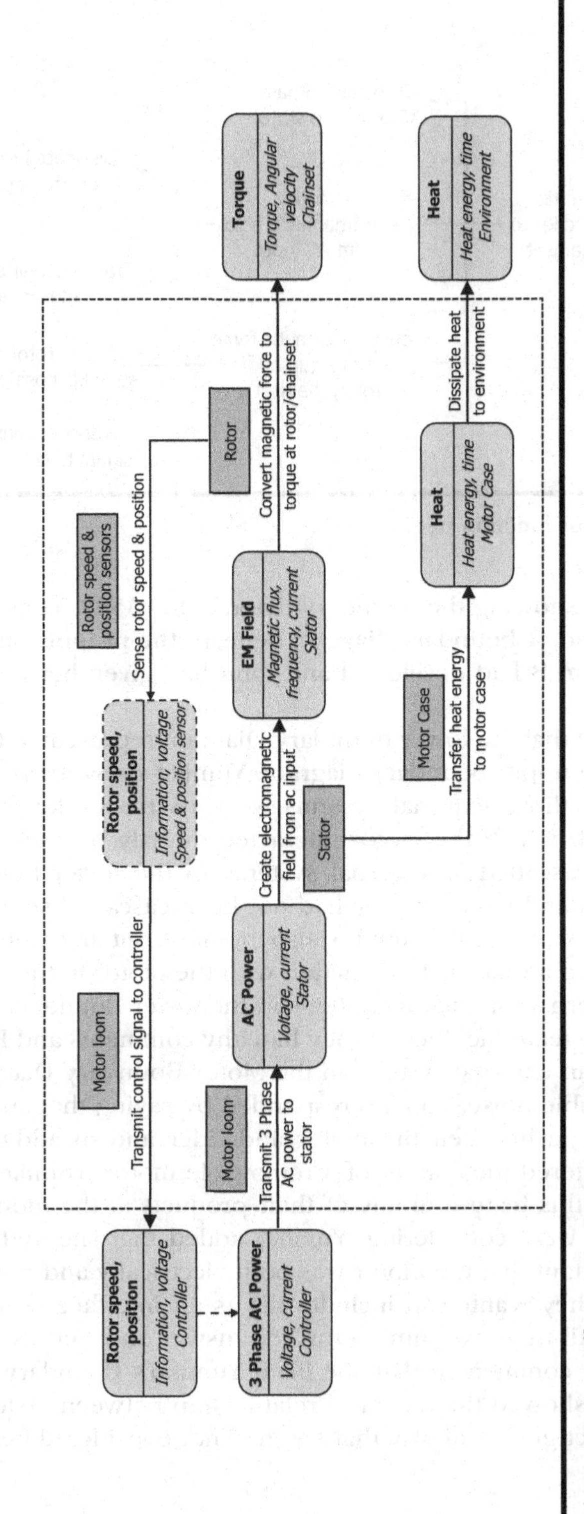

Figure 12.5 eBike motor SSFD.

(Figure 12.6) and reminded them that the top function was the one stated in the high-level SSFD.

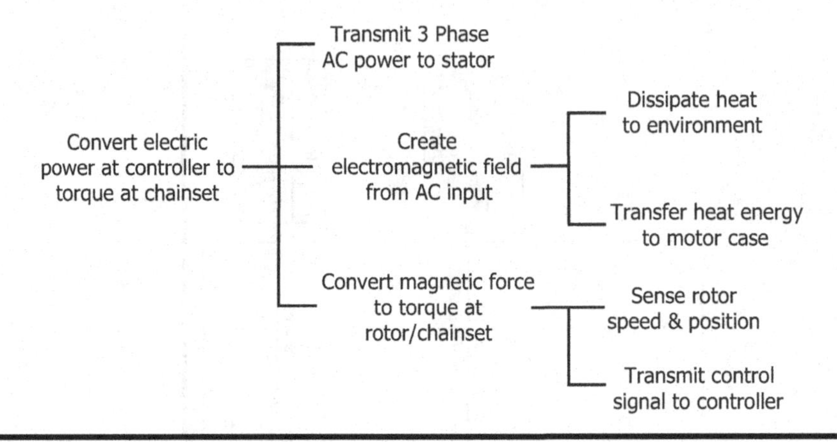

Figure 12.6 eBike motor function tree.

In sharing a slide showing the boundary diagram for A539, Yumiko said that she had positioned the motor boundary diagram beneath the propulsion system boundary diagram that Team B had developed and John had given her permission to use (Figure 12.7).

Yumiko explained that the lower Boundary Diagram represented the detail inside the motor box of the upper boundary diagram. Yumiko drew team members' attention to the fact that those external systems, such as the eBike frame and external environment with which the motor interacted directly on the upper Boundary Diagram had been cascaded as external systems to the lower boundary diagram while the handlebars and mains supply had not been cascaded since the motor did not interact directly with these. Yumiko also pointed out that the controller and chainset which formed an internal interaction with the motor in the upper boundary diagram became external interacting systems in the lower Boundary Diagram.

Yumiko asked the team members if they had any comments and Phil asked if the rider was shown as an external system on the Motor Boundary Diagram because of the exchange of audible noise. Yumiko responded by saying that audible noise was an important exchange between the motor and rider and in addition some competitor companies offered the feature of a removable motor. Yumiko explained that Agano did not offer this feature in any of their products at the moment, but it was something that they were considering. Yumiko added that she and her colleagues also felt it was important that the motor was both electrically and physically isolated from the rider and they wanted to include this as an interface requirement in the interface analysis. Phil thanked Yumiko for her answer and said that it made sense to him. Jennifer then commented that she liked Yumiko's Boundary Diagrams slide since she felt that it showed the technical relationship between Oxton and its parts suppliers in a succinct graphical way that she had not considered before.

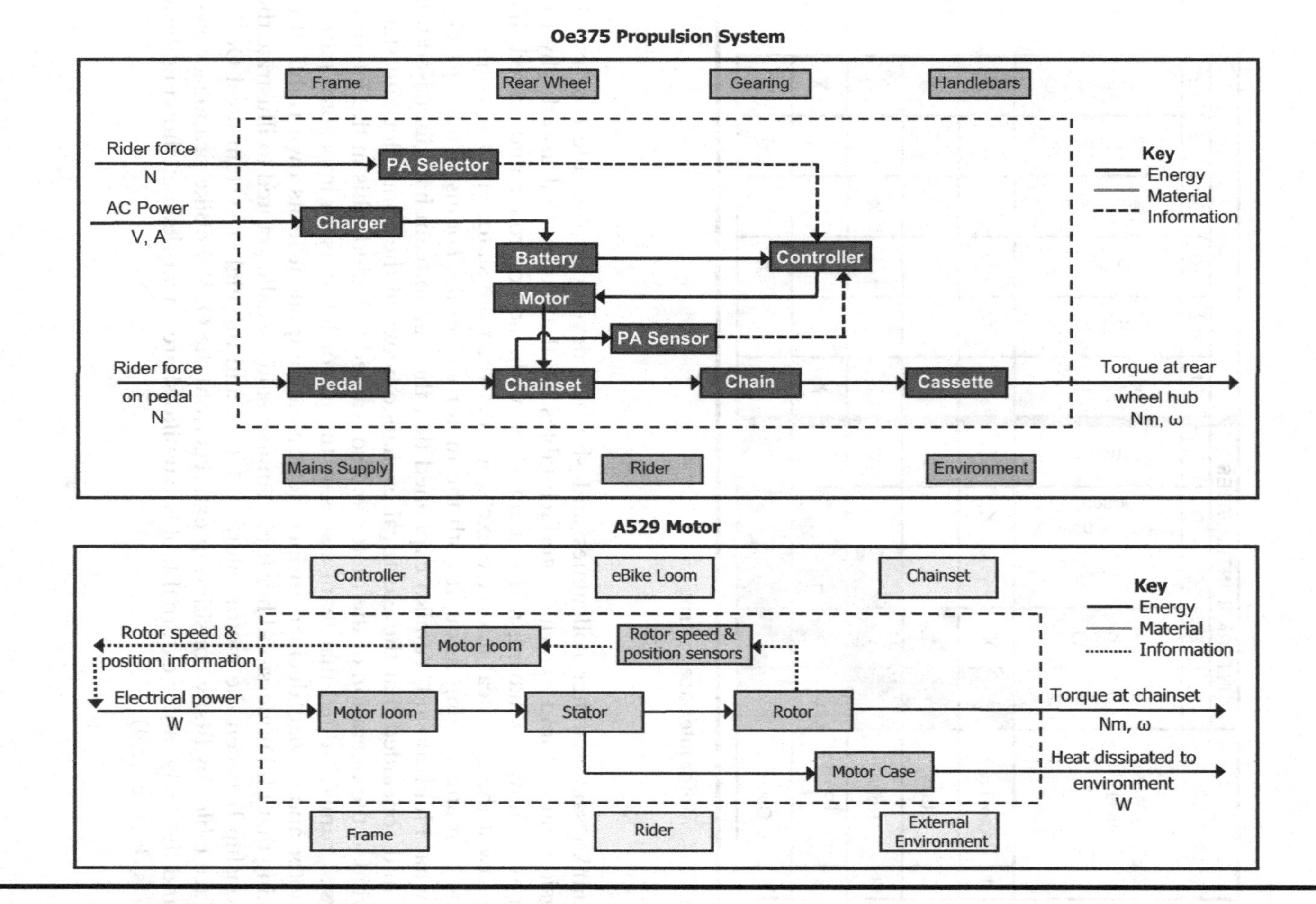

Figure 12.7 Oe375 electric drive and A539 motor boundary diagrams.

Yumiko said that the cascade of external and internal systems depicted on the boundary diagram had implications for the subsequent interface analysis and shared a worksheet showing the motor interface matrix (Figure 12.8).

	eBike Motor A539	INTERNAL INTERFACES										
		A	B	C	D	E	E1	E2	E3	E4	E5	E6
		Motor Loom	Stator	Rotor	Position & Speed Sensor	Case	Controller	eBike Loom	Frame	Chainset	Rider	Environment
1	Motor Loom		X		X	X		X				X
2	Stator			X		X	X					X
3	Rotor				X	X	X			X		X
4	Position and Speed sensor						X					X
5	Case						X		X		X	X

Figure 12.8 Motor interface matrix.

Yumiko said that her colleagues and she had developed the interface matrix alongside internal and external interface tables for the motor. Phil asked why the Stator was shown as having a direct interface with the controller and added that as he saw it the electrical interface exchange between the stator and the controller was more indirect being through both the motor and eBike Looms. While agreeing with what Phil had said, Yumiko explained that the Agano team had discussed this point and concluded that the electrical exchange between the controller and stator was vital to the operation of the motor and so wanted to include this in the interface analysis. Yumiko added that team also saw the motor loom as an important part of the motor and so had decided to include it as a component in its own right. After checking that Phil understood her logic Yumiko said that she wanted to illustrate the relationship between the motor interface tables and the higher system level Oe375 interface tables by firstly considering extracts from the Oe375 eBike propulsion system and the A539 motor external tables. Yumiko shared a worksheet showing these extracts (Figure 12.9).

Interface Ref	Interface		Type	Description of Exchange	Impact	Affected System Function
8-E1	Motor	Frame	E	Fix motor to frame so shaft can rotate freely	2	Convert controller AC power to torque at chainset
			P	Locate motor on frame so shaft can rotate freely	2	
			E	Vibration passed between frame and motor	–1	
			E	Heat from motor transferred to frame	–1	

Interface Ref	Interface		Type	Description of Exchange	Impact	Affected System Function
5-E3	Case	Frame	E	Fix case to frame	2	Convert electromagnetic force to torque at rotor/chainset
			P	Locate case on frame	2	
			E	Vibration passed between frame and case	–1	
			E	Heat transferred from case to frame	–1	

Figure 12.9 Extracts from Oe375 propulsion system and A539 motor external interface tables.

Yumiko asked team members to compare the interface 8-E1 in Oe375 propulsion system external interface table extract with the interface 5-E3 in the A539 motor external interface table extract. Yumiko asked what team members observed about the analysis for these two interfaces. Sarah said that both sets of exchanges were with the frame and were almost identical except that for the propulsion system the exchanges were with the motor while for the motor they were with the motor case. In agreeing with what Sarah had said, Yumiko said that the propulsion system exchanges had been cascaded from the frame/motor interface and allocated to the frame/case interface and added that in doing so the granularity of the detail of the exchange had increased. Yumiko asked team members to note that the frame formed an external interface exchange in both situations.

Yumiko then directed team members' attention to the interface 7-H in the Oe375 propulsion system internal interface table extract and the interfaces 2-E1, 3-E1, 4-E1 and 5-E1 in the motor external interface table extract (Figure 12.10).

Interface Ref	Interface		Type	Description of Exchange	Impact	Affected System Function
7-H	Controller	Motor	E	AC power transfer	2	Convert controller AC power to torque in motor
			E	Electromagnetic interference	-2	
			I	Motor speed and position	2	
			E	Heat transfer	-1	

Interface Ref	Interface		Type	Description of Exchange	Impact	Affected System Function
2-E1	Controller	Stator	E	AC power transfer	2	Create EM field from AC input
			E	Electromagnetic interference	-2	
3-E1	Controller	Rotor	E	Electromagnetic interference	-2	Convert EM force to torque at rotor/chainset
4-E1	Controller	Position & Speed Sensor	I	Motor speed and position	-2	Convert EM force to torque at rotor/chainset
5-E1	Controller	Case	E	Heat transfer	-1	Transmit 3 Phase AC power to stator

Figure 12.10 Extracts from Oe375 propulsion system internal and A539 motor external interface tables.

After giving team members the time to review the interface analysis, Yumiko asked them if they could explain what was happening in this analysis. Phil asked in which direction the heat was flowing in the motor interface table extract, and Yumiko said it was from the motor case to the controller and apologised that she had omitted the "from/to" columns in the extract in order to make it more legible. Yumiko suggested that team members compare the two table extracts and after a pause Jennifer said that the resolution of the description of the exchanges had increased between considering the exchanges of the Controller with the motor at the propulsion system level, and between the Controller and Stator, Rotor, Position & Speed Sensor and Case at the motor level. Yumiko agreed with Jennifer and said that the interface exchanges at the propulsion system level had been deployed to different components at the motor level and asked team members to note that the exchange of *Electromagnetic interference* had been deployed to more than one component part of the motor. Yumiko then asked what else was different. Sarah said that the controller/motor interface exchanges were internal while the controller/rotor, controller/stator, controller/position & speed sensor and controller/case interface exchanges were external. Yumiko told Sarah that she was correct and said that this was a guideline in that the internal exchanges within a system at a higher systems level are deployed as external exchanges at a lower systems level. Yumiko said that, as they had discussed, this guideline was evidenced in comparing the propulsion system and motor boundary diagrams (Figure 12.7).

Yumiko then shared an extract from the A539 motor internal interface table (Figure 12.11).

Interface Ref	Interface		Type	Description of Exchange	Impact	Affected System Function
2-C	Stator	Rotor	E	Electromagnetic torque	2	Convert electromagnetic force to torque at rotor/chainset
			E	Electromagnetic induced vibration	-1	
			E	Heat transfer	-1	
			P	Physical interference	-2	

Figure 12.11 Extract from the A359 motor internal interface table.

Yumiko said that the stator/rotor internal interface had not been considered before and these exchanges had to be identified without referencing the propulsion system analysis. Yumiko said that she and her colleagues had developed the complete internal and external interface tables for the motor corresponding to the interfaces identified on the interface matrix (Figure 12.8) and used this analysis in developing an noise and control factor table (NCFT) for the motor. Yumiko then shared an extract from this table (Table 12.1).

Table 12.1 Extract from Motor Noise and Control Factor Table

Noise and Control Factor Table MOTOR		Control Factor			
Type	**Units**	Manufacturing controls	Case Design IP Rating	Motor robustness	**Robustness Strategy**
Piece to Piece					
Rotor magnetic flux density variation	T	X			C. Reduce the manufacturing and assembly variation
Manufacturing variation	Various	X			
Assembly variation	Various	X			
Changes over time/cycles					
Bearing wear	mm			X	B. Reduce motor sensitivity to Noise using modelling
Stator heating	°C			X	
Contact resistance	Ω			X	
Sensor deterioration	V			X	
Rotor magnetic flux density loss	T			X	
Variation in sensor response	ms			X	
Customer Usage					
Excessive rider loading	kg				C. Define acceptable level of loading
Excessive road vibration	Hz				C. Define acceptable level of vibration
External Environment					
Dust ingress	mm^3		X		C Minimise ingress
Liquid ingress	ml		X		
Ambient temperature	°C			X	B. Increase motor operating temp. range
System Interaction					
EM induced vibration	Hz			X	C. Decrease level of EM vibration
Controller AC output instability	V, Hz				C. Improve controller AC output stability
Battery voltage variation	V			X	B. Reduce sensitivity to voltage fluctuation
Diverted Output					
Audible Noise (Electromagnetic source	dB			X	See EM induced vibration
Audible Noise (Mechanically sourced)	dB	X			C Remove or reduce audible noise

Robustness Strategies

A. Select the best design concept

B. Reduce the sensitivity of the design to the noise factor

C. Remove or reduce the noise factor

D. Insert a compensation device

E. Disguise the effect

F. Boost the design

Yumiko explained that as it had been some time since John had listed the different robustness strategies and their coding by letter, she was showing them alongside the NCFT extract. Yumiko said that she wanted to illustrate the depth of analysis that Agano had conducted in implementing the listed robustness strategies by outlining some of the work associated with the robustness strategy listed on the table as "Decrease level of EM vibration". She added that she remembered mentioning the two principal types of vibration in a BLDC motor of mechanical and electromagnetic when considering the chainset/motor interface in a previous meeting. Yumiko explained that Agano had reduced electromagnetic vibration by using a software model to analyse key interfaces between motor components. Yumiko continued by saying that the analysis showed that the relative geometry of the stator/rotor interface was of particular interest as a source of excitation resulting in audible noise being radiated outwards and transmitted through the motor case to the eBike rider. Yumiko said that potential frequencies of audible noise emanating from this interface were predicted using the model. The model was then validated through hardware testing which showed a close correlation between the audible frequencies forecast by the model and those found during testing. Yumiko continued by saying that the validation of the model enabled Agano to simulate the effect of changes to the geometry of the stator/rotor interface and so identify design parameter magnitudes that the model indicated would result in significantly lower levels of excitation. Yumiko explained that the design changes were then validated through hardware testing with the outcome that due to this analysis, and other such studies, A539 is widely recognised as being a very quiet motor.

Yumiko explained that while Agano had not developed an NCFT when the motor was first designed, she and her colleagues had developed the one from which the extract that she was sharing was taken and had used this along with their interface analysis to update and improve the A539 motor DFMEA. Yumiko asked team members if they had any questions or comments, and no one said anything. Yumiko asked team members if they would like to take their break and they readily accepted Yumiko's suggestion.

After the break Yumiko resumed her presentation by saying that she wanted to look next at the DFMEA that her colleagues and she had developed. Yumiko said that like the interface tables she found it interesting to compare the A539 motor DFMEA with the Oe375 propulsion system DFMEA and shared a worksheet showing corresponding extracts from both FMEAs (Figure 12.12).

Yumiko said that the extract from the A539 motor DFMEA was shown below the corresponding extract from the Oe375 propulsion system DFMEA and asked team members what they noticed about the two extracts. Phil said that the function being analysed in the A539 DFMEA extract was a lower-level function directly supporting the function being analysed in the Oe375 DFMEA extract and added that there was a similar relationship between the Oe375 and A539 Failure Modes. Phil then said that the causes on the motor DFMEA were at a lower systems level and hence more focused than the equivalent Oe375 causes. Yumiko agreed with Phil and asked if there were any further comments or questions. Jennifer asked why the causes on the Oe375 DFMEA associated with the connecting of the chainset to the rotor did not have related lower-level causes on the A539 DFMEA. Yumiko responded by saying that the equivalent causes were not feasible with the Agano design. Yumiko then asked why the two extracts bore the similarities that Phil had identified and Sam said with a smile that this was because the Agano team had copied and pasted the Oxton DFMEA and replaced

Extract from Oe375 Propulsion System DFMEA

Function	Potential Failure Mode	Potential Effect(s) of Failure	Severity	Potential Cause(s)/ Mechanism(s) of Failure	Occurrence	Current Design Controls Prevention	Current Design Controls Detection	Detection	Recommended Action
Convert controller AC power to torque at chainset	Loss in converting controller AC power to torque at chainset (greater than expected)	Increased effort required by rider	6	Insufficient torque transmitted from motor to chainset - motor power under specified	7	Mid Drive motor specification	Hardware in Loop Test	2	Review motor specification based on eBike modelling
				Motor not fully attached to chainset	1	Proven design concept	Bicycle DV test	4	Review motor spec after modelling airflow
				Motor not in full physical contact with chainset	1	Proven design concept	Bicycle DV test	4	Review fixture specification
				Motor overheats causing loss of efficiency	5	Motor specification, Motor mounting	Hardware in Loop Test	2	Review motor spec after modelling airflow
				Vibration transfer chainset-motor; loss of motor efficiency	3	Motor specification	Rig test	5	Review Motor specification
				Water ingress to motor - poor shielding	3	Motor IP specification	Water spray test	4	Review motor spec and positioning
				Water ingress to motor - inappropriate IP rating	4	Motor IP specification	Water spray test	9	Motor IP56
				Dust/dirt ingress to motor - inappropriate IP rating	3	Motor IP specification	Bicycle DV test	7	Motor IP56

Extract from A539 Motor DFMEA

Function	Potential Failure Mode	Potential Effect(s) of Failure	Severity	Potential Cause(s)/ Mechanism(s) of Failure	Occurrence	Current Design Controls Prevention	Current Design Controls Detection	Detection	Recommended Action
Convert electromagnetic force to torque at rotor/ chainset	Insufficient electromagnetic force converted to torque at rotor/ chainset	Increased effort required by rider	7	Insufficient torque transmitted from rotor to chainset	7	Motor parameter modelling	Motor eBike test stand	3	Tune motor to customer application(s)
				Rotor overheats causing loss of flux density/ motor efficiency	4	Case geometry and composition	Motor eBike test stand	3	Confirm motor robustness in customer noise space
				Magnetically induced vibration in rotor causing loss of motor efficiency	3	Geometry of rotor, stator	Motor eBike test stand	4	None
				Water ingress to rotor	3	Motor IP rating	Water jet spray test	2	None
				Dust/dirt ingress to rotor	3	Motor IP Rating	Motor dust exposure test	4	None

Figure 12.12 Extracts of Oe375 and A539 DFMEAs.

a few words here and there. Sarah smiled at Sam's response and said that the real reason was because of the influence that interface analysis had on the two documents. Jennifer asked Sarah if she would explain what she meant. Sarah said that Yumiko had just illustrated the close structural and content relationship between the Oxton and Agano Interfaces Table extracts (Figures 12.9 and 12.10). Sarah explained that since the structure and content of both the Oxton and Agano DFMEAs was derived from the corresponding interface tables the relationship between the Oxton and Agano interface table extracts had been carried across to the Oxton and Agano DFMEA extracts (Figure 12.12). Yumiko thanked Sarah for her explanation and John asked if he might show a slide that he had prepared that he felt was relevant to the current discussion. Yumiko said that of course he could and John shared the slide (Figure 12.13).

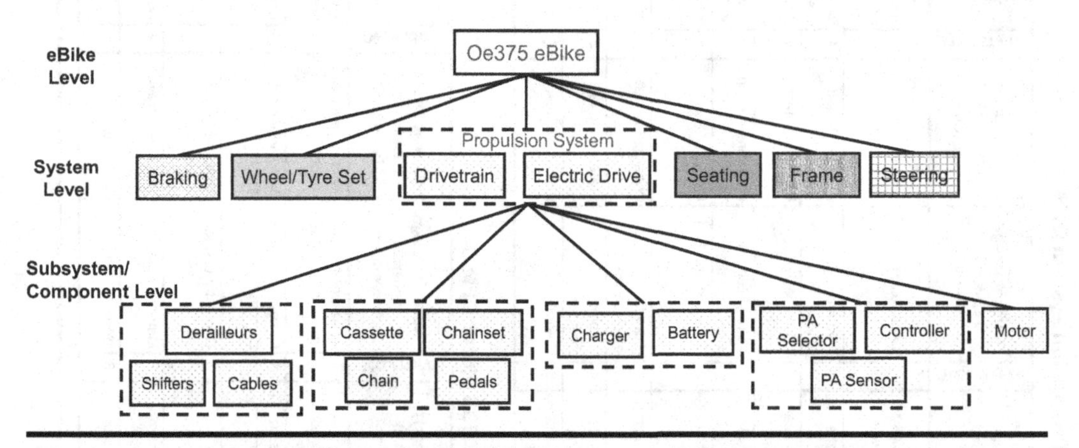

Figure 12.13 Interface analysis and DFMEA partitioning and decomposition.

John said that the Oe375 PD team had partitioned the interface tables and DFMEAs into the six architectural areas denoted on the slide at system level with the Electric Drive and Drivetrain sharing common propulsion system DFMEA and interface tables. John added that he understood from Yumiko, that Agano had developed interface tables and DFMEAs at subsystem/component level for the Drivetrain and Electric Drive in the areas denoted by the blocks towards the bottom of the slide. John continued by saying that those parts within the same dotted line boundary were a part of common Agano external and internal interface tables and DFMEA. Yumiko said that in addition Agano were analysing the braking system with the rotors, callipers, levers, and cables being a part of common interface tables and a common DFMEA. John thanked Yumiko for the additional information and said that Oxton had developed a series of six self-contained interrelated interface tables and DFMEAs. John added that the six separate, highly structured, Oxton DFMEAs for the Oe375 programme each of a few tens of pages in length could be compared to what otherwise might have been a largely unstructured single DFMEA of several hundred pages. John apologised to Yumiko for interrupting her flow and Yumiko told John that she thought the graphical representation of the systems decomposition of the interface analysis and DFMEAs was helpful.

Yumiko said that she wanted to return to the development and use of a computer-based model to tune the A539 motor to the Oe375 eBike requirements and shared an NCFT (Table 12.2)

Table 12.2 NCFT for Motor Optimisation Model

Noise and Control Factor Table Motor Torque Output				Control Factor	
Interface	**Type**	**Units**	**Noise Factor Effect**	**Motor Model**	**Robustness Strategy**
	Piece to Piece				
Tyre/Road	Rolling resistance	N	Resistive Force	X	A. Oxton specification
	Changes over time/cycles				
Tyre/Road	Rolling resistance	N	Resistive Force	X	C. Oxton specification
	Customer Usage				
	Rider mass	kg		X	C. Oxton specification
Tyre/Road	eBike mass	kg		X	B. Oxton specification
	eBike Ride cycle	Standard	Resistive Force	X	B. Optimise motor
Rider & Bike/ External Environment	Projected frontal area (U)	m^2		X	B. Oxton specification
	& Rider posture (D)	m^2		X	
	External Environment				
Tyre/External Environment	Road surface	Defined		X	B. Oxton specification
	Ambient temperature	°C		X	A. Oxton specification
	Wind speed	km/h		X	*See projected Frontal Area & Rider posture*
	Wind direction	Deg wrt eBike	Resistive Force	X	
Rider & Bike/External Environment	Ambient temperature	°C		X	
	Elevation	m		X	*See Ride cycle*
	Gradient	%		X	
	System Interaction				
Propulsion/Drivetrain	Drivetrain efficiency loss	%	Resistive Force		D. Compensate for level of loss
	Diverted Output				
Rider/Motor	Audible noise	dB	Dissatisfaction		C. Reduce level if too high

Yumiko said the NCFT summarised the sources of the resistive forces due to noise factors that the motor together with the rider had to overcome and added that team members would recognise that it was based on the NCFT that Team B had developed for the propulsion system (Table 11.6). Yumiko explained that the Noise Factors with a robustness strategy of *Oxton specification* corresponded to those for which Oxton had specified the likely magnitudes and ranges that would be experienced by the Oe375 eBike. Yumiko said that she wanted to outline the theoretical background to the strategy Agano were adopting to ensure that the A539 motor was robust to the Oe375 eBike requirements by modelling the performance of the motor. She said that although A539 was a brushless DC motor she would look first at how the performance of a brushed motor could be represented graphically and added that considering a brushed motor to which a constant voltage was applied simplified the analysis. Yumiko continued by explaining that key characteristics of an electric motor included the speed at which it operated, the torque that it generated, the power that it generated, and the current passing through the coil windings. Yumiko then shared a slide (Figure 12.14).

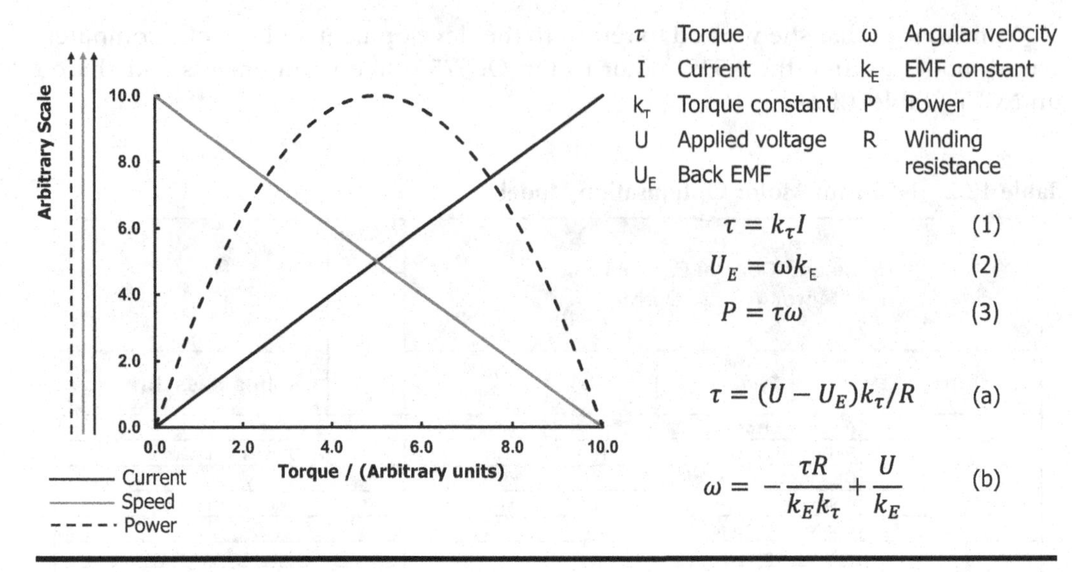

$$\tau = k_\tau I \quad (1)$$

$$U_E = \omega k_E \quad (2)$$

$$P = \tau\omega \quad (3)$$

$$\tau = (U - U_E)k_\tau / R \quad (a)$$

$$\omega = -\frac{\tau R}{k_E k_\tau} + \frac{U}{k_E} \quad (b)$$

Figure 12.14 DC motor characteristics.

Yumiko said that despite being idealised, the curves illustrated the complexity of operation of a DC motor. Jennifer asked Yumiko what she meant by being idealised and Yumiko responded by saying that, in addition to the simplification of a brushed motor with a constant applied voltage she had already mentioned, factors like the generation of heat and any internal friction in the motor had been neglected. Yumiko said that the three simple equations (1), (2) and (3) were the basis of the transfer functions used to generate the curves. Yumiko explained that, as team members could see from equation (1) the current I was directly proportional to the torque τ resulting in the solid black straight line. Yumiko said that she had used Ohm's law to replace the current I in equation (1) with the voltage across the windings divided by the winding's resistance to obtain equation (a) and asked team members to note that the voltage across the windings was the applied voltage less the Back EMF. Sam asked Yumiko if she would remind him what Back EMF was and Yumiko said to Sam that, as he knew, if a conductor was moving in relative motion within a magnetic field a current would be induced in it. Sam replied that he did know this, and Yumiko continued by saying that the stator windings were a conductor and, although stationary, the rotating magnetic field due to the rotor induced a current in the stator windings and so generated a voltage across them. Yumiko said that this induced voltage opposed the voltage applied across the stator to generate its magnetic field and was called the Back Electromagnetic Force or Back EMF. Sam thanked Yumiko for her explanation. Yumiko then continued by explaining that inserting equation (2) into equation (a) gave the relationship between speed and torque shown by equation (b) which she obtained by making the angular speed ω the subject of the equation. Yumiko said that equation (b) represented a straight line with a negative slope which was depicted by the solid grey line.

Yumiko said that while the derivation of the quadratic transfer function for power was equally straightforward, she would not bother to spend any more time on mathematics leaving it up to those team members who were interested to derive the equation for themselves. In drawing team members' attention to the scales on the axes of the graphic Yumiko explained that all the plots had been normalised to these scales. Yumiko added that while the slide was based on the analysis of a brushed motor the shape of the corresponding curves for a brushless motor were similar and she then shared a second slide showing a torque curve for a brushless DC motor (Figure 12.15).

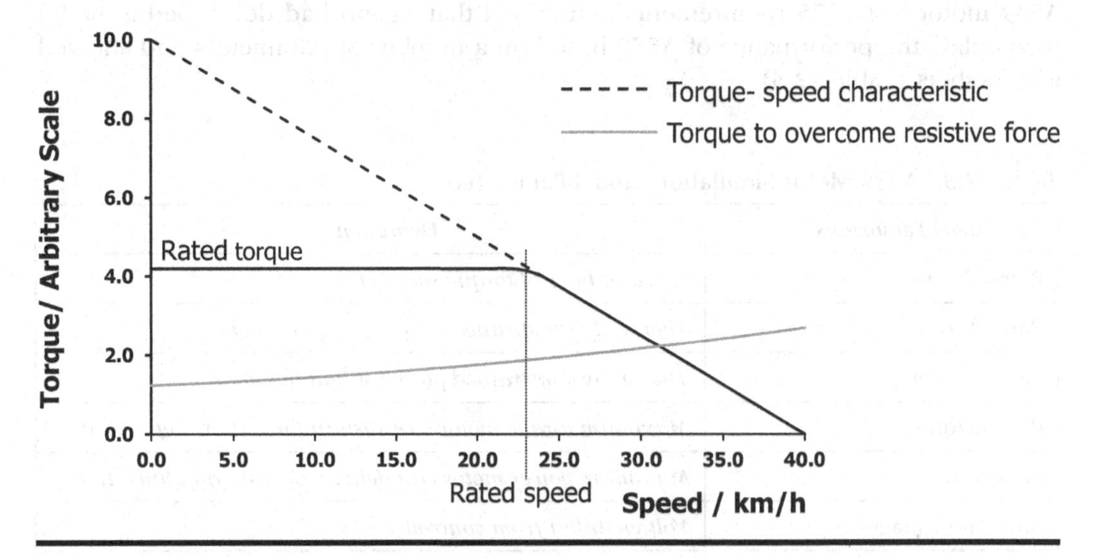

Figure 12.15 Brushless DC motor torque-speed curve.

Yumiko said that again the torque-speed plot she was sharing was idealised and asked team members to note that this time the speed of the eBike in km/h was plotted on the x axis. Yumiko said that the dotted straight line which extended into a solid line corresponded to a constant voltage. Yumiko explained that the voltage across the stator of a brushless motor was readily controlled electronically which enabled the current through the stator windings and hence torque output of the motor to be controlled. Yumiko said that the motor torque shown by the solid black line on the slide had been controlled to a steady value called the *Rated Torque* up to a speed of about 23 km/h called the *Rated Speed*. Yumiko explained that the rated torque represented the maximum torque that the motor could sustain without the current that was required to generate an increased torque causing the stator windings to overheat. Yumiko added that similarly the Rated Speed was the maximum speed beyond which the magnitude of the increased voltage and hence current required to avoid the torque decreasing would be such as to run the risk of overheating the stator to the point of burn out. Yumiko added that she had added the plot in grey of the total resistive torque based on a plot that Sam had provided for constant gradient and constant headwind.

Yumiko then asked team member if they could identify a robustness strategy based on the BLDC motor torque curve. After a pause Phil said to Yumiko that she had given the answer to her question in the NCFT extract (Table 12.1) as robustness strategy B *Reduce the sensitivity of the design to the noise factor*. Phil added that he assumed that this meant making sure that the motor torque and rider input together were higher than the effect of Noise Factors represented by the resistive torque curve over the range of speed that Oe375 would operate. Yumiko thanked Phil for his comment and said that this was the robustness strategy that Agano were using to tune the A539 motor to Oe375 requirements and added that Agano had developed a model to simulate the performance of A539 based on a number of parameters and showed a worksheet (Table 12.3).

Table 12.3 A539 Motor Simulation Model Parameters

Model Parameters	*Definition*
Rated speed	*Speed at knee of torque-speed curve*
Rated torque	*Thermally constrained torque output of motor*
Rated power	*Thermally constrained power output of motor*
Peak torque	*Maximum torque motor can sustain for a short-defined time*
Peak power	*Maximum power motor can deliver for a short-defined time*
Supplied voltage	*Voltage input from controller*
Back EMF constant	*Relationship between back EMF and angular speed of rotor*
Stator winding resistance	*Electrical resistance of stator windings*
Stator winding inductance	*Electrical inductance of stator windings*

Yumiko said that the first two model parameters were labelled features of the BLDC motor torque curve. Andy asked why the rated torque and power were described as thermally constrained and Yumiko said that operating the motor above its rated power required more current to pass through the stator windings than when the motor operated at its rated power, with the danger of the motor overheating. Yumiko added that a BLDC motor could operate for short periods of time at *Peak Torque* and *Peak Power* which were above the rated torque providing the time for which the motor operated in this way was restricted so that it would not overheat and cause damage to the motor. Andy asked for how long the motor could operate at Peak Torque or Power and Yumiko replied that it was typically of the order of a few minutes. Yumiko then said that she assumed that the remaining definitions in the table were self-evident. Jennifer asked Yumiko if she would explain the difference between Back EMF and stator inductance. Yumiko responded that the Back EMF was induced in the stator due to the rotation of the magnetic field generated by the rotor while inductance in the stator was due to the variation of the magnetic field generated by the stator itself as the current through it varied. Jennifer thanked Yumiko for her explanation. Yumiko then explained that Agano had included additional terms in their model to take account of electromagnetic, mechanical and windage losses. Jennifer asked Yumiko if she could give examples of electromagnetic and mechanical losses and explain what windage losses were. Yumiko replied that windage losses were due to air resistance in the gap between the stator and rotor as the rotor turned. She then cited losses due to heat and bearing friction as examples of electromagnetic and mechanical losses, respectively.

Yumiko continued by saying that key model parameter magnitudes were dependent on Oxton requirements for the eBike performance which was influenced by the level of the Noise Factors listed in the NCFT for the Motor Optimisation Model (Table 12.3) and the rider-selectable levels of power assistance. Yumiko asked Sarah if she would remind the team of the levels of assisted power that had been agreed. Sarah said that the Oe375 team had decided that there would be four selectable levels of power assistance of 50%, 75%, 100% and 125% additional to that generated by the rider.

Yumiko said that given the complexity of operation of a BLDC motor Agano always verified the integrity of any model they developed by comparing the performance predicted by the model with empirical data obtained in testing a prototype motor and even then, their recommendation was that customers allowed a 20% safety margin. Yumiko asked if the team members had any questions or comments. Sarah responded by saying that, as she was sure team members knew, local legislation restricted the performance of an eBike with pedal assistance with the motor power being restricted to 250 W and a maximum speed of 25 km/h in Europe. Yumiko said that the current maximum speed of a pedal assistance bicycle in Japan was 24 km/h with the level of power assistance not being greater than the rider power at the pedals. Phil said that in some states in the USA the maximum speed was 30 mph with the maximum allowed motor power being greater than in Europe. Phil said that the motor performance and control strategy for Oe375 would need to take account of differences of legislation in markets other than Europe and added that there would also need to be some future-proofing for potential changes to local legislation in Europe.

Sarah agreed with Phil and then asked John if he had anything that he would like to add to what Yumiko had spoken about. John said that he was impressed by the work that Agano had been doing and would like to put it into context within the Oe375 programme and shared a slide (Figure 12.16).

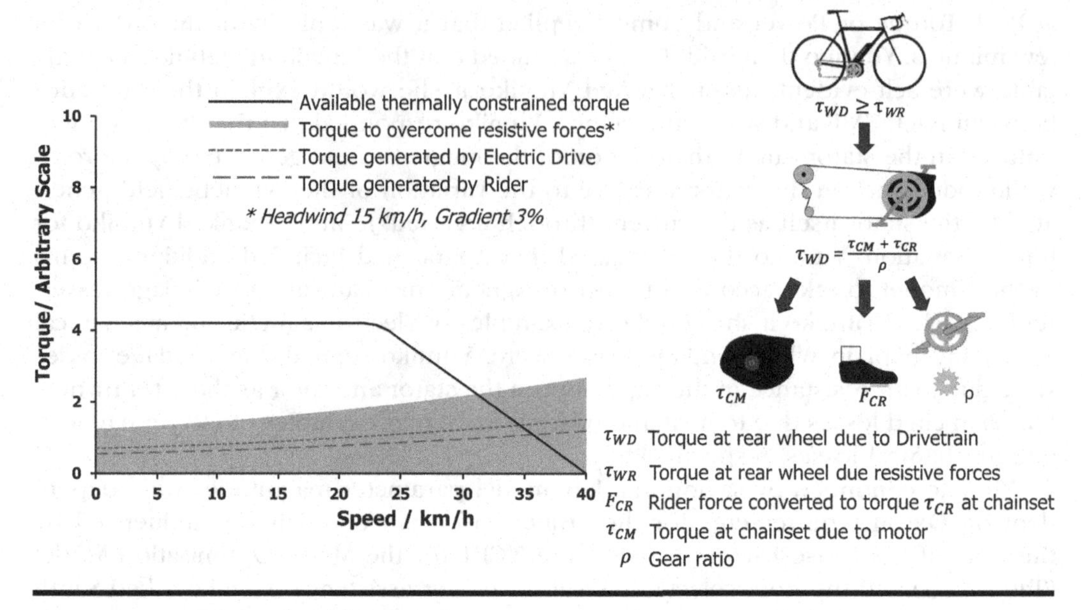

Figure 12.16 Systems cascade.

On reflecting on the slide as he shared it, while it had seemed perfectly clear to him when he had developed it, John felt that it was rather busy and needed some explanation. John said that team members would recognise the left-hand side of the slide as the brushless DC motor torque curve that Yumiko had shared along with a set of Sam's resistive torque curves. Turning to the right-hand side of the slide John said that he would explain the graphics top-down starting with the depiction of the Oe375 eBike which to move forward required that the torque at the rear wheel due the Drivetrain was greater than the torque due to resistive forces. John said that this requirement was cascaded to the propulsion system to determine the required input at the chainset due to the motor and rider taking account of the Drivetrain gearing. John added that the input requirements at the Drivetrain were subsequently cascaded to the motor, rider, chainset, chain and cassette. John said that in looking at the right-hand side slide top-down for him it depicted the journey that the Oe375 team had been on in analysing the eBike propulsive system with the eBike required performance being determined by the design of the propulsion system which in turn depended amongst other things on the performance of the electric motor.

John asked if team members had any comments and Phil said that he recalled saying in one of the team's early meetings that he had not found the Systems Vee model to be very helpful in doing systems engineering. Phil continued by saying, that in John explaining that the systems cascade depicted the path that between them the Oe375 team had been on brought the Vee model to life. Phil said that perhaps in the light of his experience of being a member of the Oe375 team he might have to modify his earlier opinion and added that this was a bit out of character for him, which brought a few laughs. John thanked Phil and concluded the discussion by saying that for him the slide illustrated the importance of Oxton cascading detailed requirements to suppliers that accurately reflected customer usage because the usefulness to Oxton of any analysis that Agano conducted using their model depended on the integrity of the data that Oxton supplied to them.

Sarah then thanked John for making what she thought was a very important point and then, in looking directly into her laptop camera, thanked Yumiko for what she described as her interesting and informative presentation and said that she thought the work that Yumiko and her colleagues had been doing deserved a round of applause. Sarah looked at the time after the applause had died down and said that in the time remaining the team could pick up where it had left off at the end of the last Team B meeting with the Steering System analysis. Yumiko said that was her signal to leave the meeting and added that she had enjoyed the meeting and hoped that her facilitation style lived up to John's standard. Sarah said that Yumiko had more that lived up to John's standard and with that Yumiko signed out of the meeting. The Oxton members of the Oe375 team duly made the best use of the remaining time and concluded the meeting with a short Warm Down in which all team members said that they had enjoyed Yumiko's presentation and agreed that the meeting had been both interesting and useful.

John returned to his office feeling refreshed and realised that he missed taking the Oe375 team through the details and nuances of the OFTEN framework, albeit on this occasion Yumiko had done most of the work. John was impressed by the way in which Yumiko had conducted the meeting and thought how far she had come from the somewhat demure individual in the early meetings who only tended to speak when spoken to. John was pleased with the way in which the work at Agano had illustrated both the applicability of the OFTEN framework at different system levels and the coherent flow of information between levels.

John broke away from his thoughts to start his day's work in earnest by opening his emails and in doing so he saw that he had a meeting notice from James Tutton inviting him to what James called a mini reunion. James was a part of the same apprenticeship intake at Blade as John and had done well for himself in progressing to a middle management role in manufacturing in the company. On reading the meeting notice further John saw that the reunion was scheduled for that Friday evening immediately after work in a small rural pub in a village some 10 miles from where John lived and was addressed to himself and Ira Grover who was also from the same Blade apprentice intake as himself. John recalled that he had caught up with both James and Ira at Mike Holmes, retirement gathering. John looked back to the meeting timing and noticed that it was scheduled for one hour and thought to himself that this did not allow for much of a reunion. John also noticed that James had sent the message from his personal email address and had stated his mobile phone number after his name. John rang James with the intent of confirming his attendance at the meeting and trying to find out why James wanted to meet. James answered the call almost immediately by stating his name in a quiet voice. John introduced himself and waited for James to respond, which he did not, and so John thanked James for his meeting notice, and after saying that he was intending to attend the reunion, he asked James if he could tell him why he was wanting to meet. James, still speaking in a low voice, said that he would be pleased to see John and could not talk now as he was in a meeting. James added that he would prefer to leave any discussion until Friday and ended the call leaving John none the wiser.

Background References

Motor characteristics Speed torque	Lynch, K. (2015) ***Brushed DC motor speed-torque curve,*** Northwestern Robotics, YouTube, www.youtube.com/watch?v=pxtRlKs0pAg&t=4s, Accessed December 18, 2022
Motor characteristics Power torque	Lynch, K. (2015) ***DC motor output power,*** Northwestern Robotics, YouTube, www.youtube.com/watch?v=drkC5P11Ch4, Accessed December 18, 2022

Chapter 13

Verifying the Design

As John drove from Oxton to the pub that James Tutton had designated as the venue for the reunion meeting, he wondered what he was going to learn. After he had received James' meeting notice and with James being reluctant to say anything over the phone, he assumed that it had something to do with the incidents in which he and someone at Blade had been falsely accused of wrong doing. John surmised that James, or perhaps Ira, was the person that Lucy Collins had referred to when she telephoned him to tell him of the second incident. While John had spoken to James at length at Mike Holmes, retirement celebration, he had only spoken to Ira in passing, and on learning that she was going to the reunion meeting, he had looked her up on the professional social media platform to which he subscribed, assuming that she would have an account. In finding Ira's details he saw that her current position in Blade was as a vehicle line director, and he thought that she must be in one of the most senior positions of anyone in his apprenticeship intake.

When John walked into the pub bar and looked around, he spotted Ira sitting alone at a table in a corner and was just going to walk over to say hello when he felt a tap on his shoulder. On looking around, he saw James standing behind him with a backpack slung over one shoulder, and John's immediate thought was that he had forgotten how tall James was. John greeted James and they shook hands with James saying that he was pleased to see John and asking him what he would like to drink. John looked towards the bar, and, spotting a golden ale amongst the row of beer pump handles, told James he would like a pint. James asked John if he had seen Ira and John told him where she was sitting. Looking at where Ira was seated James saw that she already had a drink and so ordered two pints of the beer that John had selected. While James was getting the beers, John walked across to meet Ira and, as she was about to stand, told her to remain seated and leaned over the table to shake her hand before sitting down. By the time Ira and John had exchanged initial pleasantries, James arrived with their beers which he placed on the table one in front of John and the other in front of the remaining free chair on which he sat after greeting Ira and shaking her hand in the manner that John had done.

The three former colleagues spent a few minutes catching up on family matters before John said to James that he assumed that he did not arrange the meeting

DOI: 10.4324/9781003286066-13

merely to talk about their families. James agreed with John and said that he believed that each of them had been on the end of some nasty allegations and asked John and Ira if his assumption was correct. They both confirmed that they had been falsely accused of a misdeed and John asked James why he had suspected that Ira and he had been the subjects of accusations against them. James explained that he had had a chance conversation with Lucy Collins, and, after he had mentioned what had happened to him, she had told him that John had experienced a similar thing. James said that finding out if anyone else might have been on the receiving end of untrue allegations was not as easy and added that it took a bit of arm twisting and calling in of favours from people he had promised not to name. John remembered from their days as apprentices, James always was the type of person who did not let obstacles stand in his way. Ira then said that in a funny sort of way it was reassuring to learn that she was not the only one to have been falsely accused, and the other two both agreed with her.

The three of them sat back in their chairs and sipped their drinks in silence until John asked James if he had anything else in mind in calling them together other than to confirm that they had all had allegations made against them. James replied that he wanted to discuss things in more detail and asked John and Ira what they had been accused of. John said that his supposed crime was to divulge Blade proprietary information outside of the company which of course was totally unfounded. Ira said that she had been accused of misappropriating company funds without any information or evidence to back up the claim. James then said that he had been accused of inappropriate and unprofessional behaviour towards an unnamed female colleague and like Ira no evidence was provided to substantiate the assertion.

Ira then asked James and John how the accusations against them had been made and both said that it was through an anonymous voicemail message left with HR. Ira said that the accusation against her was made in the same way. Ira asked James and John how they felt on being falsely accused and James said bloody angry, before apologising for the profanity. John said that he also felt angry, particularly because he knew the accusation was total nonsense and the person making it did not have the courage to say who they were. Ira said that she also felt anger and added that she also felt some sorrow for the poor individual who would do such a thing. John observed that while the accusation against him might have some credibility, in the sense that it was something he might have done if he was unscrupulous, the accusations made against Ira and James seemed laughable given there was nothing to back them up. While James agreed with John on this last point, he said that this did not stop the Blade rumour mill from swinging into action after rumours about the incident began to leak out and quoted the proverb "There is no smoke without fire". John asked James and Ira if there had been any serious repercussions resulting from the allegations against them. Ira said that, apart from the occasional funny look she received, most people she knew who had any knowledge of what had occurred treated the incident with the disdain it deserved. She added that it would still be good to prove once and for all the mendacity of the allegations. John and James agreed with Ira on this point.

James then asked Ira and John what the common denominator was for the three of them apart from them all having been falsely accused. Ira said that they were all from the same apprentice intake and James responded that Ira was spot on and added that they also all had successful careers. James said that he assumed because of the common modus operandi the same person had made the accusation against all three of them and suggested that the first place to look for the culprit was someone who knew them and had an axe to grind. James added that because of the diversity of their careers they should perhaps start with someone who had worked alongside the three of them and that would bring things back to the apprentice school. James then leaned down, took a folder out of his backpack and took out an A4 sized photograph which John recognised as the year group photo that they had all received in the first week of joining Blade. John was sure he still had a copy somewhere and surmised that it was probably stored somewhere at his parents' house, certainly he had not looked at it for a long time. James also had three paper copies of the photograph which he was holding beneath the original, which was still in its cardboard frame, and handed Ira and John a copy each. John's first impression was how young he and the rest of his fellow apprentices looked. He laughed out loud, and Ira asked him what was funny to which John replied that his long fringe falling over the edges of his eyes looked so dated now and he recalled how cool he thought it was at the time.

After some more giggles and reminiscing James said that he had brought the copies of the photograph, which helpfully had all their names printed on the bottom in alignment with their image, for them to identify their key suspect in what he called a secret ballot. James explained that he would like Ira and John to ring the name of the person who they felt the most likely to have perpetrated the allegations against them and added that he would like them to do it in a way so that the other two could not see who they were identifying. James said he had brought pencils with him and asked that they spend a bit of thinking time before selecting their candidate. John did not need any time to think about who he was going to select as he saw Bob Jenkins' face staring out at him the moment that he saw the photograph, nevertheless he thought it better not to make a move until someone else did. After a short while, James stood up, and using the folder as a shield and a rest, made his selection. John asked James if he might borrow the folder to lean the copy of the photograph on and, after James had given it to him, he ringed Bob's name hiding the copy of the photograph from the view of the others. Having noticed that James had put the photograph copy face down on the table in front of him James did the same and on sitting down passed the folder to Ira who was still looking at the photograph with what John took to be a blank look on her face. Seeing that James and John had made their selection Ira leaned back in her chair and made her choice placing the photograph copy face down on the table in front of her as James and John had done. James then added to the drama by saying that the moment of truth had arrived and asked John and Ira to turn their photograph copies over. John felt a mixture of surprise and reassurance to see that James had also selected Bob while Ira had selected one of the female former apprentices.

Ira said that she was surprised to see that James and John had selected the same person and James asked that the three of them explain in turn why they had made their selection. Ira said that nobody jumped out at her as being a potential culprit, and in the end, she had selected Clara because she felt that Clara had never liked her. Ira continued by saying that she felt that Clara was jealous of her because she always sailed through her exams while Clara struggled. Ira said that she could not put her finger on anything concrete except that Clara's behaviour towards her was rather catty. James spoke next and told Ira and John that he had managed to get a transcript of the voicemail message in which the accusations against him had been made and added that the call seemed more like a rant than a rational allegation. James said that, as a part of the rant, the caller had said that James had also behaved badly with respect to a woman at his tennis club. James explained that, before he had met his future wife, another male member had accused him in the changing room one day of stealing his girlfriend when the reality was that the girl had already split up from the guy and she had approached him. James explained that while the incident was a storm in a teacup it had caused quite a stir at the time. James said that this happened sometime after they had all moved on from the apprentice school and, while there were quite a few people from Blade who belonged to the tennis club, Bob was the only member of their apprentice year who was a member. John then told Ira and James about what Jane had learnt from Bob's wife at Mike Holmes' retirement celebration about Bob keeping a notebook detailing the careers of the other apprentices in their intake. Almost before John had finished relating his reasoning for selecting Bob, James said in a loud voice that what John had said had confirmed his suspicions.

After they had all taken another sip of their drink and contemplated the result of James' ballot, Ira asked what they should do next. James said that they should confront Bob to which John counselled that the evidence against him was purely circumstantial and they needed to proceed with caution. James responded to John's warning by saying that Bob was unlikely to spontaneously get a pang of remorse and admit his wrongdoing and so they need to take some action if they were to confirm that he was the culprit. Ira said that if they were to talk to Bob it needed to be in a calm manner and somewhere private and John and James agreed with her. After a pause, John said that he would think about the situation, come up with a plan and discuss it with them. John added that if they were to speak to Bob then they should do it in a descriptive manner and Ira asked him what he meant by doing things in a descriptive manner. John explained that they should not simply accuse Bob but should relate the facts to him as they knew them, allow him the time to assimilate what they were saying, and see how he reacted. James responded to John's proposition by saying that if they did get to speak to Bob, they should not "Pussy foot about". However, Ira agreed with John and said they needed to do things objectively and added that they were getting ahead of themselves because they needed to identify an opportunity to speak to Bob first. She suggested that they wait for John to come up with a plan to talk to Bob and the other two agreed with her. They sat back in silence and John thought to himself that despite Jane's reservations about bringing the matter up again he might at last get to the bottom of who had made the false accusations against him, and perhaps why. The trio then spent the rest of their short reunion catching up further with each other and finishing their drinks although John's mind kept returning to Bob Jenkins.

As John was walking into work from the car park on Monday morning his mind went back to Jane's reaction to his meeting with James and Ira when he told her what had happened. Jane had said that she had been afraid that the meeting might somehow be about the accusations against him when John first told her about James' meeting notice and the short phone call that he had subsequently made to him. When John returned from the reunion meeting on Friday evening Jane had been anxious to know what James wanted, and her worst fears were realised when John explained to her what had happened. Jane said that James sounded a bit of a hot head and that John needed to be careful, adding that he could not go around accusing people based on scant evidence. John agreed with Jane and said that he intended to approach the matter carefully, assuming that he could come up with a reasonable plan. While Jane had made it clear she would rather John left things alone she had also said that she knew that he would like to clear his name once and for all, or at least establish who the anonymous phone caller was. Jane had also added that she would support him providing he did not do anything stupid.

On reaching his office John's attention turned to business matters and he recognised that the Oe375 team would shortly need to be thinking about, and planning for, Phase 4 of the Oxton first-time engineering norm (OFTEN) framework in which they would be conducting Design Verification. On reflecting on his own experiences of working with the Oe375 team since Paul Clampin left, John recognised that his role as a coach had become blurred with that of a working member of the team. John realised that he would need to revert to his leadership role, albeit for only a short time, to introduce the team to Phase 4 of the OFTEN framework.

John was pleased with the help that the team had received from Andy and Yumiko and their respective companies which had resulted in the architecture of Oe375 maturing to the point where it could be frozen. Working with HJK through Andy the Oe375 frame geometry for an electric gravel bike was optimised for strength and comfort using an aluminium alloy and at the same time the frame mass was reduced significantly compared to the team's original specification. By using the same material for the handlebars, stem and seat post along with carbon front forks the competitive target for mass of the eBike that Phil had advised was achieved. Agano duly delivered on the Groupset and Electric Drive systems with John being both reassured and impressed by Yumiko's illustration of the depth of analysis that Agano were doing to meet the detail of Oxton requirements. In adapting the road bike Drivetrain to the Oe375 application, Agano retained the low frictional losses of the original design and optimised the Electric Drive system which included the controller, motor, battery, loom, sensors, and smart rider controller to the Oe375 application with the system having a highly competitive low overall mass. On Wolfgang's advice, which Jim Eccleston and Margaret Burrell supported, Oxton sourced the wheels from Crona, the global French bicycle parts manufacturer, with Andy using his contacts to confirm the suitability of the types of Hawk tyres chosen for Oe375.

John had agreed with Sarah that he and Yumiko would attend the forthcoming Oe375 Review/Preview meeting with a view to taking the Oe375 team through Phase 4 of the OFTEN framework in the following PD team meeting. John had spoken with Yumiko and asked her if she would be prepared to present some of the design verification work that she had been doing in Agano which she agreed to do. John had some familiarity with the work that Yumiko had been engaged in after she had asked him for some advice on the OFTEN approach to Design Verification following which they had exchanged several emails and held a couple of short virtual meetings. As he said he would, John had sent an email to Andy and Wolfgang telling them of the plan to look at Phase 4 in the upcoming Oe375 PD team meeting. John felt it was a bit like old times in walking into the Oe375 Review/Preview meeting and John was happy that Sarah and Phil had taken ownership of this aspect of the Effective Meetings process. The poster that Jim had designed on Framing Information: The Spoken Word slide and the Speaking Guidelines was still on the meeting room wall to remind John of earlier times although he thought that the humorous approach had worn thin with familiarity. After Yumiko had signed into the meeting Sarah started the meeting by saying that as John was aware from attending Oe375 team meetings the team kept a log of tasks and their completion status. Sarah then explained to John that, although the meeting they were currently attending was still called the Oe375 Review/Preview meeting, the "review" aspect of the meeting included a weekly check on the programme status against the plan as laid out by the log. Sarah added that in the "preview" Phil and she looked forward to the analysis that they projected would be completed in the forthcoming week. Sarah added that based on their review she and Phil would decide what, if any, remedial actions were necessary in the light of the OFTEN analysis work that had been done, or had not been done, during the previous week. Sarah then shared the front worksheet which summarised the detail of the log (Table 13.1).

Table 13.1 Oe375 OFTEN Analysis Status Log Summary

System	*Task*	*Status*	*System*	*Task*	*Status*
Drivetrain	SSFD	Completed	Braking	SSFD	Completed
	Interface Analysis	Completed		Interface Analysis	Completed
	NCFTs	Completed		NCFTs	Completed
	DFMEA	Completed		DFMEA	Completed
Propulsion system	SSFD	Completed	Steering	SSFD	Completed
	Interface Analysis	Completed		Interface Analysis	Completed
	NCFTs	Completed		NCFTs	Completed
	DFMEA	Completed		DFMEA	Part Comp.
Frame	Boundary Diagram	Completed	Seating	Boundary Diagram	Pending
	Function Tree	Completed		Function Tree	Pending
	Interface Analysis	Completed		Interface Analysis	Pending
	NCFT	Completed		NCFT	Pending
	DFMEA	Completed		DFMEA	Pending
Wheelset	Boundary Diagram	Completed			
	Function Tree	Completed			
	Interface Analysis	Completed			
	NCFT	Completed			
	DFMEA	Completed			

John was familiar with the log since the status of the Oe375 programme was always the first agenda item on the meetings designated as Oe375 team meetings. John was always impressed how Sarah and Phil kept on top of the programme management in keeping the log up to date and liked the fact that the Boundary Diagram and Function Tree were seen as an inherent feature of the SSFD in the log. John said that he was impressed by the depth, breadth, and rigour of the analysis that Team B had done and added that he would like to bring that work together in looking at Phase 4 of the OFTEN framework. Sarah said that they had already put Phase 4 on the agenda for the next Oe375 PD team meeting and there were another couple of aspects of the meeting that she wanted to discuss.

Sarah asked John if he wanted to discuss any People Skills in the meeting and John replied that he would like to have a brief look at the relationships between all the People Skills that the team had considered. Phil then invited John to take the Warm Up and Warm Down. John replied that he would be pleased to do this and added that since this might be one of the last Oe375 team meetings that Wolfgang and Andy would attend, if time allowed, he would like to extend the Warm Down time by ten to fifteen minutes to allow individual team members to share their experiences of being a member of the team. Sarah said that John's proposal to have a slightly extended Warm Down was a great idea. In turning to the core of the technical part of the meeting Phil asked John how much meeting time he would need on Phase 4 of the OFTEN framework. John explained that Yumiko and he would share the input on Design Verification, and he checked with Yumiko if this was still the case, to which she replied that she was happy to share the presentation with John. John asked Yumiko if she had any concerns about Andy and Wolfgang attending the meeting given that Andy in particular was from a competitive company. Yumiko replied that this should not be a problem because she did not intend to dwell on the technical details. She added that she was not intending to reveal any propriety data since, like in the meeting when John and she shared the technical presentation on robustness, she would idealise any analysis that might be sensitive and use an arbitrary scale of measurement.

John then said that he had not answered Phil's question about the quantity of information to be presented in looking at Phase 4 of the OFTEN framework and said that he thought that what he and Yumiko had to discuss would probably take the whole meeting. John added that he was hoping that there might be quite a lot of what he called lively discussion. John explained that, with Team B's permission, he would like to invite Mo Lakhani to the meeting. John said that early in the Oe375 programme he had agreed with Mo that he could be involved in the Design Verification stage of the programme due to his interest in both the subject of verification and the fact that John had explained to him that the verification process for the Oe375 programme would not focus almost exclusively on the final product as tended to be the norm with Oxton programmes. Phil said that Mo was very welcome to join them and added that he would be interested to see Mo's reaction to what John had to say, knowing Mo's views on Design Verification. John said that he had taken the liberty of getting Mo to pencil the meeting in his calendar a few days ago and said that he would forward the meeting notice to him.

Sarah said that based on what John had said they would be flexible with the meeting timings and have the Warm Up, People Skills, and Design Verification on the agenda and should there be any time left at the end of the meeting they could use that for continuing their Oe375 analysis work. Sarah confirmed with John that, since he was taking the lead in three agenda items and sharing the third with Yumiko, he would lead the meeting and John said that this would be fine. John then agreed with Yumiko that he would hand the meeting over to her after he had concluded his part of the Design Verification discussion and asked her to hand the meeting facilitation back to him when she had finished her presentation. Yumiko readily agreed to what John had asked. Sarah enquired if they had anything else to discuss and after everyone agreed they had covered everything she closed the Review/Preview meeting by saying that she was looking forward to the Oe375 meeting as it would be just like the earlier days of the Oe375 programme.

As John was walking into the conference room for the Oe375 meeting the next morning it did seem to him like he had stepped back in time as Sarah had said it would. He took his usual place and waited for Sarah to start the meeting which she did by welcoming everyone and extending an especially warm welcome to Mo Lakhani. Sarah explained for the benefit of those who did not know him that Mo was the Quality Manager at Oxton who was joining them for this meeting. Sarah said that it was good to have the whole team in attendance once again and added that John was going to take the Warm Up. John thanked Sarah and having checked that meeting communications were working shared a slide (Figure 13.1).

- Warm Up & Warm Down
- Effective Meetings
- Implicit and Explicit Knowledge
  - Visualisation
- Communication Skills
  - Listening Skills
  - Questioning for Clarification
  - Questioning for Information
  - Framing Information
    - The Spoken Word
    - The Written Word
- Feedback in Meetings
  - Explicit Feedback
  - Descriptive Feedback
- Attitudes
- Attribution

Figure 13.1 Oe375 team People Skills.

John explained that the slide was essentially the same as the slide he had shown in the first Oe375 team meeting (Figure 3.2) with a bit more added detail and represented all the People Skills that he had shared with the team. John asked team members to review the list individually and identify which single People Skill they thought was the most important, before discussing their choice with a team member and agreeing between them the most important People Skill. John added that team members could only choose one of the Communication Skills or one type of feedback as a single People Skill. John said he would then ask team members to discuss their choices as a team. After telling the team that he would designate the pairs, John turned towards Mo and said that he assumed he would rather not participate in the Warm Up exercise. John was both surprised and pleased when Mo said that he was happy to participate. Mo explained that while his perception of some of the People Skills might be different to that of Oe375 team members, he was familiar with the vast majority of those listed on John's slide. John thanked Mo, designated the pairs ensuring that Mo paired with Sarah, set up the breakout rooms to include the external team members and asked team members to start the exercise.

As John expected, it soon became apparent in the subsequent team discussion that no one skill stood out as the clear choice of a majority of members with team members agreeing that it was difficult to choose any one skill as being more important than any other. John felt that Jennifer summed up the situation when she said that all the People Skills on the list were important with them supporting each other. Phil asked John which People Skill he would choose as the most important and John replied that he would cheat and choose Descriptive Feedback as a powerful interpersonal skill and Attitudes as a self-awareness skill. Mo then directed a question to team members asking them if they had found the People Skills to be useful. In responding to Mo's question there was a consensus amongst that team members that they had found the People Skills to be beneficial. Sarah added that she thought that the People Skills were as, if not more, useful than the OFTEN framework Technical Skills in influencing the way in which the Oe375 team had worked together effectively.

John thanked team members for their participation and then stated the meeting purpose and outlined the agenda. John moved directly on to the next agenda item by saying that he was going to pick up the point that Jennifer made in the Warm Up exercise when she had said that the People Skills supported each other and shared a slide (Figure 13.2).

John explained that he saw the People Skills that the team had considered in their meetings forming an overall system, just as the component parts of a bicycle formed an overall system. John added that the People Skills act together synergistically, in the same way as the bicycle components do, to form a whole which was greater than the sum of the parts. John cited the example of tacit knowledge being shared as explicit knowledge through visualisation with the explicit knowledge then being clarified and extended through the application of listening and questioning skills. In looking back to the Warm Up exercise John said that no one People Skill stood out as being more important than the others in the same way that he found it difficult to identify the most important component of a bicycle in isolation from other components. Andy said that the frame was obviously the most important component since everything else hung on it, which drew a few smiles, and Phil retorted that he would not want to ride a bike without pedals, handlebars or a seat, producing more smiles.

John asked team members if they could give examples of the synergistic use of People Skills in their meetings. Sam said that when he and Phil were developing the propulsion system noise Simulator he had found Questioning for Clarification and Restatement to be very helpful. Phil nodded his head in agreement with what Sam was saying. Andy said that he had found the Listening and Speaking Communication skills very useful in working with his colleagues in HJK. Sarah said that she saw the Effective Meetings process as a combination of People Skills and said that she now appreciated the benefit of planning a meeting before it occurred and reviewing it after

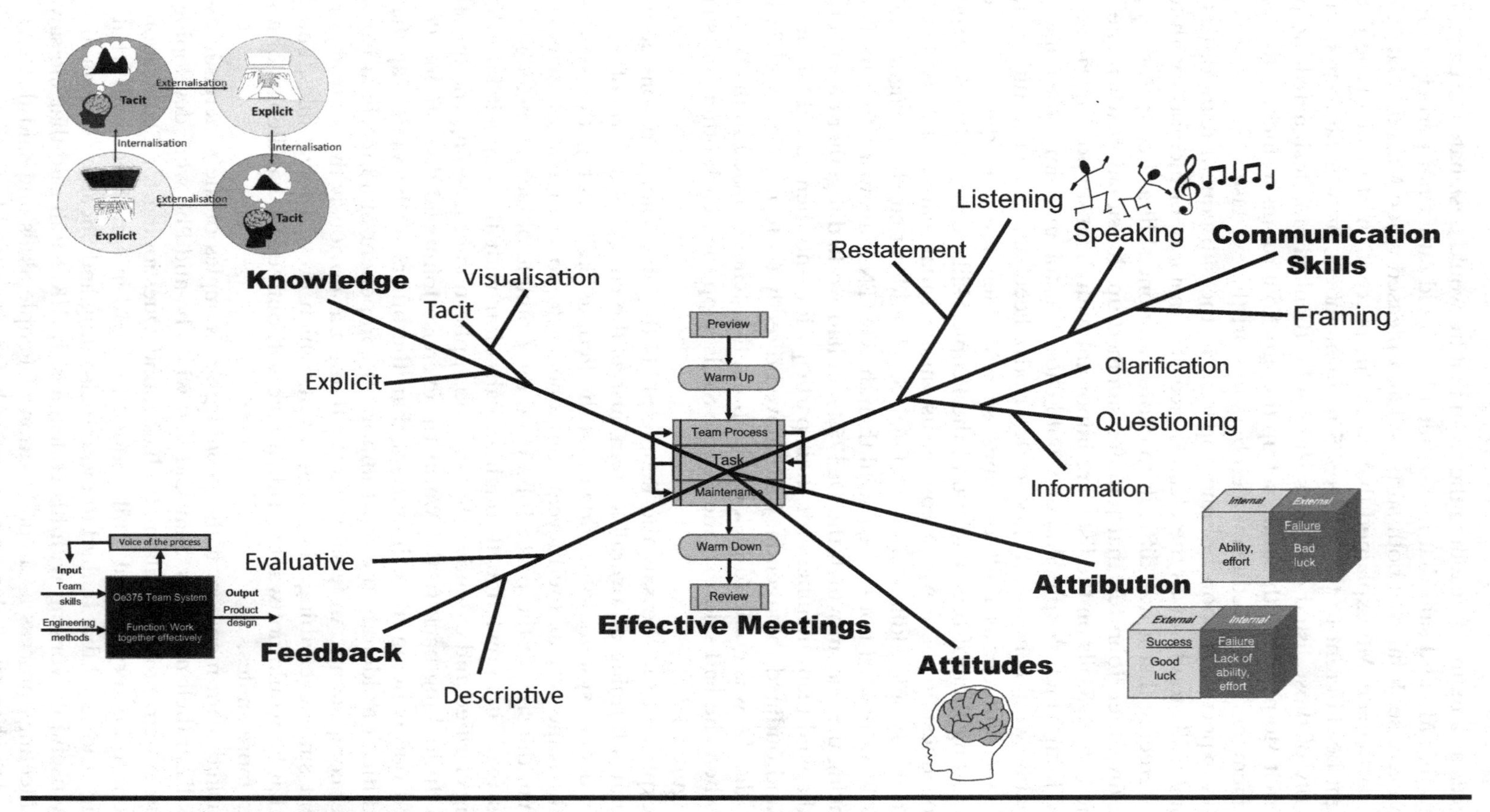

Figure 13.2 Oe375 People Skills mind map.

it had happened as well as taking a break during the meeting. Jennifer said that having a better understanding of how her attitudes influenced what she thought, along with the framing of verbal information, helped her in taking clearer ownership of her opinions. Yumiko explained that Descriptive Feedback had helped her in interacting with people in Agano and added that she saw Restatement as a form of Descriptive Feedback which had helped her in Team B meetings. Not to be left out Wolfgang said that he had found the framing of information useful in talking to people. Mo said that he was impressed by the enthusiastic way in which Oe375 team members spoke about the People Skills and John used this as a cue to conclude the discussion by thanking all team members for their examples. John then moved on to the next agenda item.

John said that since the Oe375 team had almost completed their Phase 3 analysis for all Oe375 systems he wanted to look at Phase 4 of the OFTEN framework entitled *Design Verification* and shared a slide (Figure 13.3).

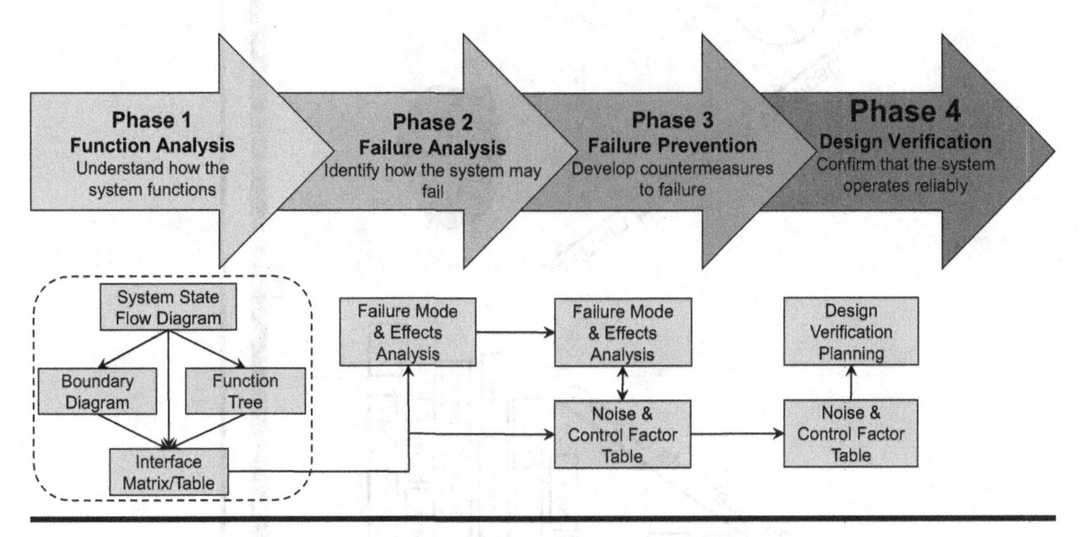

Figure 13.3 OFTEN FMA Framework – Phase 4.

John said that the purpose of Phase 4 was to verify the reliability of a design by using a verification process that was effective and efficient. John explained that the interface analysis conducted in Phase 1 of the framework facilitated the identification of the noise factors documented in the noise and control factor tables (NCFTs) developed in Phase 3 of the framework with design verification subsequently needing to ensure that a design was robust to the effects of these noises. In confirming that a design was robust to noise, a verification test evaluated the effectiveness of the countermeasures to failure summarised as robustness strategies in the NCFT and/or as recommended actions in the DFMEA with the latter document having been developed over the previous two phases of the framework.

John explained that effective Design Verification should be a good indicator of the reliability of a design in customer usage while efficient Design Verification used the least number of tests in achieving effective verification. John shared another slide (Figure 13.4) which he said team members would recognise and added that the right-hand side of the Systems Vee depicted the integration of the motor into the propulsion system and the propulsion system into the eBike.

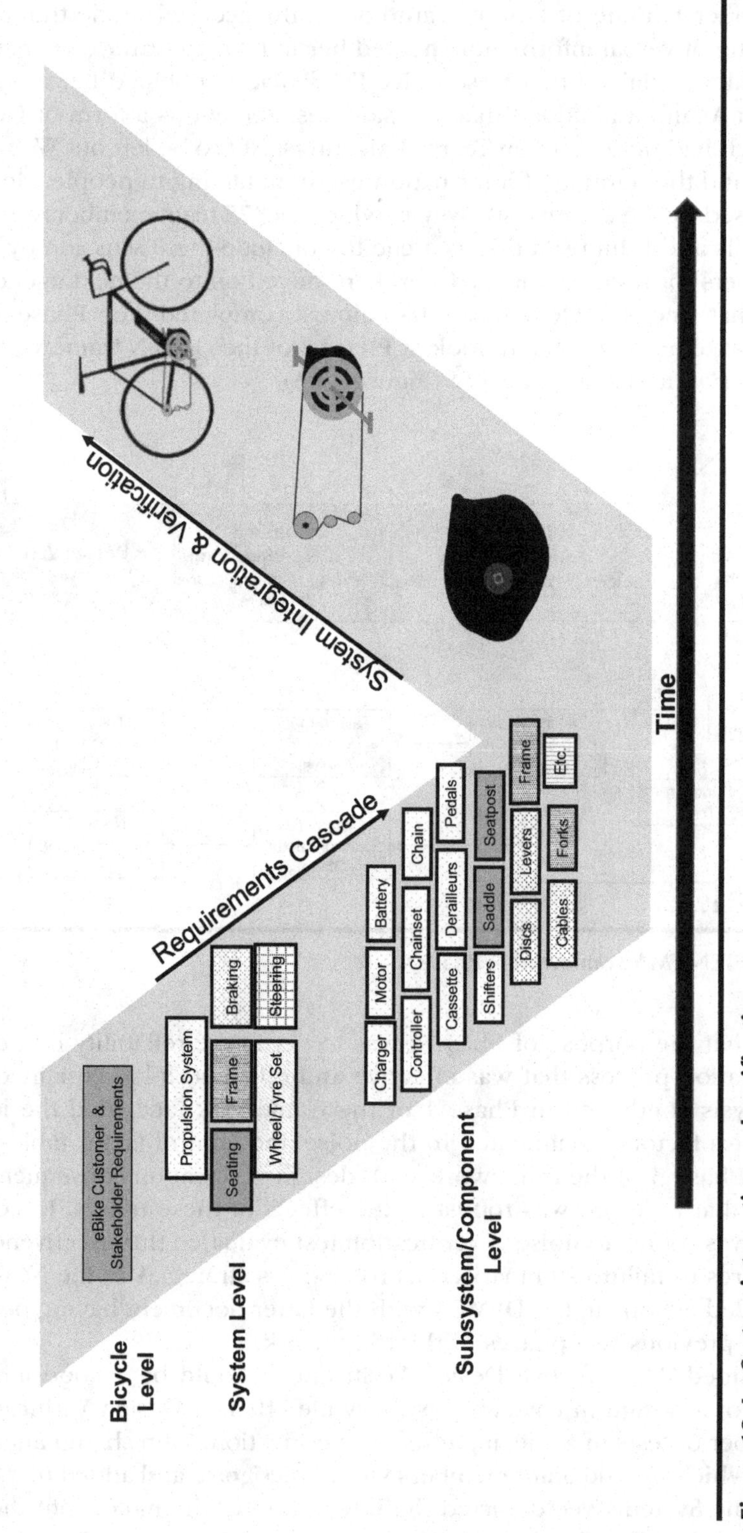

Figure 13.4 System integration and verification.

John said he wanted to consider the verification process illustrated on the right-hand side of the slide by conducting a quick vote and explained that he wanted team members to send him the answer to the question that he had just sent to them on meeting Chat (Figure 13.5).

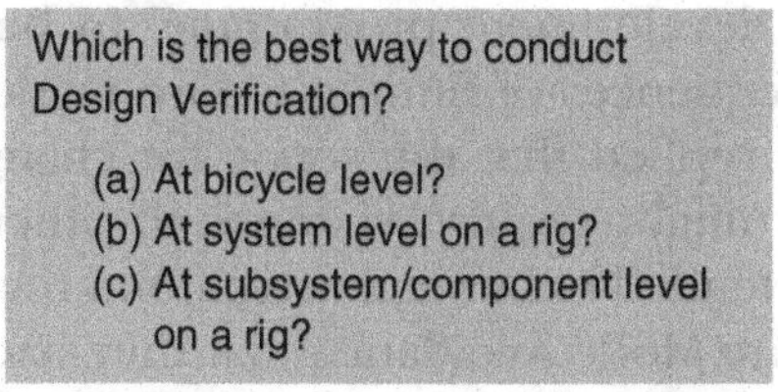

Which is the best way to conduct Design Verification?

(a) At bicycle level?
(b) At system level on a rig?
(c) At subsystem/component level on a rig?

Figure 13.5 Chat question.

John explained that he would like all team members to respond (a), (b) or (c) and added that he would collate the answers into a bar chart. John said that team members could choose any one option on its own, any two options, or all three. John asked team members to send him their answers after they had given the question some thought. John invited Mo to participate, and he responded that he intended to do so.

After receiving answers from all team members John quickly collated the responses and shared the bar chart that he had developed (Figure 13.6).

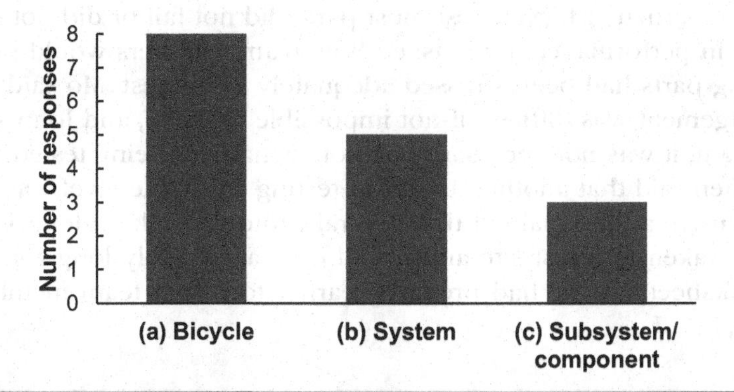

Figure 13.6 Responses identifying the best level to conduct Design Verification.

John said that he wanted to understand the results better and asked team members to use the meeting software *Raise your hand* in answering his next few questions. John asked team members to 'raise their hand' if they had chosen all three options and Sarah, Jennifer and Yumiko indicated that this was their selection. John then asked team members to indicate if they had chosen one option only and John noted that Wolfgang, Sam and Mo 'raised their hand'. John said that since everyone had chosen the bicycle level this must be the single option choice. John worked out that Phil and Andy must have both chosen options (a) and (b), although in the spirit of openness, he asked that those team members who had chosen two options to 'raise their hand', and Phil and Andy duly obliged. John explained that he wanted to discuss why the results might be as they were.

John asked team members to give him reasons why someone might choose option (a) as the best way to conduct Design Verification and Mo said that a test based on riding the eBike reflected the way a rider would use it. John asked if there were any drawbacks in conducting verification testing at bicycle level when compared to testing at the other levels and Sarah said that it can only be done when you have bicycles, which will be late in the development programme. In building on Sarah's answer, John asked what the consequences of finding failure modes late in the development programme was, and Phil replied that the scope for change was restricted because changing one component might require changes to other components which also meant that fixing a failure tended to be expensive. John remarked that this was a fundamental rational for Failure Mode Avoidance that they had identified in one of their early Oe375 team meetings. John then asked what was being tested in a verification event conducted at bicycle level, and Sam responded by saying that everything was being tested. John queried Sam's answer by asking him to imagine that the power loss of a bicycle under test gradually increased and asked him if it would be obvious what was causing it. Mo intervened to say that in his experience sometimes the cause of power loss was obvious, like a brake pad rubbing but on other occasions, particularly with an intermittent fault, it was difficult to locate the cause. In building on Mo's comment John asked him if he meant it was sometimes difficult to understand what had been stressed in a bicycle test and Mo said that John could draw this inference from what he had said. John said that when this was the case it meant that it may not be obvious what was being tested. John then asked team members to imagine that at the end of a particular bicycle test most parts did not fail or did not seem to have any decrease in performance. John asked how team members would judge whether the non-failing parts had been stressed adequately in the test. Mo said that in many cases this judgement was difficult if not impossible to make, and John said that consequently, again, it was not necessarily obvious what was being tested.

Jennifer then said that another aspect of testing at bicycle level was that the testing was expensive and explained that several prototype bikes along with their riders had to be taken to a test site and tested over a relatively long route. John then shared a worksheet that he had prepared earlier to reflect team members' answers (Figure 13.7).

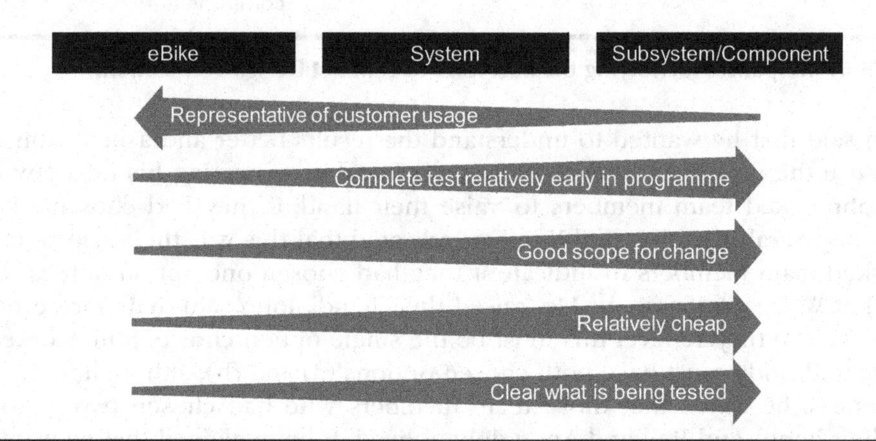

Figure 13.7 Weighting of features of verification testing in context of system levels.

John explained that he had divided the graphic into the three system levels indicated by the black boxes at the top with the direction and width of the arrows representing the systems level towards which the feature that was written on the arrow best corresponded. By way of example John said that the top arrow on the graphic indicated that verification at the eBike level was most representative of customer usage while the next arrow down indicated that testing at subsystem/component level could be completed relatively early in the programme. John said that he believed the graphic covered all of the points made by the team although since he had developed the graphic prior to the meeting the listing of aspects was not in the same order as the team had identified them.

Andy observed that John's graphic was stated in the positive sense while the team had largely identified the features in a negative sense. John asked Andy if he would give an example of what he meant, and Andy said that Jennifer had said that testing at eBike level was expensive, yet John's graphic identified testing at component level as relatively cheap. John thanked Andy for what he said was an interesting observation. John said that he was probably guilty, explaining that firstly he framed his original question in a negative way in asking if there were any drawbacks in conducting verification testing at bicycle level and secondly, he must have been in a positive frame of mind when he developed the slide which drew smiles. John added that this was a good example of the way in which a question is asked affecting the answer that the question promotes.

Yumiko said that the statement *Clear what is being tested* may be right when Oxton was testing a motor as a part of a propulsion system but might not apply when Agano was testing a motor as a system since failure of the motor might be due to a failure of any of the component parts that comprise the motor. John told Yumiko she had made a good point in that a company's perspective on the systems hierarchy depicted in the Systems Vee slide, and hence what he called the arrow graphic, depended on the products corresponding to their top and bottom levels. John said that they were developing the table from Oxton's perspective and from this perspective the motor was treated as a single entity and explained that if a motor was to fail during a verification test, Oxton would probably not investigate the root cause of this by identifying which component of the motor had failed. John continued by saying that he assumed that Agano ran a series of component-level tests before and after assembly of the motor such as testing the electrical integrity of the motor loom and stator windings. Yumiko confirmed that this was the case and said that while these focused tests were important this did not mean that an overall motor performance verification test was not just as important and valuable. John asked Yumiko if what she had just said was the reason that she had answered the Chat question by selecting "all levels" and Yumiko confirmed that this was the reason behind her choice. John asked Sarah and Jennifer if they had chosen "all levels" for the same reason as Yumiko and Sarah said that like Yumiko, she saw testing at all levels as important. Jennifer said that she agreed with Yumiko and Sarah.

John thanked team members for their input and said that he agreed that testing at all levels was important. John asked how testing at lower levels might be made more representative of rider usage given the significant relative advantages of testing at these levels. John then immediately asked a follow-up more focused question by asking what it was that made a test representative of rider usage. After a pause,

Wolfgang said that a good lower-level test would include the same noise factors that the component or system would encounter in customer usage as a part of the eBike. John said that he agreed with Wolfgang and told him that he would add one more word "effects" to what he had said and restated Wolfgang's statement with the word included – *A good test would include the same Noise Factor effects that the component or system would encounter in customer usage when a part of the eBike.*

John said that the NCFT provided a good input to verification testing because it listed the significant noise factors, and their effects, that a design would be subjected to. John reminded team members that the NCFT had been used in Phase 3 of the OFTEN framework in selecting a robustness strategy to improve design robustness. John added that once a robustness strategy had been implemented its effectiveness needed to be verified. John explained that the robustness of a design could only be verified if it was exposed during testing to the Noise factor effects to which it had been made robust. He added that the aim of verification testing was to demonstrate that the design performed as expected despite being exposed to the Noise factor effects. John said that, as Wolfgang had implied, this meant that all the Noise factor effects that a design had been made robust to must be included in any testing regime, and the NCFT helped to ensure that this happened. John shared an updated version of the propulsion system NCFT that the team had considered in Phase 3 of the OFTEN framework that now included columns for Design Verification tests (Table 13.2).

Table 13.2 Propulsion System NCF Table Including Design Verification Tests

Noise and Control Factor Table PROPULSIVE FORCES				Control Factor			DV Tests			
Interface	**Type**	**Units**	**Noise Factor Effect**	**Type of tyre**	**Tyre pressure**	**Propulsion system model**	**Simulation**	**HIL Rig**	**Wind Tunnel**	**Robustness Strategy**
	Piece to Piece									
	Tyre composition	Mfg-Hawk								
Tyre/Road	Tyre tread pattern	Pattern	ΔC_r	X	X	X	X	X		A. Select most robust tyres
	Tyre-road contact patch	mm^2								
	Tyre width	ISO		X	X					
	Changes over time/cycles									
Tyre/Road	Tyre wear	Tread	-							*Minimal effect*
	Tyre pressure	bar	ΔC_r	X	X	X	X	X		C. Specify most robust pressure
	Customer Usage									
	Rider mass	kg	Δm	X	X	X	X	X		B. Increase robustness to rider mass
Tyre/Road	eBike mass	kg		X	X	X	X	X		B. Reduce mass
	Dirt on tyre	mm^3	ΔC_r							*Neglect*
	eBike Ride cycle	Standard	ΔC_r, $\Delta\theta$			X		X		B. Optimise propulsion system
Rider & Bike/ External Environment	Projected frontal area U	m^2	ΔAC_a			X			X	B. Streamline frame
	& Rider posture D	m^2				X			X	
	External Environment									
Tyre/External Environment	Road surface	Defined	ΔC_r	X	X	X	X	X		B. Tyre pressure to suit surface
	Weather (rain)	mm/hour	-							*Ignore*
	Ambient temperature	°C	ΔC_r	X	X		X	X		A. Select most robust tyres
Rider & Bike/ External Environment	Wind speed	km/h	Δv_a			X			X	*See projected Frontal Area & Rider posture*
	Wind direction	Deg wrt eBike	Δv_a			X			X	
	Ambient temperature	°C	$\Delta\rho$			X			?	
	Elevation	m	$\Delta\rho$			X				*See Ride cycle*
	Gradient	%	$\Delta\theta$			X	X	X		
	System Interaction									
Propulsion/Braking	Resistive force/torque	N, Nm	(ΔF_p)							C. Eliminate non intended braking
	Diverted Output									
Propulsion/Drivetrain	Drivetrain efficiency loss	%	(ΔF_p)							C. Reduce frictional losses

John explained that the Oe375 propulsion system was at system level in the Oe375 eBike systems hierarchy and added that he had hidden the columns corresponding to the Noise factor ranges to make the table more legible. John said that as team members could see the NCFT now included three verification tests. John added that one of the verification tests was a virtual test based on the simulation that Sam and Phil had developed, another was the hardware-in-loop (HIL) test that they had discussed previously and the third a wind tunnel test. John said that the table showed which Noise factor effects could be included in each test by the "Xs" in the test columns and so the NCFT could be used as a checklist to ensure that all the key Noise factor effects that had been identified were included in at least one test. Sam asked why there was a question mark against ambient temperature in the wind tunnel test and John replied that he did not know if the ambient temperature could be varied in the wind tunnel that Oxton used. Sarah said that she believed the facility did not have this ability. John said that he had updated the propulsion system NCFT to include verification tests on his own and added that the table would need be reviewed and edited where necessary by the Oe375 team.

John explained that as team members knew all the noise factors listed in the NCFT affected the resistive forces acting on the propulsion system and added that the combined effect of the noise factors could be included in the HIL rig test by increasing the level of electromagnetic braking that was built into the testing equipment. Sarah asked if the combined effects of the noise factors included in a verification test was found by adding them together and John told Sarah that she had asked an interesting question. John said combining Noise factor effects could be tricky because some may tend to increase a design parameter while others might decrease the same design parameter and added that not all noise factors were necessarily going to act simultaneously or be at their extreme levels at the same time. John continued by saying that he tended to take a pragmatic approach since, as they had discussed before, it was normally the case that a small number of noise factors predominated and tended to swamp the effect of the other noise factors. John said that he relied mainly on engineering judgement in combining Noise factor effects. John continued by saying that a key aspect of any verification test was what was measured or how the robustness of the system or component was judged. John then shared a slide (Figure 13.8).

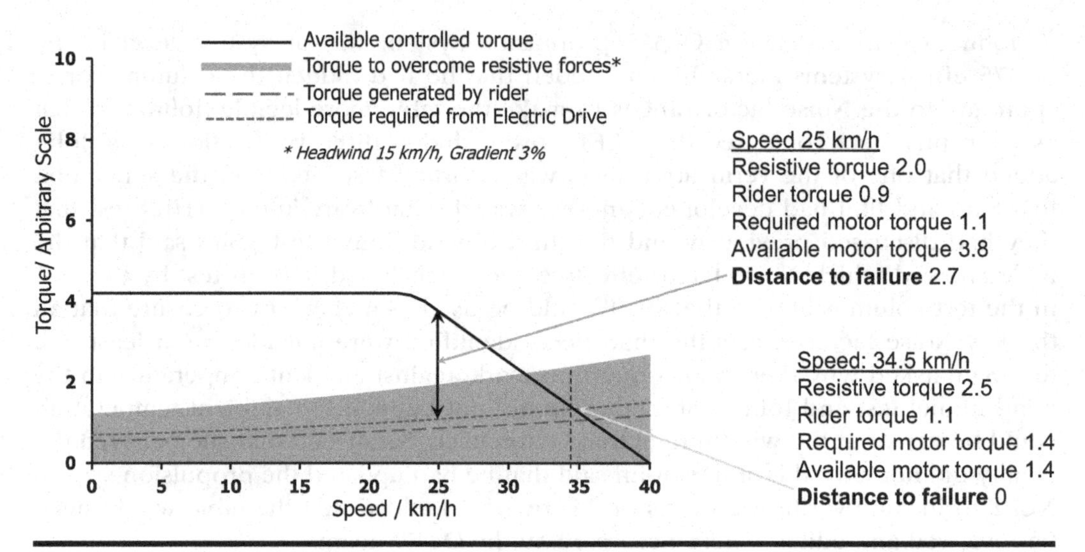

Figure 13.8 Distance to failure for 125% power assistance.

John said that team members might recognise that the left-hand side of the slide that he was sharing now was based on a slide (Figure 12.16) that he had shown in the Oe375 team meeting in which Yumiko had presented the work that she and her colleagues had been doing on the A539 motor. John reminded team members that this part of the slide was based on the brushless DC motor torque-speed curve that Yumiko had developed along with a set of Sam and Phil's resistive torque curves from the Oe375 propulsion system model. John said that the graphic was idealised and explained that the graphic related to the available output torque, neglecting losses, at a level of Electric Drive assistance of 125% with the constant headwind and gradient as stated on the slide. In bringing team members' attention to the data shown on the right-hand side of the slide John explained that this data related to two speeds of the eBike. John reminded team members that the torque was measured in the arbitrary units given on the y-axis of the plot and said that as team members could see that at 25 km/h the available motor torque given by the torque-speed curve, along with the torque provided by the rider, was significantly greater than that required to overcome the total resistive torque. John added that the difference between the available motor torque and the required motor torque at 25 km/h was 2.7 units and said that he had marked this on the plot by what he termed "*Distance to failure*".

John asked team members to compare the situation at 25 km/h with that at 35 km/h when the torque available from the motor together with the torque generated by the rider was only just sufficient to overcome the total resistive torque. John said that the situation represented the distance to failure being zero with the motor being unable to meet any further demands put on it. John said that distance to failure represented the best measure of system robustness or what was called the "Response" in a Design Verification test. John added that while other types of robustness metric are used in verification testing such as testing to bogey and testing to failure, distance to failure was the most information rich metric and shared a slide (Figure 13.9).

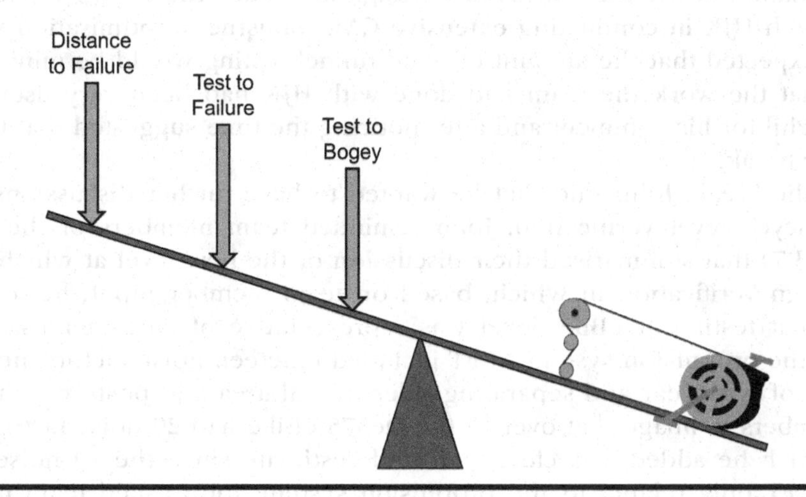

Figure 13.9 Measures of system robustness.

John said that, as team members would know, in measuring to bogey a part is tested until it reaches a predetermined target of a given time or number of cycles. John said that in bogey testing, a part is deemed robust if it does not fail during the test which leaves unanswered questions such as how long it would take for the part to fail. John continued that while the answer to this question is realised in testing to failure, in such a test it might not be clear which Noise factor effects caused the part to fail. John explained that, as shown for the propulsion system verification test response, the distance to failure related the performance of a system directly to key noise factor effects documented in the propulsion system NCFT that influence system robustness. John added that distance to failure, being the most useful, gave the most leverage in quantifying system robustness in a verification test.

John asked team members if they had any questions or comments. Jennifer observed that the Simulation and HIL tests were used in Phase 3 of the OFTEN framework to investigate and improve the propulsion system robustness and again in Phase 4 as verification tests. John thanked Jennifer for her comment and said that, as they had considered briefly in Phase 3 of the OFTEN framework, this was often the case and added that whether a particular test was used as a part of the design process or as a verification test depended on the phase of the OFTEN framework in which it was used.

Phil then reminded John that in Oxton Wind Tunnel testing was reserved for the high-end bikes because of the expense of hiring the facility. John said that he was aware of this situation and added that he had discussed this with Jim Eccleston and Mo who had agreed that Oe375 could piggy-back on the wind tunnel testing of the current high-end bike programme code named O549. John said that he had told Jim that because Design Verification testing was intended purely as confirmation of the robustness of a design and since the Oe375 PD team had worked with HJK in conducting extensive CAE robustness optimisation work on drag he expected that the amount of wind tunnel testing would be minimal. Phil agreed that the work the team had done with HJK had been very useful. John thanked Phil for his comment and after noticing the time suggested that the team take their break.

After the break, John said that he wanted to have further discussions on the role of bicycle-level verification. John reminded team members of the graphic (Figure 13.7) that summarised their discussion of the best level at which to conduct Design Verification in which, based on team member input, he had documented that testing at eBike level was representative of customer usage. John said that the propulsion system NCFT included nineteen noise factors, neglecting the effect of tyre wear and separating rider frontal area and posture. John asked team members to image that overall, the Oe375 eBike had 20 noise factors acting on in, which he added was clearly an underestimate since the 19 noise factors in the NCFT only related to the propulsion system. John asked team members how many possible combinations of Noise factor levels including all noise factors there would be, taking each Noise factor at both its minimum and maximum level. Andy quickly replied that there would be two to the power twenty combinations and after a brief pause Sam said that this was over a million. John said that this many tests would be required to experience all noise factors at their maximum and minimum levels and to know what Noise factor was causing which effect by changing one Noise factor at a time between bicycle road tests. Sam interjected to say that there were ways of significantly reducing the number of tests through statistically designed experiments which would allow multiple noise factors to be changed at a time between tests and still be able to tell which Noise factor caused which effect.

John thanked Sam and said to him that, while he was right, they were still looking at a large number of tests. John said that in addition to the number of tests required there was the difficulty of manipulating some of the noise factors to their maximum and minimum levels and cited the example of road testing with a need for the wind to blow at 25 km/h on some test days with no wind on other days. John added that it was generally much easier to manipulate Noise factor effects in system and subsystem/component-level tests. John said that he was not decrying bicycle-level testing as he saw it as an important part of an overall Design Verification regime. Looking at Mo, John said that he would only add a few words to Mo's well-known mantra and quoted ***There is no substitute for real-world testing with the final*** *product as a part of an efficient regime of testing at all system levels.* John said that of course while most, if not all, of the verification testing of subsystems/components would be done within the supply base he would like to see more testing at system level than was currently the case in Oxton. John then asked Mo, who had been quiet during most of the technical part of the meeting, for his perspective on what the team had just discussed.

After thanking John, Mo said that he had found the discussion very interesting. He admitted that he might have to modify his views on the role of what John had called "testing at the bicycle level" and added that, like John, he still saw this as very important. Mo said that the point that had struck him the most was the one that John had just made, which was that it was easier to include Noise factor effects in lower system-level testing than at bicycle level. Mo added that while he and John had met regularly to discuss the Oe375 programme before what he called the Oxton reorganisation, they had both been very busy lately and consequently, while he had discussed the functional approach to DFMEA with John, the NCFT was new to him. Mo continued by saying that, from what he had gathered in the meeting, the NCFT seemed a useful tool in planning effective Design Verification. Mo added that, as he has promised John originally, as far as time permitted, he was happy to help the Oe375 team in conducting their Design Verification. John was pleased that Mo was still prepared to support the Oe375 team despite his increased workload and that he had viewed the NCFT positively. John asked team members if they had any more comments and Mo said that he assumed that the Oe375 team would be documenting the schedule of verification testing in a Design Verification Plan and John agreed that this would be the case. Since there were no more comments John asked Yumiko if she would like to share with the team what she had being doing in verifying the design of the A539 motor.

Yumiko thanked John and said that she had found what they had discussed very interesting. Yumiko reminded team members that the design of the A539 motor had been adapted by herself and two Agano colleagues for use on eBikes like Oe375. Yumiko reiterated that although Agano had not applied the NCFT in the original design of A539, her colleagues and she had used this tool in adapting the A539 design during their application of Phase 3 of the first-time engineering design (FTED) framework. Yumiko said that she wanted to illustrate the use of the NCFT in Design Verification and she thanked John for the advice that he had given her about using the NCFT in this context. John said that he was pleased to help and said that the help he gave was minimal. Yumiko then shared an updated version of the NCFT for the motor that she had shown the team before (Table 12.1) which now included Design Verification test columns (Table 13.3).

Table 13.3 Motor Noise and Control, Factor Table Including Verification Tests

Noise and Control Factor Table MOTOR		Control Factor			DV Tests				Robustness Strategy
Type	Units	Manufacturing controls	Case Design IP Rating	Motor robustness	Virtual Motor Test	HIL Rig	Electrical Integrity Test	Accelerated Thermal Test	
Piece to Piece									
Rotor magnetic density variation	T	X							C. Reduce the manufacturing and assembly variation
Manufacturing variation	Various	X			X	X			
Assembly variation	Various	X							
Changes over time/cycles									
Bearing wear	mm			X		X			B. Reduce motor sensitivity to Noise
Stator heating	°C			X				X	
Contact resistance	Ω			X			X		
Sensor deterioration	V			X			X		
Rotor magnetic density loss	T			X	X	X			
Variation in sensor response	ms			X	X	X			
Customer Usage									
Excessive rider loading	kg				X	X			C. Define acceptable level of vibration
Excessive road vibration	Hz				X	X			
External Environment									
Dust ingress	mm^3		X		X	X			C. Minimise ingress
Liquid ingress	ml		X			X			
Ambient temperature	°C			X	X	X			B. Reduce motor sensitivity to Noise
System Interaction									
EM induced vibration	Hz			X	X				B. Reduce motor sensitivity to Noise
Controller AC output instability	V, Hz				X				C. Improve controller AC output stability
Battery voltage variation	V			X	X				B. Reduce motor sensitivity to Noise
Diverted Output									
Audible Noise	dB			X					See EM induced vibration

In referencing the table, Yumiko said that she had spoken about the development of the motor simulation model when she had presented the robustness work that Agano were doing on the A539 motor, and this model formed the basis of the Virtual Motor Test. Yumiko explained that in addition, the HIL rig test was like the test that Oxton members of the team had been conducting on the Oe375 propulsion system and so she would not go into any further detail about this test either. Yumiko said that she would summarise the other two verification tests documented in the NCFT starting with the Electrical Integrity Test. Yumiko said that this test included a check of electrical continuity and electrical resistance of the motor loom, stator windings and stator winding insulation as well as a check of the integrity of the rotor position and speed sensors. Yumiko explained that because this test was a simple automated test it was used both in Design Verification of the adaptation of the A539 motor and as an end-of-line test in the manufacturing of the motor. Yumiko added that, having heard what John had said about robustness metrics, she appreciated that the metric in this test was a simple pass/fail and added that since the test was highly focused it was a relatively simple matter to find out where any electrical discontinuity had occurred.

In describing the Accelerated Thermal Test, Yumiko explained that the stator was hung beneath a fume hood with a controlled variable airflow passing over it and a variable current passed through the stator windings. Yumiko said that the maximum voltage across, and hence maximum current through, the stator windings was greater than that experienced in normal usage so as to heat the stator winding insulation above its stated operating temperature range. Yumiko added that to further increase the thermal stress on the stator windings' insulation, the stator was subjected to repeated thermal slew and variation in airflow across it. Yumiko added that the resistance of the winding insulation was measured every 12 hours using a megohmmeter with these readings being plotted against time until the resistance started to drop, indicating the beginning of a breakdown in the insulation. Yumiko said a projection of the results from the test indicated that the winding insulation would retain its electrical integrity for a period significantly longer than the target life of the motor and added that the test response was time to failure induced by the noise factor of heat. Yumiko added that providing the level of insulation breakdown which was judged to be unacceptable was clearly defined in terms of the insulation resistance then this metric of time to failure was in effect a distance to failure.

Yumiko asked if team members had any questions, and Wolfgang asked what a megohmmeter was, to which Yumiko replied that it was essential an ohmmeter which applied a relatively high voltage across the winding insulation and measured any leakage current through the insulation. Sam then asked what an accelerated test was, and Yumiko said that one in which the stress applied to a part in a test is significantly larger than the stress the part would experience in normal usage, to speed up any degradation in the part characteristic being measured. Yumiko said that in the case of the stator winding her colleagues agreed with her that the estimated time to failure was reduced by several orders of magnitude in the test. Sarah said that care had to be taken in conducting an accelerated test to ensure that the magnified stress level did not cause failure modes that would not occur when the part was subjected to a normal level of stress. Yumiko thanked Sarah and said that Sarah had made an important point. Andy then asked what thermal slew was and Yumiko told him that it was an increase or decrease of temperature. Since there were no more questions Yumiko handed the meeting back to Sarah.

Sarah thanked both John and Yumiko for their presentations and said that she was sure she spoke for the Oe375 team when she said that she was looking forward to putting into practice the aspects of Design Verification they had discussed in the meeting. John asked Sarah if he might share an observation and, after Sarah readily agreed, said that one thing that struck him in listening to Yumiko was that Yumiko and he had given their presentations in reverse order. John explained what he meant by this was in that in practice Design Verification of the motor should occur before the Design Verification of the propulsion system. Sarah thanked John for his observation and handed the meeting over to him.

John thanked Sarah and said that he thought now that the team was getting into Phase 4 of the OFTEN framework, it might be a good time for team members to take a look back at the journey that they had taken within the Oe375 team. John then shared a slide (Figure 13.10).

John said that the journey that the Oe375 team had taken was directed by the four-Phase OFTEN framework represented in the middle of the slide and the use

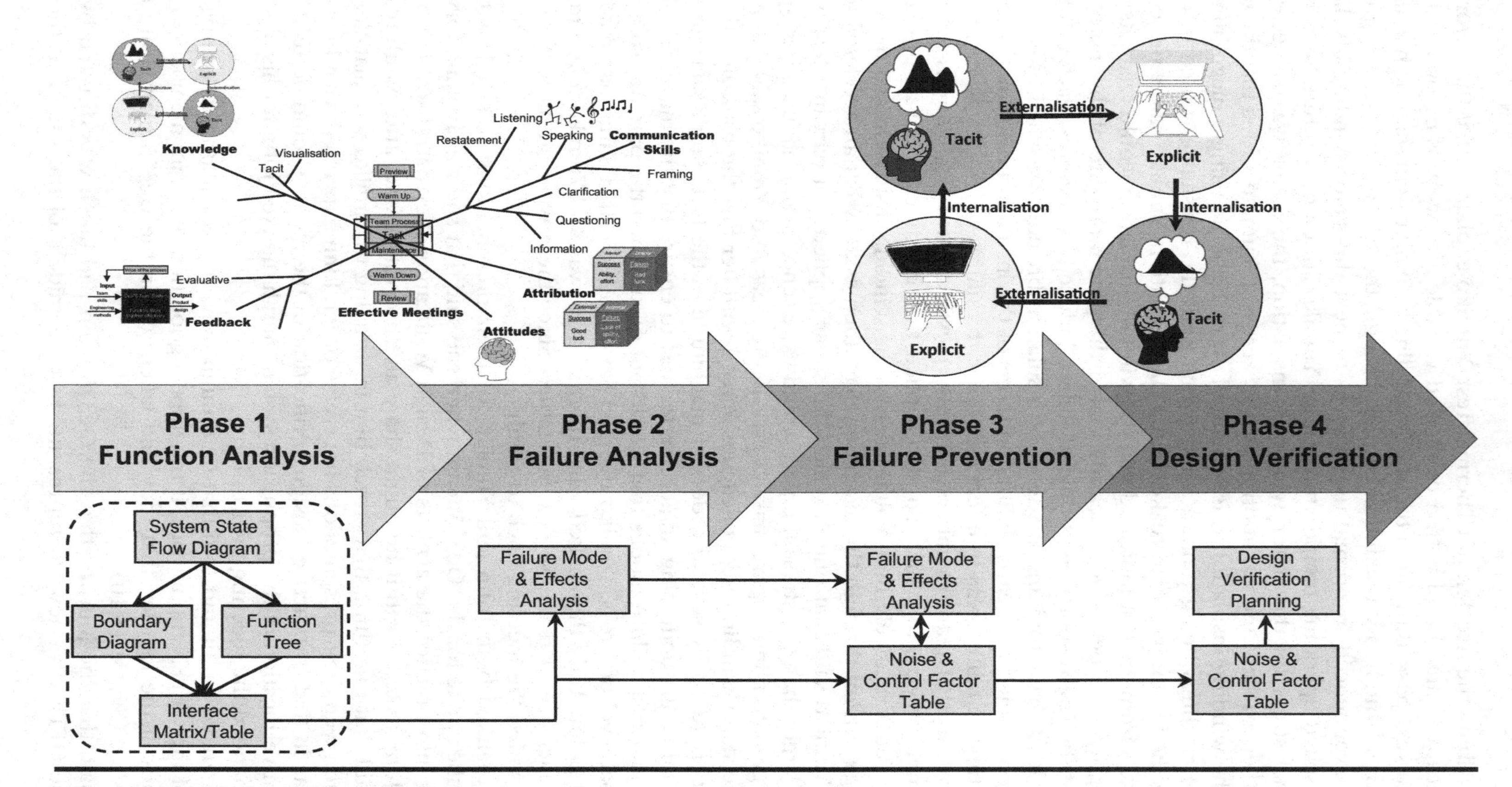

Figure 13.10 The Oe375 team journey.

of the People Skills portrayed at the top of the slide and the engineering tools and methodologies depicted at the bottom. John apologised if the People Skills graphic was illegible and said that team members would recognise it as the mind map that he had shown earlier in the meeting (Figure 13.2). John explained that he had deliberately picked out the Tacit to Explicit knowledge transfer cycle on the top right of the slide as he hoped that the increased use of tacit knowledge in the Oe375 team represented a key legacy from the journey that they had all been on. John said that, as team members knew, it was important to view the OFTEN framework in the context of the overall product development process and shared a slide (Figure 13.11).

John reminded team members that they had seen a different version of the slide in one of their early meetings. John explained that the slide depicted the overall context in which the team had been applying the OFTEN framework and added that in particular the OFTEN framework had been used within the System Design, Detailed Design and Prototype Testing Phases of the Oxton Product Development Process. John said that the placing of the People Skills mind map in the Manufacturing Readiness and Ship Product phases of the OPDP did not mean that People Skills use was reserved for these phases only, adding that it was the only space on the slide that was available, which drew a few smiles.

John said that the aim of applying the OFTEN framework was to ensure that the design of the Oe375 eBike resulted in robust and reliable product in the hands of the customer after its manufacture. John added that a key test of the successful application of the enhanced Technical and People Skills used within the OFTEN framework was that the launch of the eBike within the Manufacturing Readiness phase of the OPDP would be problem-free. John said that he was also looking forward to the time when Oe375 began being shipped to customers, since customer reaction to using the product in earnest was the ultimate validation of the effectiveness of using the OFTEN framework as a key part of the product's development. John added that he was getting ahead of himself, since the team had only just begun Phase 4 of the Framework. Aware that he was both in danger of getting into what team members might see as "training mode", and the fact that he wanted to spend more time than usual on the meeting Warm Down, John concluded what he saw as a "mini-presentation" and thanked team members for their attention.

John then explained that as this might be one of the last times that the full Oe375 Product Design team met he was going to conduct an extended Warm Down. In beginning the Warm Down John said that he believed that the Oe375 team were making a significant contribution to increasing Oxton Bikes' corporate engineering knowledge on the design and development of failure-free product. John said that he was interested to hear how team members viewed their experience of being a member of the Oe375 team and added that he would also like to hear team members' individual perspective on how their knowledge and skills might have benefitted from being part of Oe375 programme.

John then initiated what he called an extended pass-the-pen exercise and after giving team members a few minutes to collect their thoughts, he asked for a volunteer to start the process of sharing them. Sarah said that she would go first. Extracts from John's notes on the statements that individual team members made are reproduced below (Table 13.4).

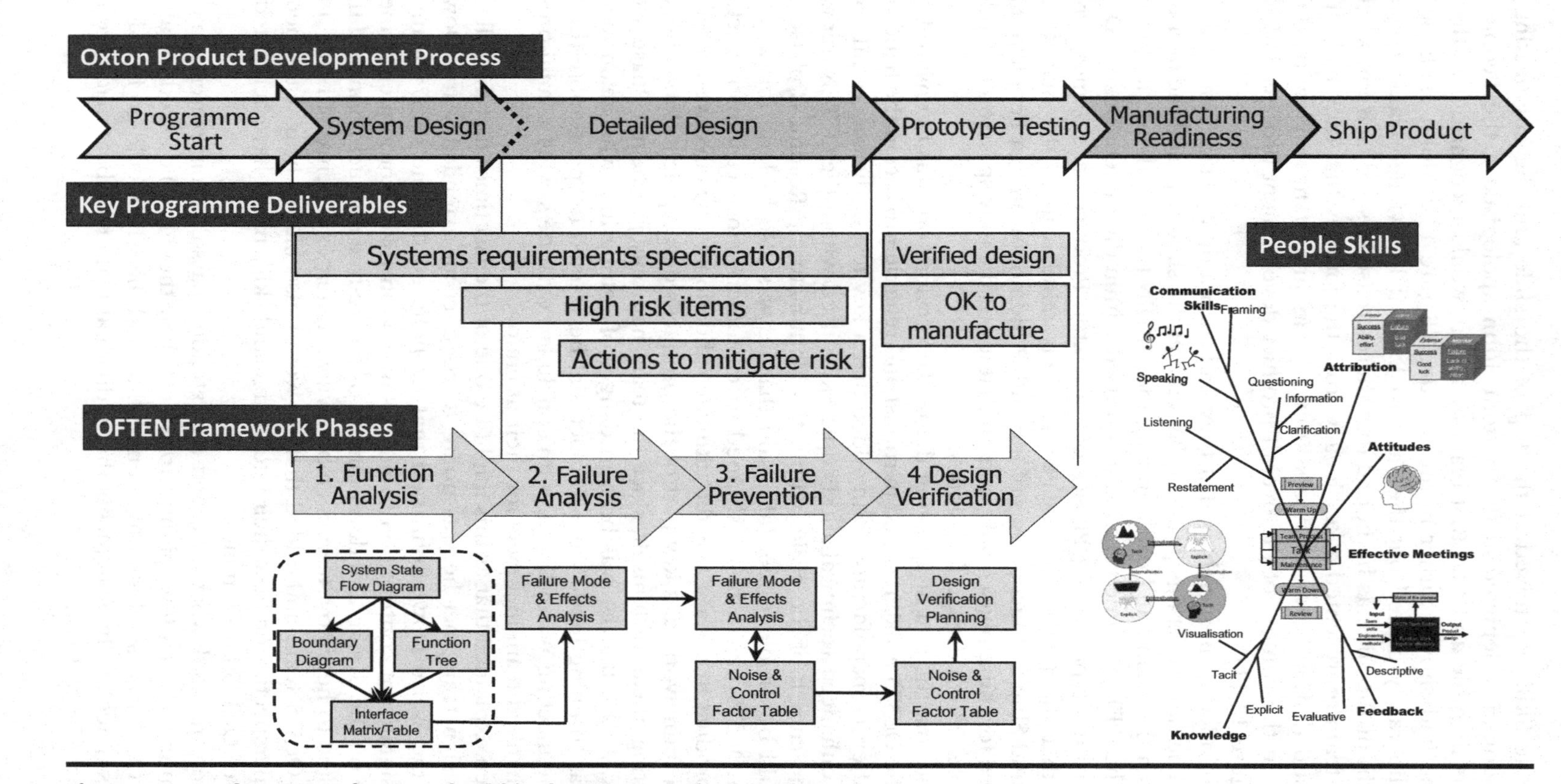

Figure 13.11 The OFTEN framework within the Oxton product development process.

Table 13.4 Verbatim Notes on Oe375 Team Member Warm Down Comments

Team Member	*Verbatim Extracts*
Sarah	*I have used the People Skills both in the Oe375 team and generally in the company and find them very useful. Interface Analysis is very time consuming but being the core of OFTEN facilitates the development of the other tools and gives them structure. I now see the importance of DFMEA and developing it in a team is so much better than doing it on your own. I like the way that the NCFT distils the knowledge gained in applying the OFTEN framework to facilitate the designing and verifying of failure-free product. All in all, the Oe375 programme has been the best programme I have worked on in Oxton.*
Sam	*My initial reaction to what we were doing on Oe375 was not good and I felt that the People Skills and OFTEN were just theoretical and a waste of time. Once we got into model-based engineering, however, I realised the worth of what we had been doing and it all became interesting and exciting. Once I took the People Skills more seriously, they made me think and I realised things about myself that helped me in working with others.*
Jennifer	*If I am honest, I was sceptical at first about both the People Skills and OFTEN as to me they felt like some of the peripheral stuff I learnt at Uni – while you could not argue with the logic it seemed to have little practical day to day use. However, I soon became convinced that the OFTEN framework was a good way of doing things. I think the People Skills are just common sense and are very helpful when working in a team.*
Phil	*I am too long in the tooth to be taken in by the latest management fad and usually just wait for the next one to come along. I have been pleasantly surprised by OFTEN as it genuinely seems a good approach and I'm looking forward to the proof that all the time and effort that the team has put into the first three phases of the Framework has been worthwhile with many fewer problems being found at or after Design Verification. I have also taken to the People Skills which have helped me both as an individual and I hope I am a better at communicating with people now.*
Andy	*I have enjoyed my time so far with the Oe375 team and believe it or not I actually look forward our meetings. I don't think that I will use the OFTEN framework in HJK as we have our own way of doing things never the less there are ways of working embedded in it that might prove useful. Similarly, I do not see myself promoting the People Skills in HJK although like Phil, I think that they have probably helped me as an individual.*
Wolfgang	*I have found it interesting and enjoyable to work with everyone on Oe375. I do not see Teutoburg using the OFTEN framework as we do not do any design work. I'm looking forward to Oe375 going on sale and seeing how customers react to it. I found some of the People Skills useful, but I think that others are a waste of time. I can't see the People Skills being adopted in Teutoburg.*

(Continued)

Table 13.4 (Continued)

Team Member	*Verbatim Extracts*
Yumiko	*I recently had a performance appraisal and my manager told me that she had noticed that I have grown in confidence since I started to work closely with Oxton. I put this down to several things and thank everyone in the Oe375 team for being supportive. I think that the logical way in which the OFTEN framework is structured has helped me to see how what we are doing fits together. I enjoyed using our version of the Framework in Agano and was pleased by the way in which my two colleagues that I have worked with on the A539 motor took to it. Although there are cultural differences between Oxton and Agano I have also found the People Skills have helped me in my job in Agano as well as in working with the Oxton teams.*
Mo	*As you know this is the first time that I have attended a Oe375 PD team meeting and so I can't reflect back on your journey. I have been impressed by you as a team and the way that you work together. From what John has told me about the OFTEN framework and what I have seen today it certainly challenges the way in which we have been working in Oxton and I believe this can only be a good thing. I am looking forward to working with you all on Oe375 Design Verification.* *Having heard more about the People Skills and seeing them in operation has convinced me of their potential.*
John	*When I started to work with the Oe375 team I was apprehensive as to how what became OFTEN, and the People Skills would be received. Overall, I was pleasantly surprised how you took to them. I also thought that there would be resistance to the methodologies since I was coming from automotive into a different industry, but I detected little opposition at least from within the team. I have been very impressed with your willingness to listen and apply new concepts and have learnt a lot from you all. I have enjoyed my time on the Oe375 programme and like others I am looking forward to seeing the fruits of our labour when the eBike is launched.*

John was very encouraged listening to team members' positive reflections and felt that although Wolfgang's comments were more negative, they reflected the reality of his situation. John was also reassured by Mo's renewed commitment to work with the team on Design Verification and his positive feedback on the meeting, including both the Technical and People Skills content, as well as the way in which the meeting was conducted. John thanked team members for their contributions in the meeting and said he looked forward to working with them on Phase 4 of the OFTEN framework. John handed the meeting back to Sarah who declared the meeting closed.

John returned to his office pleased with the way in which the Oe375 meeting had gone and in reflecting on what Sarah had said in the Review/Preview meeting he agreed that it did seem like old times. John found himself looking forward to working with the Oe375 team and Mo in developing and implementing the programme Design Verification test plan although he was getting worried about his workload, given the O254 programme would also soon get to the stage of conducting Design Verification testing. His mind was directed away from this line of thought when, on checking his emails, he saw he had a meeting notice from James Tutton. John saw that James was inviting Ira Grover and him to another get-together on Friday evening after work in the same pub in which they had met the last time. The only information John could glean from the notice was that the meeting was entitled "Agree Plan of Action". John's attention then switched again when he realised that he had a meeting with the O254 team in a few minutes. In comparing the meeting that he had just had with his meetings with the O254 team he realised that working with O254 was never going to be the same as working with Oe375. While John felt that he had developed a good working relationship with the members of the O254 team, and that together they were making positive inroads into the O254 programme, to him O254 team meetings often seemed more like problem-solving than design because of the need to overcome the less than adequate work that had been done under Paul Clampin's direction.

Although he had yet to discuss this with Mo Lakhani, John was hoping that the O254 team could take advantage of the work that the Oe375 team would be doing in Phase 4 of the OFTEN framework in the same way as they had made use of the analysis that the Oe375 team had completed in the other three phases. John realised that although the O254 programme would have to place more reliance on bicycle-level design verification testing than he would have liked he was reassured that following discussion with Jim Eccleston and Margaret Burrell, the frame and groupset for the hybrid bike were being sourced from HJK and Agano respectively, albeit in the form of off-the-shelf components. Given Oxton's good working relationship with both companies John was sure that, if necessary, he would be able to get detailed information about the components over and above those quoted in the literature available on the companies' websites.

John met with Mo the next day to discuss the Design Verification needs of both the Oe375 and O254 programmes. Mo said that while he was happy to support the Design Verification of both programmes, he was constrained in what he could do because of his current workload. After John had explained the synergies between the programmes Mo suggested that John and he form what he called a "DV Core team" with representation from both programme teams with the team planning and implementing the Design Verification of both programmes in parallel with each other. John was pleased that Mo and he were thinking along the same lines in enabling the O254 programme to continue to make best use of the detailed analysis that the Oe375 team had completed. John also realised that Mo's suggestion had the added bonus of reducing the overall administrative workload associated with organising and running tests. John told Mo that he thought his proposal was a good idea and added that while he had people in mind that he would like to see on the Core Team he would like to speak to both programme teams before finalising the team membership. In the event, with Jim Eccleston's blessing, both programme teams agreed that forming a Core DV team was in the interest of the company and, as John hoped, Sarah and Phil were nominated from Oe375 with Peter Andersson and Emma Stead joining them from the O254 team.

Towards the end of Friday afternoon, and at the end of another tiring week, John was driving to meet James and Ira. He was looking forward to his weekend with Jane and while he saw the impending meeting as somewhat of an unwanted distraction, he was keen to share his idea for how they might meet with Bob Jenkins. John had found little time to think about a plan to confront Bob since he last met with Ira and James but had developed an idea that morning while walking the dogs. James and Ira both pulled into the car park as John was walking from his car to the pub door. John bought the drinks and they sat at the same table as last time. After introductory pleasantries James asked John if he had come up with any ideas about confronting Bob. John said that he had, and added that he was proposing to ask Bob to meet in order that he might learn something to his advantage. James said that Bob might not want to attend a meeting with them and while John agreed with James that this was the case, he explained that by saying that Bob might learn something to his advantage this might be sufficient of a carrot to persuade him to go. Ira said that while John's idea sounded interesting, she did not see how any meeting would be to Bob's advantage. John said that if Bob had made the voicemail calls to Blade HR it was better for him to discuss the matter with the three of them rather than them pursuing the matter directly with Blade without involving Bob. James said that he did not give a hoot how they invited Bob to a meeting as long as the invitation got him to go.

Ira said that she was prepared to go along with John's suggestion and cautioned against contacting Bob through company channels. John said that he agreed with Ira and explained that he proposed to contact him through the free internet-based messaging network that everyone seemed to use these days. John added that the only problem was that he did not know Bob's personal mobile number which he needed to be able to contact him. James said that this was not a problem since all departments in Blade now kept a log of everyone's personal contact details for emergency purposes. Ira said that she understood that such information was confidential to each department and James, in touching the side of his nose with his index finger, replied that working in Blade relied as much on who you knew as what you knew. John said that he hoped Bob would not learn that any of his personal details were being accessed and James assured John that this would not be a problem.

Ira asked John where and when he intended that they meet with Bob and John responded that his thinking was to leave it for Bob to choose the date, time, and venue as this would make it seem that he was not being pressured into the meeting. Ira then reminded John that he had said the information that they confronted Bob with should be descriptive. John said that he had been thinking about that as well, and his proposal was to present Bob with a factual list of what might be called his misdemeanours. John suggested that if they could add to the list over and above the three voicemail messages that would be good, and James responded by saying that he would use his contacts to see what he could find out on the quiet. John said that he would collate the list and added that he thought that in approaching Bob it might be better if, at least initially, only one of them spoke with him to reduce the chance of scaring him off. Ira said that she thought that John's proposal was sound and suggested that this person be John as he no longer worked for Blade. John confirmed that he would be happy to do this, and James said he concurred with the proposition. Ira added that she thought that any further contact between the three of them should be through their personal rather than company mobile phone and email accounts and they duly shared the details that they did not already have.

After concluding the discussion on his proposal, John did not feel like spending much more time with Ira and James in friendly conversation, and it seemed that they felt likewise as they both finished their drinks and got up to leave. In leaving, John suggested that the three of them continue their discussions through the messaging app and email. As John expected, Jane did not greet the plan to meet with Bob with wild enthusiasm when he got home, and he told her of the outcome of his meeting. However, John was pleased that, while he knew Jane was not at all happy with what he was proposing to do, she again gave him her support providing that he was careful and did not do anything silly.

Background References

Design Verification	Henshall, E. and Campean, F. (2010) Design Verification as a Key Deliverable of Function Failure Avoidance, *SAE International Journal of Materials and Manufacturing*, 3(1), 445–453, https://doi.org/10.4271/2010-01-0708

Chapter 14

Reviewing Achievements

John was pleased with the way his plan to meet with Bob developed. James had reminded Ira and him of the time during their apprenticeship when Bob had been accused by fellow apprentices of taking more than his fair share of credit in a group project by claiming to have done work that others had done. While the event had caused quite a stir at the time within the apprentice intake, John had quickly forgotten about it. John had decided to include this occurrence on the list he was compiling of what he called "Bob's Misdemeanours". James had passed on Bob's personal mobile number, and John had sent Bob a message. In messaging Bob, John had taken a leaf out of Mo Lakhani's book and invited him for a chat to discuss something in their mutual interest. Bob seemed to take the bait asking John why he wanted to meet and where and when the meeting would take place. After John had asked Bob to select the time and venue and said that he would discuss the purpose of the meeting when they met, they had agreed to meet on a Friday after Bob finished work in a small café in the town a few miles from the Blade Engineering Centre. In the meantime, John had agreed with Ira and James on the content and style of a one-page document summarising the incidents that they wanted to discuss with Bob. The document was not headed and comprised five bullet points which included details of the voicemail accusations against James, Ira and himself, preceded by the fact that Bob kept a log of the careers of the apprentices in their intake and the incident in which Bob had been accused of taking more credit than he deserved in the apprentice group project. John kept the description of events brief and factual and included Ira's, James' and his names at the bottom alongside the date on which he had arranged to meet Bob. John had agreed to meet Ira and James at Blade before they travelled in the same car to the café to meet Bob.

John left work early on the Friday afternoon that he had arranged to meet with Bob and parked his car in the Blade visitor's car park and called Ira on his mobile before walking through the main entrance doors and across the foyer to the reception desk. After informing the security guard that he was there to meet Ira Grover, John signed in and picked up his visitor's badge and lanyard to find Ira waiting to greet him on the other side of the turnstile which the security guard freed enabling John to walk through and shake Ira's hand. After initially greeting him, Ira walked with John in silence up the flight of stairs and along the corridor to the canteen. They had

DOI: 10.4324/9781003286066-14

arranged to meet James in the “Drink ‘n’ Snack” area which was clearly distinguished from the rest of the canteen by its carpeting and more secluded location. John saw that James was already seated in one of three comfortable chairs arranged around a low table. James stood up and asked Ira and John what they would like to drink, and both settled on a latte. James duly returned with three coffees on a tray which he set down on the table and indicated to Ira and James to take their coffees with the taller lattes being clearly distinguished from his flat white. As a quick reminder, John then ran over the plan that they had agreed for approaching Bob and gave Ira and James printed copies of the one-page document. John said that although the café in which he had arranged to meet Bob had a small car park, he thought they should park the car in a side street a short walk from the café in case the sight of Ira and James alarmed Bob before he had met with John, given that John was meeting Bob alone initially. John told Ira and James that he would text them at an opportune moment so that they could join Bob and him. Having finished their coffees, James said that they should make their way to Ira’s car in which they had agreed to travel. The three former colleagues travelled to the agreed venue in silence, and John assumed that, like him, Ira and James were rehearsing the impending meeting in their mind. Ira found a convenient parking spot and, although it was still ten minutes before the time at which John had agreed to meet with Bob, John said that he would get seated in the café at a table for four in as secluded an area as he could, before Bob arrived, to avoid Bob selecting a less convenient table. John picked up his backpack, got out of the car and walked to the café.

As it turned out John had just got seated at what he considered to be the best available empty table and told the waiter that he would wait to meet someone before ordering, when he saw Bob walk through the café door. After standing to shake Bob’s hand, John asked him what he would like to eat and drink. Bob replied that he would have a latte and toasted teacake, which John duly ordered along with a tea for himself. John started to ask Bob how he was keeping when Bob interrupted him and asked what it was that John wanted to discuss. John delved into his backpack and handed Bob a copy of the one-page document placing a second copy on the table in front of himself. Having taken the sheet of paper, Bob appeared to scan the document quickly before calmly asking John what he intended by the document. John replied that he assumed that Bob would recognise the events summarised in the document. Just then the waiter arrived with the drinks and teacake, and Bob stirred his latte before returning to the document which he appeared to read more carefully. After taking a sip of his coffee, Bob said that the situation was ridiculous as the events described in the document had nothing to do with him and added that he needed to get home. John suggested to Bob that it was probably in his interest to stay a bit longer, and Bob started to butter his teacake. John said that he was sure that Bob would remember as a matter of fact the event that occurred in respect of the apprentice team project. Bob said that he had no recollection of any such event and added that whatever John believed was of no consequence to him. At this point, John sent the pre-written text to James and Ira requesting that they come and join them in the café.

John directed Bob’s attention to the bullet describing the log that Bob kept, and Bob asked John how he knew about that. John explained that Bob’s wife had mentioned it to his wife at Mike Holmes’ retirement celebration. Having glanced down at the document, Bob said that the voicemail messages had nothing to do with him. John explained that he had spoken to Ira and James, and they had helped him compile the information described in the document. Bob said that John was wasting his time and quickly finishing off his teacake and coffee, stood up and started to walk to the door when Ira and James walked through it. James stood to his full height in front of Bob and asked him, in what John took to be James' polite voice, where he was going. Bob replied he was going home. Ira calmly told Bob that she would like to speak with him and said that it was in his interest to hear what she had to say. Bob hesitated for a short while and then turned around and returned to the table at which John was still sitting, with James and Ira following behind him.

James asked John how things were going, and John replied that Bob and he were just chatting. The waiter came over to the table and asked Ira and James what they would like. Ira said that she would have a latte and James settled on a glass of water. James asked if Bob had admitted to any of the voicemail messages described in the document, pointing to the copy still on the table in front of John. John replied to James's question by saying that Bob and he had just started to discuss these events. James said in a raised voice that he knew that Bob had left the voicemail accusing himself because the message had included details about an occurrence at James' tennis club some time ago that Bob would know about. John noticed that other customers had turned to look at what was going on. Bob replied that he had not made any accusation against anyone either anonymously or by name and, after a pause, he added that he could understand why someone might do such a thing. John was intrigued by Bob's statement and asked him why he believed someone might falsely accuse people like James, Ira and himself. It seemed to John that Bob had realised that he had said too much as he responded breezily to John's question by saying that there was no reason. Bob's remark seemed to trigger James' anger and in banging his fist on the table, he asked Bob why someone would make unfounded accusations. Bob, whom John thought seemed remarkably calm, responded to James by telling him that he had answered his own question by his demeanour, which only served to make James even angrier as he banged his fist on the table again, repeated his question and at the same time called Bob a little rat. By this time most of the other customers in the café were watching and listening to the proceedings. Bob calmly told James that he was a bully and added that he had only got to where he was in the company by bullying people. John suddenly thought he understood the underlying reason why Bob had made the accusations and asked him if he was jealous of the way in which James, Ira and he had progressed in their careers. Bob responded to John's question by asking John why he would be jealous of three of his former apprentice colleagues who had smarmed their way up the greasy pole.

In changing the subject, Ira calmly told Bob that John, James and she were going to write a joint letter to Blade HR setting out their suspicions about him. This seemed to have an impact on Bob, and John thought for the first time in their conversation that Bob looked worried as he requested Ira not to do what she had threatened. Ira asked Bob why they should not write to HR and Bob replied that he had recently received a written warning about, what he called, an unrelated matter, and even the suspicion of something else might go against him even if he strenuously denied it, as he would. As James seemed about to say something, Ira intervened and told Bob that what James, John and she would like to happen was for HR to receive a communication in which it was made clear firstly that that person making the communication was the same person who had made the accusations, and, secondly to explain that the accusations were totally false. Bob asked how he could make this happen when he did not know either who had made the accusations in the first place or their nature. John decided that it would not be productive to extend the conversation with Bob any further and merely told him that the ball was in his court and picking up the document in front of him returned it to his backpack, stood up and wished Bob goodbye with Ira and James following suit. John walked over to the café counter and paid the bill before joining Ira and James outside.

John, Ira and James walked quickly back to Ira's car without speaking, not bothering to look around to see if Bob had left the café. As Ira was driving the car away from the kerb, James asked John and Ira how they thought, what he called, the confrontation with Bob had gone. Ira said that she thought it had gone as well as they could have expected, and John said he agreed with Ira and asked James for his opinion. James said that he was disappointed as he was hoping for a confession from Bob. Ira said that a confession was never likely as Bob was probably always going to deny everything. John agreed with Ira and said that at least Bob now knows that we know it was him. Ira said to John that he sounded as if he was sure Bob was the perpetrator, and John said that having met Bob again, he was even more sure than he was before they met. John added that he thought Ira's line about writing a letter to HR was a masterstroke which, judging from the expression on Bob's face had struck home. Ira said that the idea had just come to her and added that she did not think they would carry it out due to the circumstantial nature of the evidence against Bob. John said that, while he agreed with Ira that writing a letter would probably not reflect well on them, Bob could not be sure of their intentions.

On returning to Blade, Ira dropped John off first near the entrance to the visitor's car park. Before John got out of the car, James, Ira and he agreed that they would wait to see what happened next and arranged that each would let the other two know if they learnt of any developments. When John got home and recounted the encounter with Bob to Jane, she said that her overwhelming reaction was one of relief that the confrontation had not become physical. John agreed with Jane, although admitting that James appeared to get close on a couple of occasions. Jane advised John to try to forget the whole incident and said that they should not speak about it again unless something positive happened. While John agreed with Jane that this was the best course of action, he could not help himself replaying the encounter in his mind several times over the weekend.

As John hoped that it would, the DV Core Team proved to be a godsend with the team being able to manage the development and implementation of both Oe375 and O254 Design Verification Plans in a synergistic manner. Again, as John expected most of the new knowledge gained from the Interface and Noise Factor analysis in the Oe375 programme that influenced the verification testing migrated from Oe375 to the O254 programme rather than in the opposite direction. In the event, the system-level testing for Oe375 identified a few failure modes that had not been identified on the design failure mode and effects analysis (DFMEA). On inspection of the interface analysis, it was found that, while all but one of the exchanges associated with the failure modes had been identified in the interface analysis, the learning from this information had not been transferred to the respective DFMEA and noise and control factor table (NCFT). One potentially significant safety-critical issue was identified at bicycle-level verification on Oe375 which was associated with the locking of the quick release wheel skewer which turned out to be the result of a design modification to the part by the manufacturer which had been communicated to Oxton during the Oe375 programme but had not been picked up in the design analysis. The O254 programme was not as fortunate as Oe375 with several failure modes including a couple of a safety-critical nature being identified during subsystem/component level testing, including fatigue cracks appearing in the seat post and pedals becoming loose due to the difficulty of installing them. Further failure modes, while not being of a safety-critical nature but likely to cause annoyance to customers, were discovered during the bicycle-level testing of O254. Mo's expertise in bicycle-level verification proved invaluable as on the occasion when bottom bracket creak was incorrectly diagnosed as due to dry bearings and was subsequently identified by Mo as the simpler, and more easily fixed, assembly fault of an incorrectly torqued crank bolt. John was pleased that no failure modes associated with the excitement features that they had incorporated into the design of Oe375 were identified during the Design Verification.

Following the implementation of countermeasures to the failure modes discovered during Design Verification both Oe375 and O254 programmes moved into the manufacture of the product that would be used for marketing purposes. John was thrilled to see all four of the Oe375 primary build variants in the three different colour options lined up ready for the press pack photo shoot. Oe375 had been given the public name of Oxton RGV 5.2e with the O254 being designated Oxton HYC 5.1 with the abbreviations standing for road/gravel and hybrid commuter respectively. The numbers associated with the bike names appeared to John to be plucked out of the air by the Marketing Department as they did not seem to fall into any numerical sequence of Oxton past or present products. John still thought of Oe375 and O254 by their programme designations, however. O254 also included different colour options as well as two build options of the base bike and, what was termed by Oxton Marketing Department as, "the fully loaded" option having accessories such as mudguards, pannier rack and lights as standard. John was pleased that he had persuaded Jim Eccleston to introduce a selection of two-tone paint schemes in vibrant colours for both models which contrasted with Oxton's conventional more conservative single-colour product that had been available up to then.

Having seen prototypes of both Oe375 and O254, Jim Eccleston decided that both products should be included in the press launch that he was organising for the new model year upmarket road bike O549 which had been designated SRP 5.8 as its public name. In view of the inclusion of Oe375 and O254, Jim decided to invite more of a mix of distributors and trade customers alongside the cycling press representatives. Working with the Marketing Department, Jim had settled on holding the launch in a 4-star hotel in a small market town on the fringe of the Yorkshire Dales. The hotel, which had been a graceful country house in its heyday, specialised in weddings and so had good conference and guest facilities and was set in its own expansive grounds with the hills of the Yorkshire Dales as a backdrop. Having hosted the Grand Depart of the Tour de France, the Yorkshire Dales provided excellent on- and off-road cycling country, and although some distance from a major conurbation the location was served reasonably well by both road and rail. Jim asked John to introduce both the Oe375 and O254 products at the launch with Sarah Jones' and Peter Andersson's help. Jim also requested that John outline the role of the Oxton first-time engineering norm (OFTEN) framework and People Skills in the design of both products. Jim said that he would set the OFTEN framework and People Skills in the context of what he called the Oxton turnaround strategy in his introductory presentation, and he asked John to keep his presentation of the use of the framework and People Skills short and at a high level, focusing more on the aims rather than the mechanics of the approach.

The launch started formally with a lunch which allowed those participants travelling on the day to get there with shuttle buses being laid on from the local rail station and Manchester Airport. Rooms were booked for the participants for the two-day duration of the event with the formal proceedings taking place in a large conference room which had been set out with a stage and rows of chairs. Following lunch and before the presentation of the new product, Jim gave his introductory presentation. In looking around the room from his vantage point on the stage, John was shocked to see Paul Clampin sitting as bold as brass in the second row of participants. John had not seen Paul at lunch and so assumed he had come straight to the conference room. John thought Paul's appearance strange since he had not seen his name on the delegate list which he had carefully studied several times in looking at which organisations had accepted Jim's invitation to attend. John wondered what Paul's motive was in attending the launch since he was not involved in the bicycle industry anymore. John hoped that Paul was not there to cause trouble but thought this was the most likely explanation for his presence. Jim had also seen Paul and in introducing those participants that he knew directly, rather than welcome Paul, Jim merely said that he was surprised to see him, to which remark Paul smiled but did not reply. After his brief introductory presentation, Jim moved directly into an explanation of the key design features of the SRP 5.8 road bike. Jim was ably assisted by Isla Johnstone, the Technical Specialist working on the O549 programme, who pointed at appropriate times during Jim's presentation to specific parts of the example bike that was stood on the stage. John thought that Jim did an excellent job in presenting the virtues of the bike without making it seem like an overly enthusiastic sales pitch. At a couple of points in his presentation, Jim said that of course participants did not have to take his word for what he was saying since they would get plenty of chance to ride the bike for themselves around the beautiful countryside in which the launch was taking place. Jim and Isla then took questions from participants. John was impressed with the direct way the questions were answered even where there was an implied criticism and compared this to the politicians that he had heard on the radio recently.

Jim then handed over to John who led the presentation and Q&A session on the HYC 5.1 ably assisted by Peter. It was during the Q&A session that John thought he began to understand Paul's motive for being at the launch. Paul introduced himself by name and said that he had led the design and development of HYC 5.1 with John taking over towards the end of the programme when Paul said he had left Oxton for what he described as pastures new. Paul then asked John if he thought the pricing of the bike was competitive, given that it was higher than that originally projected at the start of product development. John replied that the bike represented good value for money and Jim followed up on John's response by saying that as Paul knew well Oxton as a company was founded on offering high-quality product at competitive prices. This then spawned an observation from another participant whom John recognised as being from the cycling press that Oxton product quality had taken somewhat of a nose-dive recently. Jim replied that John was going to explain shortly what Oxton was doing in rectifying this situation. John wondered if the comment on Oxton quality would have been made if Paul had not asked his question and moved on swiftly by taking the next question.

At the end of the Q&A session John said that it was time for a refreshment break and directed participants to the room in which the refreshments were being served saying that proceedings would recommence in 20 minutes. After the last participant had left the room, Jim came across to John and asked in an exasperated voice how Paul had got in and what he was playing at. John said that he assumed Paul was there to disrupt proceedings as best he could while at the same time basking in any reflected glory afforded by the O254 programme. Jim said that he was going to throw him out, but Paul was nowhere to be seen and John thought that he had probably made himself scarce deliberately. The hotel had offered to pick up any late comers to the press conference after it had started to allow Oxton staff to focus on other things and so Jim asked Peter if he would check with the hotel reception, if they remembered Paul. Jim added that Peter should find out how Paul had signed in. Peter joined Jim and John in the refreshment area a few minutes later and pulled Jim aside from the group of participants to whom he was chatting. John overheard Peter tell Jim that Ann on the hotel reception desk remembered Paul, and had said that he registered about half an hour after the lunch had started saying that he had been delayed by traffic. Peter added that Ann had said that Paul looked at the four identity badges that were placed on one end of the reception desk, picked up one, and said he was the person named on the badge. Jim asked Peter to see if he could find Paul, and ask him to leave.

After the refreshment break John noticed that Paul was back sitting in the same seat. Since Jim had not yet returned to the room, John turned to Sarah, who was sitting next to him on the stage, and asked her if she would get hotel security to quietly ask Paul to leave without disrupting the proceedings. John then started his presentation by taking the participants through the approach that the RGV 5.2e (Oe375) Product Design team had taken in using both the OFTEN framework and People Skills and was shortly re-joined by Sarah. As John continued his presentation, he noticed a hotel security guard speaking softly to Paul who shook his head and sat steadfastly in his seat. Having concluded his brief overview of the use of the OFTEN framework and People Skills with Sarah's help, John demonstrated the features of RGV 5.2e. John began by speaking of the general impression that the eBike made using both the eBike itself and a selection of photos that had been taken for the marketing brochure. John then stressed what he called the rideability of the eBike, and like Jim asked participants to judge for themselves when they got a chance to ride the launch product. John suggested that riders might mentally rather than physically close their eyes for a short while and judge whether they were riding an electric bike or a light weight gravel bike. Without calling them as such, John then ran through the performance and excitement features of RGV 5.2e including the highly competitive power assistance range and smart phone connectivity including fingerprint locking.

John then asked for questions from the audience and was pleased to see what he took to be interest in both the eBike itself and the OFTEN framework and People

Skills. Aware of what Jim had said, John and Sarah were careful not to share too much detail on the OFTEN framework answering questions at a general level and directing the questioner to John's published material on FTED for more detailed information. Building on a question on the use of People Skills, Paul stood and asked John about what he called the significant resistance to People Skills within the RGV 5.2e development team. Somewhat taken aback by the forthright nature of Paul's question, John bought himself time by thanking Paul for an interesting question and then quickly decided to answer the question directly by saying that resistance to change was not at all unusual in a group of people particularly when they were asked to do things in a significantly different way. John then cited the example of him insisting that the Oe375 Product Design Team took a break in their meetings. Realising that he had used the Oxton coding, John explained that Oe375 was the Oxton coding for the eBike now designated RGV 5.2e. John said that some in PD team found taking a break strange when their previous manager had insisted that the team spent every minute of every meeting in addressing the task in hand. John hoped that this reference to Paul was sufficiently subtle as not to be noticed by audience members while acting as a signal to Paul that he could give as good as he got. John then reminded his audience that Sarah who was sitting next to him was a member of the RGV 5.2e Product Design team and asked Sarah to comment on the use of breaks in a team meeting specifically and People Skills generally.

Sarah thanked John and said that she was sceptical when John first introduced the People Skills but had quickly changed her mind adding that their use had helped the RGV 5.2e team to be more efficient. Sarah added that based on her experience taking a short break mid-way through meetings helped them to get more done, not less. John thanked Sarah and continued with the Q&A session until the questions appeared to have dried up. As John was handing the discussion back to Jim, Jim whispered to him to not let Paul out of his sight. Jim then introduced what he said was the most interesting part of the launch in which participants could ride each of the three bikes that they had been introduced to. As John made his way round to position himself at the side of the room just behind Paul, he heard Jim outline the procedure for participants to put their names down for bike riding adding that those that were not staying at the hotel could use the hotel swimming pool changing rooms to change into their riding gear. Jim then directed participants to the area used for signing into the event before lunch, to put their names down for rides adding that refreshments would be available on a rolling basis during the riding which would continue the following morning. Jim concluded by saying that dinner was served at 7:30pm in the hotel main restaurant and that Oxton staff would be available to assist and answer any further questions.

As the participants started to leave the room, John saw Paul exit by the side door and John followed at a distance behind him into the hotel grounds where Paul sat on a bench in a quiet area. John did not attempt to hide from Paul but kept a sufficient distance where he hoped he appeared to be out for a stroll rather than being interested in him. After about five minutes, when John was just starting on the third lap of the circular route that he had taken, John saw Jim emerge from the hotel with the same security guard that had spoken to Paul in the conference room. John waved and seeing Jim respond pointed in the direction where Paul was sitting. As Jim and the security guard strolled quickly in Paul's direction, John walked across the grass towards

Jim and on reaching him asked if he should join him. Jim told John that this would be fine, adding that there was strength in numbers as the three men quickly approached Paul who had remained sitting on the bench. Without any preamble, Jim asked Paul what he was doing, bringing his attention to the fact that the name on the badge hanging around his neck on a lanyard was not his name and that he was attending the launch under false pretences. Paul got up from where he was sitting but did not speak and the security guard then asked Paul to leave the hotel premises. Paul said that he was leaving anyway and added that his car was parked in the hotel car park. Without any further discussion with Paul, Jim and John left it to the security guard to see Paul back to the car park and off the premises and strolled back to the hotel for some refreshment. On their way back, Jim's only comment on their brief face-to-face encounter with Paul was to say that he was glad that Paul's departure had been swiftly executed without fuss and he hoped that none of the other participants had noticed.

The rides seemed to John to go very well with participants giving good feedback, although John was aware that what a journalist says to your face can be different to what they write in black and white. The O549 road bike had the greatest demand for riding but since Jim had said that number of bikes of each type should equate to just over one third of the number of proposed participants with a small number of bikes kept in reserve in case of punctures or mechanical problems the Oe375 and O254 got their fair share of rides. As John expected the O549 got the most enthusiastic reception with the Oe375 being the next well received. John was pleased, however, with the generally positive reception of the O254.

Jim gave a well-received after-dinner speech having told John that he was not going to pay for a speaker when he could do the job himself. John told Jim that he had enjoyed the speech liking both his personal anecdotes from his time as a professional cyclist and his jokes. Jim replied that he had heard so many cycling-related after-dinner speeches that he knew most of the standard jokes off by heart, as well as others which he was not prepared to repeat in public.

In the event, the launch proved to be a good investment with the cycling press, distributors and trade customers all being enthusiastic about Oxton's new products. John was pleased that the Oe375 gained a good overall critique, although as he expected the O549 received the majority of the favourable coverage. A number of reviews commented on the sleek looks of the Oe375 and described the eBike as easy and comfortable to ride with almost all reviews commenting on the lightweight nature of the bike. Several reviews said that the bike was fun to ride with other commenting on the responsiveness of what was seen as a powerful but well-matched motor particularly when accelerating from low speeds on steep inclines. John was pleased that the Oe375 team had taken Yumiko's advice on including a torque sensor in the Electric Drive which enabled the motor to react quickly to any change of rider force on the pedals. As John expected, there were contradictions between reviews with one saying that the eBike felt as if it almost had too much torque. John was amused that two reviews said the single chainring, and hence no front derailleur, lowered maintenance with other reviews picking up on the flip side of this of the reduction in gearing options. John was also pleasantly surprised that the O254 also got reasonable overall reviews with a number saying that the bike represented good value for money and was an ideal commuter bike. One review said that the braking felt soft and somewhat ridiculously to John's mind another said that the external exposure of some of the cabling on the bike was more prone to dirt than the internal cabling which John thought was a statement of the obvious and he wondered what the reviewer expected from a budget bike. In addition to the cycling press, the Oe375 and the O254 also featured in the national press albeit, in passing in an article on the O549.

Several reviewers mention that the good variety of cycling routes for the launch provided by the scenic Yorkshire Dales proved a good testing ground for Oxton's new products. John was interested in the coverage of what was described as Jim Eccleston's strategy for turning around the company and the positive mention of the OFTEN framework and People Skills in this context. John was amused that one journalist mentioned Jim ushering out Oxton's old guard as a part of his turnaround strategy which John took to be a reference to Paul Clampin's somewhat embarrassing departure from the event. John reflected that Paul's attempt at sabotaging the launch had failed since most of the cycling press reviews agreed that Oxton's new products demonstrated that the company was on track in its effort to recover its reputation for high quality product.

Oe375 was subsequently accredited as "eBike of the Year" by one of the prestigious cycling websites with an article entitled "Oxton is Back" accompanying the technical review describing the Oe375 as Oxton's return to the level of build and ride quality for which it had been renowned. John was particularly proud of this accolade since the eBike was not one of the high-end more expensive bicycles on which Oxton had originally earned its reputation. Oe375 also featured in Oxton's first ever TV campaign which ran during the final week of Giro d'Italia Grand Tour. One of Jim's former team mates, who was just starting his professional career when Jim retired and was still riding professionally, fronted up the advertisement and was seen riding both the O549 road bike and the Oe375.

The launch of Oe375 and O254 proved a turning point in Oxton's financial position with the company returning to profitability by year end. The first-year warranty figures were very encouraging with Oe375 warranty being 15% of the outgoing model warranty and O254 warranty being reduced by 60% compared to the former model. John thought that this reflected well on the depth and integrity of the failure mode avoidance analysis that was implemented on both programmes. John was annoyed, however, at the need for a recall campaign 12 months after the launch of O254 when a small number of customers complained that bolts securing the optional factory fitted rear pannier rack to the frame had sheared. Although no injuries were reported, Jim took the decision to replace all of the bolts due to the potential for a loose pannier strut to interfere with the rear wheel. On reviewing the interface analysis for O254, John was disappointed to find that this failure mode had not been picked up. The design of the pannier rack fixing had been carried over from the previous model, and, even though the carry-over decision had been agreed before John took over managing the development of O254, John felt responsible for the lack of vigilance which had allowed the failure mode to occur.

Jim Eccleston was keen to fill the vacancy left by Paul Clampin as soon as company finances would allow to ease his own heavy workload. After discussing the matter with Oxton senior management, and agreeing it with the Board, Jim reassigned John to take on Paul's Engineering Director role in addition to retaining his oversight of teamwork in the company. John was pleased that his previous position, after being advertised as a Programme Manager, was awarded to Sarah who stood head and shoulders above the external candidates. Phil was then given the status of a Technical Specialist left vacant by Sarah's promotion. Jim recruited a graduate engineer to bring the company headcount to the level it was at before Paul Clampin's departure. Peter Andersson was rewarded with a significant pay rise for his work on the O254 programme with Mo Lakhani and Margaret Burrell benefitted from Oxton's improved financial position in their healthy year-end bonuses. Jim was also keen to ensure that the OFTEN framework and People Skills were used in all future Oxton programmes. After discussing the matter with John, Jim was persuaded that, while the development of OFTEN documentation from scratch, as had happened in Oe375, was a thing of the past, it would take considerable discipline and effort to ensure that learning from one programme was made best use of in subsequent programmes. Based largely on his discussion with John, Jim decided to reorganise his senior management team so that they worked across all programmes rather than within fixed programmes. Working with John, Phil and Jennifer, Sarah led the development of a manual detailing the use of the OFTEN framework with the manual including both a case study based on the Oe375 design experience and People Skills.

Several weeks after James, Ira and he had met with Bob Jenkins, John received a phone call at work from Ira. Ira asked John if he had heard anything since their meeting with Bob, and John responded that he had heard nothing. Ira then explained that both she and James had separate meetings with the Blade HR Director, Ayleen Macar, in the previous week in which they had learnt that new information had been received and that HR were now convinced that they both were innocent of the accusations that had been made against them. Ira told John that she had asked Ayleen why HR now thought this way and Ayleen had told her that information had been received in the same way that the original accusations were made. Ira added that she had also asked Ayleen why she was treating this new information with any more credibility than the original information and Ayleen had told her that the person calling appeared to be the same person as had made the original accusations and had quoted aspects of the original messages. John asked if Ayleen had mention him at all and Ira said that she had brought up his name in asking Ayleen if she intended to inform James and John of the new situation. Ira said that Ayleen seemed surprised that she knew that James and John had also been the subject of accusations and said that she had spoken with James but was non-committal about informing you. John quickly thought over what Ira had told him and told her that he was glad that Bob had done, what he called, the "right thing", and added that he would not be surprised if Blade HR did not contact him preferring instead to sweep things under the carpet for fear of it getting out into the public domain. In agreeing with John, Ira said that she would also not be surprised if Blade HR took no further action in John's case. When John told Jane of Ira's phone call, she said that she was pleased that it looked like the matter was over at last.

As things transpired, John did not hear anything further from Blade or from Jim Eccleston about the allegation. Several weeks after Ira's call, John took Sarah and Jennifer into his confidence and told them that Blade had withdrawn the accusation against him. Sarah and Jennifer told John that they never doubted him. John hoped that informing Sarah and Jennifer would be the final act in the sorry story, which it proved to be.

Appendix

Key Graphics

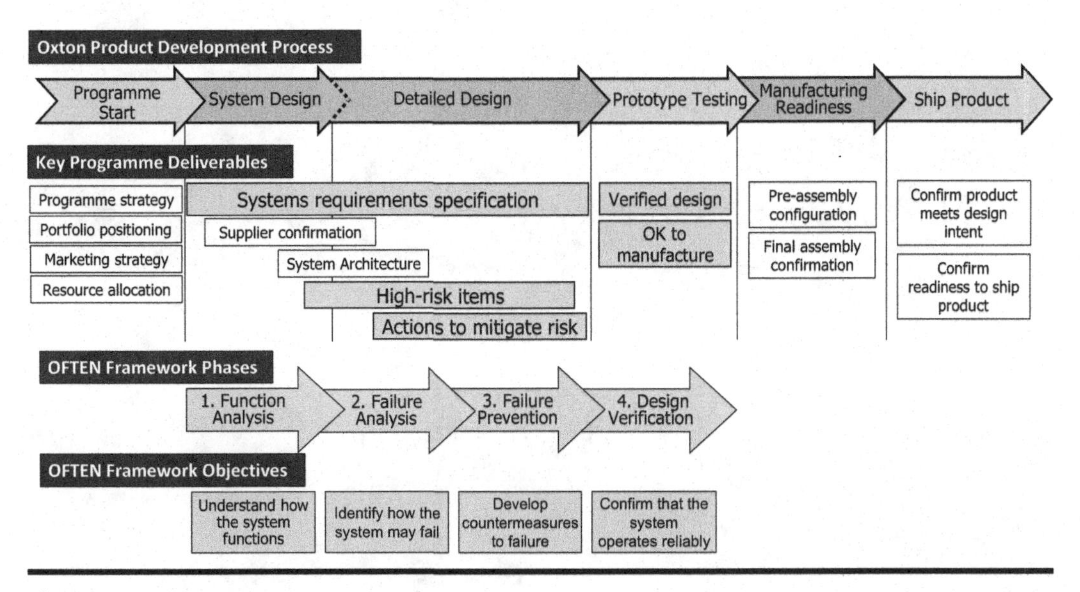

Figure A.1 Oe375 programme roadmap.

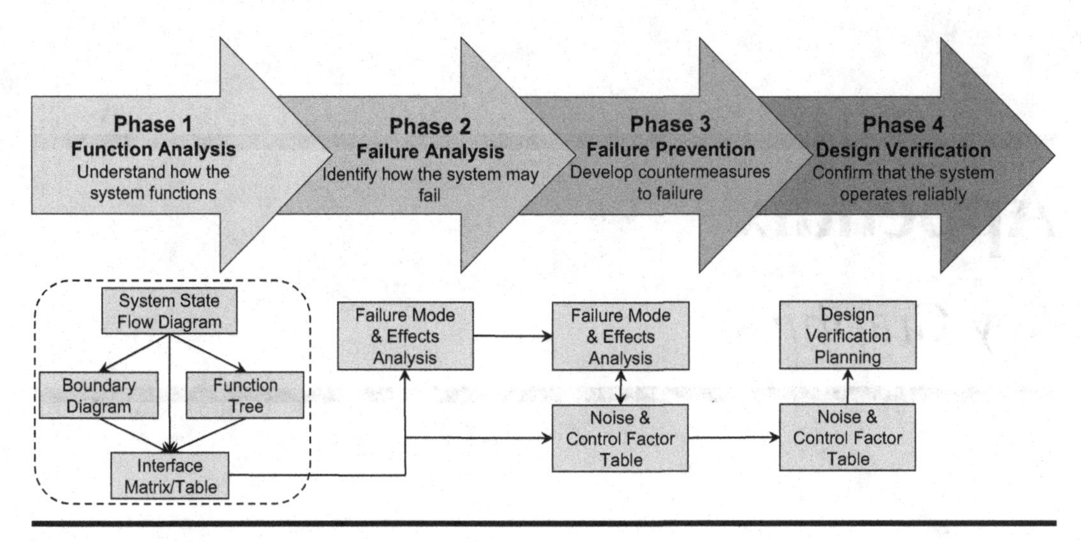

Figure A.2 OFTEN framework.

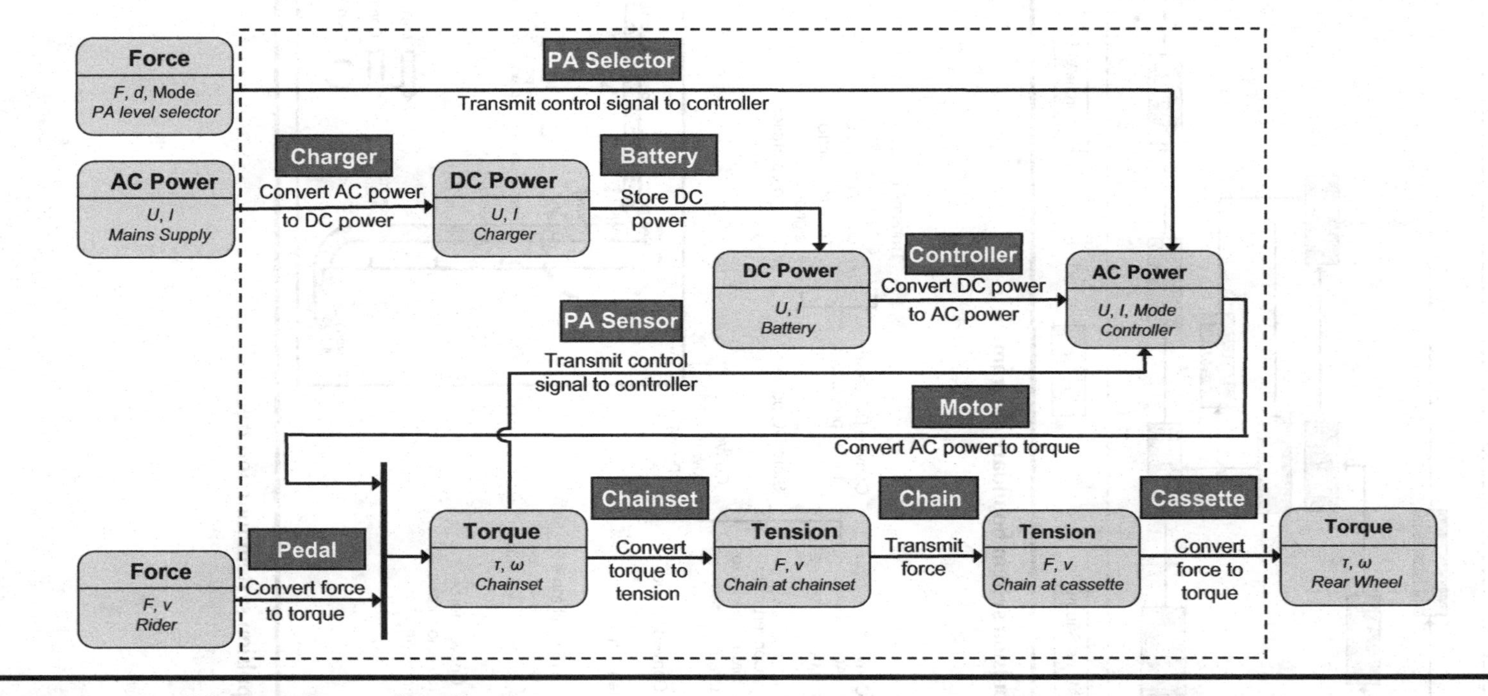

Figure A.3 Propulsion system state flow diagram.

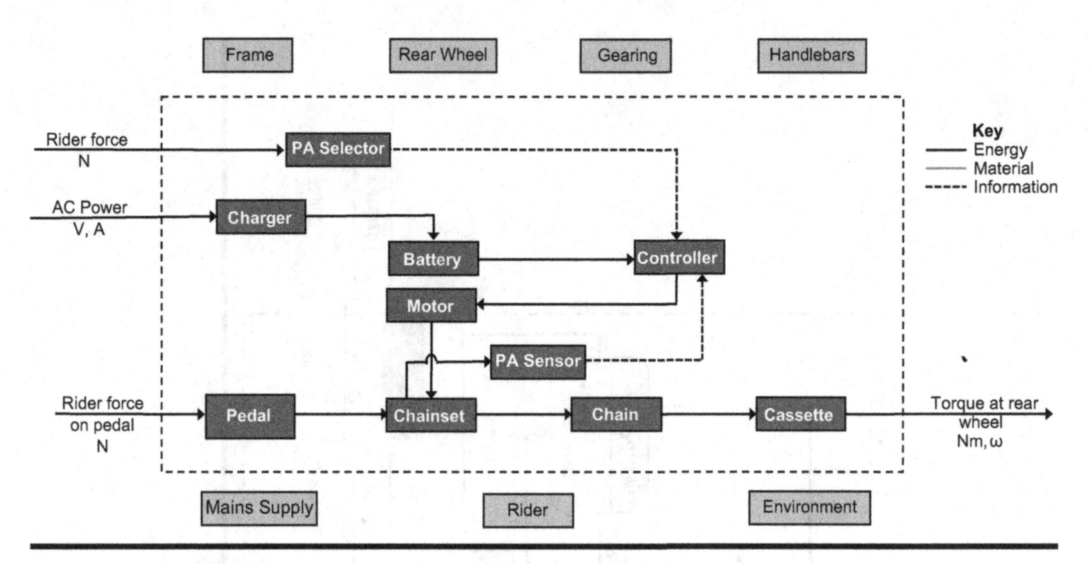

Figure A.4 Propulsion system boundary diagram.

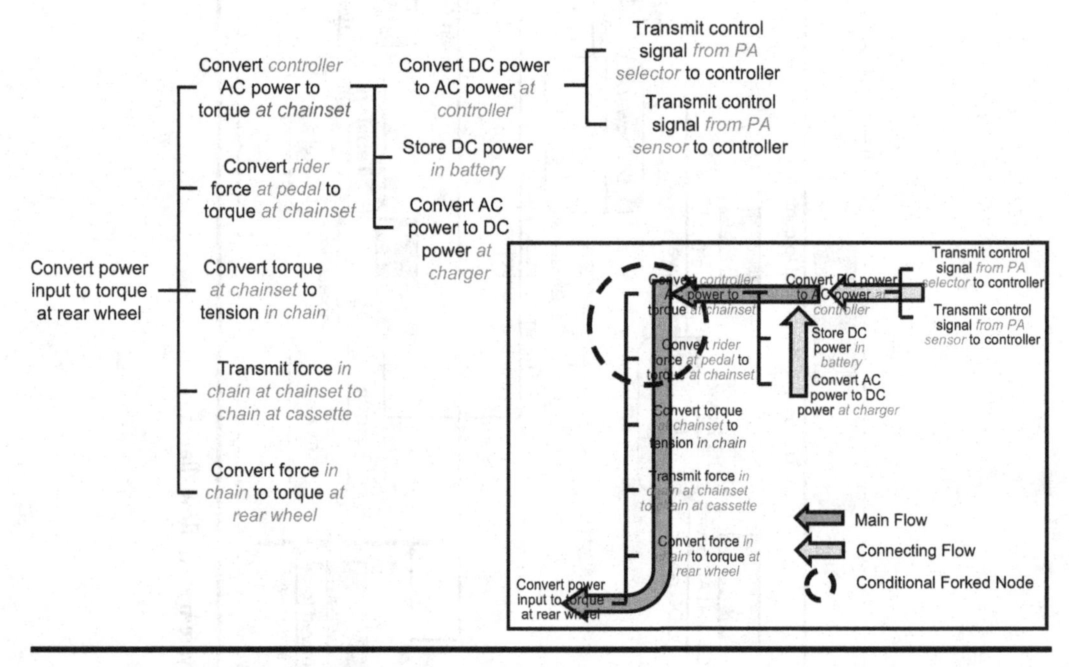

Figure A.5 Propulsion system function tree.

Oe375 Propulsion System		INTERNAL INTERFACES										EXTERNAL INTERFACES						
		A	B	C	D	E	F	G	H		I	E1	E2	E3	E4	E5	E6	E7
		Pedal	Chainset	Chain	Cassette	Charger	Battery	Controller	Motor	PA Selector	PA Sensor	Frame	Rear Wheel	Derailleur Gearing	Handlebars	Mains Supply	Rider	Environment
1	Pedal		X														X	X
2	Chainset			X					X		X	X					X	X
3	Chain				X							X	X	X			X	X
4	Cassette												X				X	X
5	Charger						X	X								X	X	X
6	Battery							X	X			X					X	X
7	Controller								X	X	X	X					X	X
8	Motor											X					X	X
9	PA Selector														X		X	X
10	PA Sensor											X					X	X

Figure A.6 Propulsion system interface matrix.

Glossary

Note that the word bicycle is taken to include eBikes

Active Listening Listening carefully to a speaker to understand what they are saying in order both to respond and retain the information for later.

Affected System Function The system function identified on a state flow in a system state flow diagram and/or on a system function tree that is directly affected by particular exchanges at an interface.

Air Resistance Resistance to the motion of an object through the air caused by the frictional force between the object and air, sometimes called Drag.

Attitudes A learned tendency to evaluate and respond to things in a certain way based on experience.

Attribute A quantifiable characteristic assigned to a state that distinguishes it from other states in a system state flow diagram.

Attribution The act of regarding something as being caused by a person or thing.

Back EMF A voltage that is generated in an electric motor by the relative motion of the stator and rotor which opposes the voltage applied across the windings.

Battery A group of cells that store and provide electrical energy.

Basic Event A root cause event at the lowest level of a fault tree.

Boundary Diagram A graphical representation of the component parts of a system along with the external entities that interact with the system, which defines the scope of the system and depicts the flow of energy, material and information through the system from input(s) to output(s).

Branching Flow A flow that branches out of the main flow in a system state flow diagram.

Brushless Direct Current Motor (BLDC) A direct current electric motor in which commutation is achieved electronically.

Cassette The cluster of sprockets located on the hub of a bicycle rear wheel that converts tension in the chain into rotation of the rear wheel.

Cause The factor(s) that contribute to a failure.

Chain The component on a bicycle that transfers the rotational force at the chainset to the cassette and hence rear wheel hub.

Chainring A sprocket that is part of the chainset onto which the chain engages. The rotation of the chainring causes the chain to move (see Chainset).

Chainset The subsystem on a bicycle that converts the force exerted by the rider on the pedals into tension in the chain. It comprises the chainrings, the right and left crankarms and the spindle upon which chainset rotates. Also known as the Crankset.

Charger A device for charging a battery.

Closed Questions Questions which give the respondent limited options from which to choose in answering a question with the most limiting being questions with the answer yes or no.

Commutation The reversal of current in the windings of an electric motor.

Component A part that along with other such parts makes up a larger subsystem.

Conditional Fork Node Two or more flows in a system state flow diagram combining into a single flow under defined conditions or a single flow diverging into two or more flows under defined conditions.

Connecting Flow A flow directed into the main flow to facilitate particular conversions within the main flow in a system state flow diagram.

Control Factor A design parameter that can be manipulated to improve the robustness of a system (see Robustness).

Controller (eBike) The system that provides power to an eBike motor and monitors and/or controls key aspects of the eBike propulsion such as pedal movement, battery state of charge and required level of pedal assistance.

Crankarm The levers to which the pedals attach which cause the chainring to rotate (see Pedal and Chainring).

Current Controls Actions that are currently used in designs similar to the one being analysed to prevent or detect failure modes.

Deletion Including certain aspects of a topic and excluding others in what is said or written.

Derailleur The subsystem that changes the gearing of a bicycle by moving the chain to different sprockets on the chainset and cassette.

Descriptive Feedback Feedback based solely on what an individual has observed and/or heard and given without judgement.

Design Controls Measures that mitigate the effect of a failure mode.

Design Failure Mode and Effects Analysis (DFMEA) A disciplined process to identify failure modes inherent in the design of a product along with their effects and causes in order to assess failure risk and prioritise actions to mitigate that risk.

Design Solution An entity that generates a function within a system and is identified on a system state flow diagram.

Design Verification Confirmation through a test or other event that a system performs in the way it has been designed to perform.

Design Verification Plan A regime of Design Verification events/tests conducted to ensure that all aspects of a product perform in the way they have been designed to perform (see Design Verification).

Detection Control An event/test that seeks to detect a failure mode in a product before the product design is released for production.

Detection Rating A rating which indicates the relative effectiveness of an event/test at detecting a particular failure mode.

Distance to Failure The amount of a variable entity that is available for use before failure occurs due to a lack of robustness (see Robustness).

Distortion Including inaccurate or vague information in a spoken or written statement.

Diverted Output Non-useful or undesired output generated by a system. Also known as Error State.

Drag Coefficient A measure of the effect of air resistance to the forward motion of a bicycle.
Drivetrain The part of a bicycle responsible for propulsion comprising the pedals, chainset, chain, cassette and derailleur.
Duty Cycle The load cycle(s) to which an electric motor is subjected.
Effect The impact of a particular functional failure mode on system performance as perceived by the customer.
Effective Frontal Area The product of the projected frontal area and the drag coefficient (see Projected Frontal Area and Drag Coefficient).
Effective Meetings A process by which meetings can be conducted efficiently and effectively by paying attention to Warm Up, Team Process, Task, Team Maintenance and Warm Down (see separate items).
Electric Drive Those components that contribute to propelling an electric bicycle with electric power.
Evaluative Feedback Judgemental feedback based on the opinion of the person giving the feedback.
Exchange A transmission or translation of energy, material or information between different parts. The presence, or where desirable, the absence of physical contact between parts is also considered an exchange.
Explicit Knowledge Knowledge that is documented in a tangible form such as in an email, written report or on presentation slides and can be readily shared between people.
Failure Mode The way in which failure of system function can occur.
Failure Mode and Effects Analysis (FMEA) A disciplined process to identify failure modes, their effects and causes in order to assess failure risk and prioritise actions to mitigate that risk.
Failure Mode Avoidance A strategic approach to product design in which failure modes are found and countermeasures applied during the design process.
Failure Mode Types The four types of functional failure modes are total function loss, partial function loss, intermittent function loss and unintended function.
Fault Tree Analysis (FTA) Graphically based top-down root cause analysis used to establish the potential or actual causes of a single failure of a system with causes occurring either in isolation or in combination.
Framing (Information) Conveying information in a way that helps the person receiving the information to assimilate it.
Function The purpose or specific action of a system, subsystem or component.
Function Fault Tree Analysis A form of fault tree analysis where the causal events on the tree are function failure modes.
FTED An acronym for First Time Engineering Design.
Generalisation Taking one example of a situation to represent the entire situation in what is said or written.
Gravitational Force The force due to gravity. For a bicycle, the component of gravitational force acting down an incline opposes the forward movement of a bicycle travelling up the incline.
Groupset The components on a bicycle that propel the bicycle through rider input on the pedals and cause the bicycle to brake.
Impact of Exchange A coded indicator of the beneficial or detrimental effect of an exchange on the system function being analysed.

Interface The boundary between one entity and another across which exchanges of energy, material and/or information occur. Physical contact between entities may also occur.

Interface Analysis An investigation of the exchanges that occur both between the constituent parts of a system and between system parts and external systems.

Interface Function/Requirement The function or requirement that is required to generate or manage an exchange at an interface.

Interface Matrix A rectangular matrix used to visualise the existence of exchanges between both the internal design elements of a system and between the internal design elements and systems external to the system being analysed.

Interface Table A table which details the exchanges that occur at interfaces.

Intermediate Event A causal event on a fault tree on a failure path that bridges between the top event and the root cause events at the base of a fault tree (see Top Event).

Kano Model An approach for identifying and prioritising product features that will maintain and enhance customer satisfaction by distinguishing those which are expected from those that are not.

Loom The harness of wires associated with the Electric Drive on an eBike.

Main Flow The flow of states from input(s) to output(s) on a system state flow diagram that enables a system to achieve its overall function.

Maintenance (Team) Paying attention to individual team members' personal needs.

Mid-Drive Motor A motor which is housed close to the lower centre of an eBike frame and propels the rear wheel through the Drivetrain.

Noise and Control Factor Table (NCFT) A table listing noise factors acting on a particular system along with their effect, control factors and design verification tests in which the noise factor effects can be incorporated (see Control Factor and Noise Factor).

Noise Factor A factor that an engineer cannot control or chooses not to control in designing a product such as the wear of a bicycle tyre.

Noise Factor Types The categories of noise factors are piece-to-piece, changes over time/cycles, customer usage, external environment and subsystem interaction.

Object A thing that can have physical and/or cerebral existence.

Occurrence Rating A rating which indicates the relative probability of occurrence of a particular failure mode.

OFTEN An acronym for Oxton first-time engineering norm.

Open Questions Questions that allow the respondent to answer in their own frame of reference rather than the frame of reference of the questioner.

P-Diagram A high-level graphical representation of a system in which system inputs and outputs are identified along with noise factors that affect system performance and control factors that can be used to mitigate the effect of noise (See Control Factor and Noise Factor).

Pedal The part of a bicycle that the rider pushes with their foot to propel it.

People Skills Behaviours that reflect an individual's strengths and weaknesses and can be enhanced both at an individual level and in groups of individuals working together.

Personal Assistant (PA) A person who provides a senior manager with administrative support by conducting secretarial work and performing such duties as answering phone calls, managing the manager's diary, answering emails and other correspondence.

Prevention Control A design action taken to prevent a failure mode occurring in a product in customer use.
Preview (Meeting) Planning for a meeting by deciding what is to be accomplished and the way the accomplishments will be achieved.
Priority Matrix A matrix that helps to identify the level of risk associated with particular failure modes quantified as high, medium and low risk.
Programme Roadmap A high-level graphical representation of the major steps, and their sequence, in developing a product.
Projected Frontal Area The area that a cyclist and bicycle present to the forward direction of motion.
Propulsion System The system on a bicycle or electric bicycle that generates propulsion. On an eBike, this comprises the Drivetrain and Electric Drive.
Propulsive Force The force that drives a bicycle forward.
Question Delivery The way in which a question is asked in terms of the words that are used, the manner in which the question is asked and the question structure.
Questions for Clarification Questions that the listener asks in an attempt to minimise misunderstanding or ambiguity.
Recommended Action Action taken in developing a countermeasure to mitigate the effect of a failure mode.
Repulsive Force A force which acts to slow the motion of a bicycle.
Restatement Listening to a person speaking and reflecting back to the speaker what the listener has heard in order to confirm to the speaker that the listener has understood what the speaker has said.
Review (Meeting) A meeting to assess if the meeting under review achieved its purpose and identify how any shortcomings might be overcome before, or in, the next meeting.
Robustness An in-built property of a product that allows it to perform as expected despite the presence of noise factors (see Noise Factors).
Rolling Resistance The energy required to keep a bicycle tyre moving over a surface by overcoming the energy lost by the tyre through its movement and friction.
Root Cause The lowest level event or combination of events that initiate a failure. Known as a basic event on a fault tree.
Rotor The permanent magnet-based component in a brushless direct current motor which revolves within the electromagnetic field generated by windings of the stator (See Stator).
Severity Rating A rating scale which indicates the severity of the effect of a particular failure mode.
Shifter A lever used to change gear.
Speaking Guidelines A series of straightforward pragmatic guidelines to enhance spoken communication in teams.
Sprocket A toothed gear wheel on the chainset or cassette that engages with the chain on a bicycle.
State An object represented in a system state flow diagram that is measured in terms of attributes/values and its location.
Stator The static part of an electric motor that generates an electromagnetic field within which the rotor revolves (see Rotor).
Subsystem (1) A group of interdependent components forming a unified whole.
Subsystem (2) A system that with other such systems makes up a larger system.

System (1) A group of interdependent subsystems forming a unified whole.
System (2) A term used in a general sense to refer to a component, subsystem or system.
System Level The level, system, subsystem or component, at which analysis is being conducted in a system.
System State Flow Diagram A graphical method based on the analysis of the state transitions associated with the flow of energy, material and information through a system to support solution-neutral functional decomposition and architecture modelling.
Systems Vee Model A model of systems engineering which considers the whole product development life cycle.
Tacit Knowledge Knowledge residing in people's heads based on experience which is difficult to share.
Task (Team) The core task(s) to be completed in a meeting in order that a meeting achieves its purpose.
Team Process The manner in which team members interact in a meeting.
Top Event The single system failure event that is placed at the top of a fault tree and is the subject of the fault tree analysis.
Triad (System State Flow) A focused graphical model of state flow including input and output states, the function required to achieve the state flow and the Design Solution that generates the function.
Use Case Diagram A graphical representation of the ways in which a user can interact with a system.
Verbal Communication One person speaking to another person or other persons. Key elements of verbal communication are the words that are used, the speaker's body language, and the speed, tone and flow of what is said.
Visualisation (1) A technique in which individuals can share their tacit knowledge of a system by imagining that they are a small component inside the system.
Visualisation (2) Making shared mental images visible as a model of a system by, for example, developing a system state flow diagram.
Warm Down Closure to a meeting when the meeting accomplishments are reviewed, and individuals are able to share their experience of the meeting.
Warm Up The introduction to a meeting in which team members can attune mentally to the meeting and (re)acquaint themselves with others in the team before confirming the meeting's purpose and agenda.
Wheelset The combination of the front and back wheels on a bicycle along with the wheel axles.
Windage The resistance due to air between the stator and rotor in an electric motor (see Stator and Rotor).
Windings The coils of wire that form the stator in a brushless direct current motor.

Bibliography

Albers, A., Turki, T. and Lohmeyer, Q. (2012) Transfer of Engineering Experience by Shared Mental Models, International Conference on Engineering and Product Design Education, Antwerp, URL: https://www.designsociety.org/download-publication/33170/ Accessed 22 October 2022.

Anand, G., Ward, T. and Tatikonda, M. (2009) Role of Explicit and Tacit Knowledge in Six Sigma Projects: An Empirical Examination of Differential Project Success, *Journal of Operational Management*, 28(4), 303–315, https://doi.org/10.1016/j.jom.2009.10.003

Campean, F., Henshall, E. and Brunson, D. (2010) Failure mode avoidance paradigm in automotive engineering design. *International Congress on Automotive and Transport Engineering - CONAT, Conference Proceedings. Brasov*, 25th–27th Oct 2010, 207–214.

Campean, F., Henshall, E., Yildirim, U., Uddin, A. and Williams, H. (2013a) A Structured Approach for Function Based Decomposition of Complex Multi-disciplinary Systems. In: Abramovici, M. and Stark, R. (editors) *Smart Product Engineering. Lecture Notes in Production Engineering*, Berlin Heidelberg, Springer, https://doi.org/10.1007/978-3-642-30817-8_12

Campean, F., Williams, H., Uddin, A. and Yildirim, U. (2017) *Robust Engineering Systems Analysis through Failure Mode Avoidance*, Short Course Lecture Notes, University of Bradford, UK.

Campean, F. and Yildirim, U. (2020) Functional Modelling of Complex Multi-disciplinary Systems Using the Enhanced Sequence Diagram, *Research in Engineering Design*, 31(4), 429–448, https://doi.org/10.1007/s00163-020-00343-8

Campean, I., Henshall, E., Brunson, D., Day, A., McLellan, R. and Hartley, J. (2011) A Structured Approach for Function Analysis of Complex Automotive Systems. *SAE International Journal of Materials and Manufacturing*, 4(1), 1255–1267, https://doi.org/10.4271/2011-01-1268

Campean, I., Henshall, E. and Rutter, B. (2013b) Systems Engineering Excellence Through Design: An Integrated Approach Based on Failure Mode Avoidance. *SAE International Journal of Materials and Manufacturing*, 6, 389–401, https://doi.org/10.4271/2013-01-0595

Campean, F. and Henshall, E. (2022) Fault Tree Analysis in the Context of Systems Engineering Design Analysis. In: El Hami, A., Delaux, D. and Grezskowiak, H. (editors) *Reliability and Physics-of-Healthy in Mechatronics*, 15, 139–197. https://doi.org/10.1002/9781394186068.ch3

Davis, T. (2006) Science, Engineering, and Statistics, *Applied Stochastic Models in Business and Industry*, 22(5–6), 401–430, https://doi.org/10.1002/asmb.643

Davis, T. (2022) Turning Hindsight into Foresight How DFMEA Can Foster a Proactive Approach, *Quality Progress*, 55(7), 36–43.

Henshall, E. and Campean, F. (2010) Design Verification as a Key Deliverable of Function Failure Avoidance, *SAE International Journal of Materials and Manufacturing*, 3(1), 445–453, https://doi.org/10.4271/2010-01-0708

Henshall, E., Campean, F. and Rutter, B. (2014) A Systems Approach to the Development and Use of FMEA in Complex Automotive Applications, *SAE International Journal of Materials and Manufacturing*, 7(2), https://doi.org/10.4271/2014-01-0740

Henshall, E., Campean, I. and Brunson, D. (2011) *Robust and Reliable Teamwork within Engineering Projects*, SAE Technical Paper 2011-01-127, https://doi.org/10.4271/2011-01-1273

Henshall, E., Rutter, B. and Souch, D. (2015) Extending the Role of Interface Analysis within a Systems Engineering Approach to the Design of Robust and Reliable Automotive Product, *SAE International Journal of Materials and Manufacturing*, 8(2), 322–335.

Mackin, D. (2007) *The Team-Building Toolkit*, American Management Association.

Nonaka, I. and Konno, N. (1998) The Concept of "Ba": Building a Foundation for Knowledge Creation, *California Management Review*, https://doi.org/10.2307/41165942

Saxena, A., Davis, T. and Jones, J. (2015) A Failure Mode Avoidance Approach to Reliability, Proceedings. *Annual Reliability and Maintainability Symposium (RAMS)* January 2015, https://doi.org/10.1109/RAMS.2015.7105062

Syer, J. and Connolly, C. (1996) *How Teamwork Works: The Dynamics of Effective Teamwork*, McGraw-Hill.

Yildirim, U., Campean, F. and Williams, H. (2017) Function Modeling Using the System State Flow Diagram. *Artificial Intelligence for Engineering Design, Analysis and Manufacturing*, 31(4), 413–435, https://doi.org/10.1017/S0890060417000294

Zhou, J. (2005) *Reliability and Robustness Mindset in Automotive Product Development for Global Markets*, SAE Technical Paper 2005-01-1212, 2005, https://doi.org/10.4271/2005-01-1212

Index

Pages in *italics* refer figures and pages in **bold** refer tables.

E

F

W

Printed in the United States
by Baker & Taylor Publisher Services